MEANINGFUL PASTS

T0179914

Historical Narratives, Commemorative Landscapes, and Everyday Lives

In *Meaningful Pasts*, Russell Johnston and Michael Ripmeester explore two strands of identity-making among residents of the Niagara region in Ontario, Canada.

First, they describe the region's official narratives, most of which celebrate the achievements of white settlers with a mix of storytelling, rituals, and monuments. Despite their presence in local lore and landmarks, these official narratives did not resonate with the nearly one thousand residents who participated in five surveys conducted over eleven years. Instead, participants drew on contemporary people, places, and events. Second, the authors explore the emergence of Niagara's wine industry as a heritage narrative. The book shares how the survey participants embraced the industry as a local identifier and indicates how the industry's efforts have rekindled the residents' interest in agriculture as a significant element of regional heritage and local identities.

Revealing how the profiles of local narratives and commemorations become entwined with social, cultural, economic, and political power, *Meaningful Pasts* illuminates the fact that local narratives retain their relevance only if residents find them meaningful in their day-to-day lives.

RUSSELL JOHNSTON is an associate professor in the Department of Communication, Popular Culture, and Film at Brock University.

MICHAEL RIPMEESTER is a professor in the Department of Geography and Tourism Studies at Brock University.

Meaningful Pasts

Historical Narratives, Commemorative Landscapes, and Everyday Lives

RUSSELL JOHNSTON AND MICHAEL RIPMEESTER

UNIVERSITY OF TORONTO PRESS
Toronto Buffalo London

ISBN 978-1-4875-2873-7 (cloth)   ISBN 978-1-4875-2875-1 (EPUB)
ISBN 978-1-4875-5042-4 (paper)   ISBN 978-1-4875-2874-4 (PDF)

Library and Archives Canada Cataloguing in Publication

Title: Meaningful pasts : historical narratives, commemorative landscapes, and
     everyday lives / Russell Johnston and Michael Ripmeester.
Names: Johnston, Russell T. (Russell Todd), 1967–, author. | Ripmeester, Michael,
     author.
Description: Includes bibliographical references and index.
Identifiers: Canadiana (print) 20230551580 | Canadiana (ebook) 20230551602 |
     ISBN 9781487528737 (cloth) | ISBN 9781487550424 (paper) |
     ISBN 9781487528751 (EPUB) | ISBN 9781487528744 (PDF)
Subjects: LCSH: Collective memory – Ontario – Niagara (regional municipality)
Classification: LCC FC3099.N53 J64 2024 | DDC 971.3/38 – dc23

Cover design: Will Brown
Cover image: Sketch of the Lundy's Lane battlefield and Village of Drummondville,
Canadian Illustrated News, 15 July 1876. From the Archives & Special Collections, Brock
University Library.

We wish to acknowledge the land on which the University of Toronto Press
operates. This land is the traditional territory of the Wendat, the Anishnaabeg, the
Haudenosaunee, the Métis, and the Mississaugas of the Credit First Nation.

This book has been published with the help of a grant from the Federation for the
Humanities and Social Sciences, through the Awards to Scholarly Publications
Program, using funds provided by the Social Sciences and Humanities Research
Council of Canada.

University of Toronto Press acknowledges the financial support of the Government of
Canada, the Canada Council for the Arts, and the Ontario Arts Council, an agency of
the Government of Ontario, for its publishing activities.

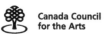

Contents

List of Maps, Illustrations, and Tables   vii

Acknowledgments   ix

Introduction   3

Part One: Resident Responses to "Official" Mnemonic Products

1 Assessing Public Engagement with Historical
   Narratives in Niagara   19

Part Two: The Private Alexander Watson Monument

2 A War Memorial as a Mnemonic Device   57
3 The Watson Monument through Time   85
4 Residents Engage the Watson Monument   111

Part Three: The Niagara Grape and Wine Industry

5 Viticulture as a Mnemonic Product   147
6 Residents Engage the Niagara Wine Industry   180
7 Conclusion   205

Notes   214

Index   283

Maps, Illustrations, and Tables

Maps

0.1. The Niagara region   5
1.1. Plaques in the Niagara region, by location   31
6.1. Niagara vineyards and wineries, 2017   191

Illustrations

0.1. Salem Chapel and four mnemonic products, St. Catharines   4
1.1. Commemogram: plaques in the Niagara region, by decade of event commemorated   28
1.2. Plaques in the Niagara region, by memory entrepreneur   29
1.3. Commemogram: plaques in the Niagara region, by memory entrepreneur and decade   32
1.4. Commemogram: participant opinions on Niagara identity, 2005   35
1.5. Commemogram: participant opinions on Niagara's contribution to Canadian identity, 2008   38
1.6. Commemogram: participant opinions on Niagara's identity, 2016. "What is the first thing that comes to mind when you hear the word 'Niagara'?"   49
2.1. "Part of Funeral Procession"   75
2.2. Lieutenant A.W. Kippen's grave monument, Elmwood Cemetery, Perth   78
2.3. Watson Monument as depicted in a 1907 booklet, *The Garden City of Canada*   79
2.4. Unveiling the Watson Monument, 1886   81

3.1.   St. Catharines veterans of Ridgeway and Red River receive their medals, 1900   91

3.2.   The Soldiers' Cross and Watson Monument, ca. 1919   94

3.3.   Great War Veterans Association Club with War Shrine   99

4.1.   Watson Monument, former Lincoln County Courthouse, and city hall, 2021   112

Tables

1.1.   If you had to choose one thing that identifies Niagara, what would it be? (2009)   40

1.2.   If you had to choose one thing that identifies Niagara, what would it be? (2012)   42

1.3.   Is wine part of Niagara's local heritage?   42

4.1.   Number of unique social media accounts and posts regarding the petition and motion to remove the Watson Monument   128

5.1.   Points of interest promoted through pamphlet kiosks in Niagara (2010)   170

5.2.   Themes manifest on winery websites (2010)   171

6.1.   If you had to choose one thing that identifies Niagara, what would it be?   189

6.2.   Are wine or wine activities part of your daily routine?   190

6.3.   How do wine or wine activities fit into your daily routine?   192

6.4.   In what ways do you participate in wine and wine-related activities?   192

6.5.   Resident engagement with local agriculture and opinions regarding local agricultural heritage   193

6.6.   Residents' sources of knowledge about wine and wine-related activities   195

6.7.   To what extent do you learn about wine and wine-related activities from advertising?   197

Acknowledgments

This book would have been inconceivable without the participation of Niagara residents. Over the course of five surveys, 932 individuals took time to share their thoughts with our research team. These individuals' spontaneous, insightful, and often surprising responses drove our research and challenged many of our original assumptions. Their generous contributions to these pages remain anonymous, but we are grateful to each one individually. We would be delighted to hear from any participant who spoke with us. Perhaps you still have the "thank-you" card you received when you finished the questionnaire.

Our crack team of research assistants conducted most of the interviews. Their interest in the project was palpable and their awareness in the field leant depth to our interpretation of the data. Most importantly, their professionalism and meticulous attention to detail ensured the participants' thoughts were captured accurately. Our sincere thanks to Zorianna Zurba, Sarah Bradley, Erin Kaipainen, Laura Visan, Richard Lagani, Sarah Middleton, Kirby Calvert, Sierra Sheppard, Andrea Tirone, Kim Randall, Michael Daleo, and Stephanie Murray. Further thanks are due to Zorianna Zurba, Sarah Bradley, and Hannah Johnston for their tenacious archival research. Our book also benefits from the excellent cartographic and image-manipulation skills of Loris Gasparotto, Sharon Janzen, and Nick Ripmeester.

When exploring the legends and lore of Niagara that matter to residents, look to the region's local historians. We received tremendous support from Dennis Gannon and John Burtniak; Arden Phair and Kathleen Powell at the St. Catharines Museum; David Sharron and Edie Williams at Brock University Archives and Special Collections; Tiffany Tifan at the Grimsby Museum; and all the staff at St. Catharines Public Library.

Funding for the surveys came from two sources. From 2005 to 2010, we shared a SSHRCC standard research grant with colleagues in the

Popular Culture Niagara Research Group at Brock University. In 2012 and 2016, we received two separate grants from Brock University's Council for Research in the Social Sciences, under the purview of the Faculty of Social Sciences. Our thanks to both agencies for their support.

Thanks are also due to friends and colleagues who provided their feedback on drafts of this book: Brian Osborne, Ross Fair, James Opp, John C. Walsh, Matthew Rofe, Phillip Macintosh, Chris Fullerton, Joan Nicks, and Barry Grant, as well as the late Alan Gordon and Charles M. Johnston. We also acknowledge the invaluable input of the many anonymous referees who reviewed earlier drafts of these pages. Thanks are also due to groups who invited us to share our research: the Cultural Business Group at Glasgow Caledonian University, Scotland; the Greenlines Institute headquartered in Barcelos, Portugal; the Centre for Military, Strategic, and Disarmament Studies at Wilfrid Laurier University and the Lincoln and Welland Regiment (Niagara), who co-hosted a Remembrance Day symposium in St. Catharines; the Regional Municipality of Niagara Cultural Committee; the City of St. Catharines Culture Committee; and the Cool Climate Oenology and Viticulture Institute at Brock University. Last but not least, we greatly appreciate the enthusiastic support of our editor, Jodi Lewchuk, who shepherded us through the publication process at University of Toronto Press.

To our families, who lived through this rather lengthy project: thank you for your enduring patience and occasional field trips to look at weed-strewn relics of Niagara's past. Russell dedicates this book to his wife, Sara, and daughter, Hannah. Mike dedicates this book to his wife, Wendy, and children, Nick, Jill, and Mollie.

MEANINGFUL PASTS

Introduction

A small white church sits on a busy thoroughfare on the margins of downtown St. Catharines. Local historians believe it is home to the oldest African Canadian congregation in Ontario.[1] Erected in the 1850s, Salem Chapel was built for those who escaped slavery by following the Underground Railroad to its terminus in Upper Canada. White residents were not overly welcoming of these Freedom Seekers.[2] Both the church and its congregation were segregated in the city's "Coloured Village." As we write these words, five historical plaques adorn the property while a sixth is located a short block away (figure 0.1). They reflect the priorities of their sponsors, as all plaques do. The City of St. Catharines placed two of them, one to mark the local history of the Underground Railroad and one to celebrate the origins of a local African Canadian community. The provincial Government of Ontario memorialized famed conductor Harriet Tubman's residency, while the federal Government of Canada honoured the church's role in a nation-building narrative that highlights multiculturalism. Two further plaques celebrating multiculturalism were added by the city and by the congregation. The Underground Railroad is also the focus of a permanent exhibit at the St. Catharines Museum and the city's annual celebration of Black History Month. A downtown public school was named in Tubman's honour. This multi-scalar commemoration effort makes Salem Chapel an intriguing example of how people, places, events, and things become marked as worthy of remembrance.

The story of the Underground Railroad is one of many local historical narratives marked in the Regional Municipality of Niagara (map 0.1). Governments, the private sector, and voluntary organizations pour their energy and resources into the commemoration of white settlement, military engagements, pioneering industry, and other noteworthy individuals, events, and places. This work will never be complete. Some

(a)

(b)

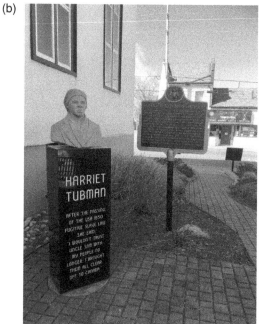

Figure 0.1. (a) Salem Chapel and (b) four mnemonic products, St. Catharines

Source: Michael Ripmeester.

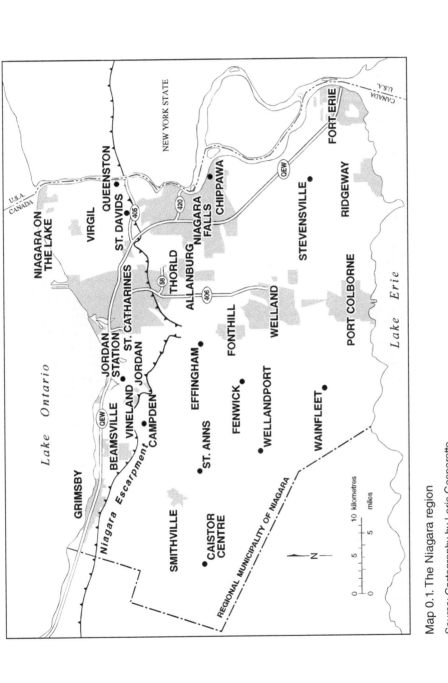

Map 0.1. The Niagara region

Source: Cartography by Loris Gasparotto.

stories endure, their narratives seemingly timeless. Other stories, however, are re-conceived and retold to ensure their relevance for multiple, shifting audiences. In some cases, sponsors may add or delete certain aspects from a story as its cultural currency waxes or wanes. Thus, in the case of Salem Chapel we might ask the following questions: What do residents know about the chapel and its story? Does this story make an impact on contemporary lives? And does it help to foster a sense of collective identity among city residents?

The relationships among a narrative, individual engagement, and group identity is often termed "collective memory." Its nature is the subject of active and lively debate. A key point of contention addresses the relationship between individual and sociocultural memory. Most individuals can, for example, recall details of particular events they experienced during their childhood, be they deeply personal moments or world events. Similarly, individuals may speak knowledgably of the Underground Railroad even though no one living today met Harriet Tubman or witnessed the arrival of the first Freedom Seekers. Both sets of stories are remembered, and both sets of stories may be mediated by the individual's social, cultural, economic, and political contexts. These facts suggest the two types of memory are related. They are also distinct.[3] Individuals can recall both, but one is lived and the other is learned. Recognizing this difference, Maurice Halbwachs – a pioneering scholar in the field of memory studies – was careful to differentiate between autobiographical memory (the ability to recall the moments one has witnessed) and historical memory (the ability to recall things learned via secondary sources such as a book or museum exhibit).[4] Our interest lies in the second type. Jeffrey Olick provides a useful definition:

> collective memory really refers to a wide variety of *mnemonic products and practices*, often quite different from one another.... Mnemonic practices – though they occur in an infinity of contexts and through a shifting multiplicity of media – are always simultaneously individual and social. And no matter how concrete mnemonic products may be, they gain their reality only by being used, interpreted, and reproduced or changed.[5]

In similar fashion, Laurajane Smith argues that using mnemonic products in a social setting constitutes a cultural performance.[6] Memories may consist of narrative details, but they are made and remade through every family gathering, every lesson in a classroom, or every visit to a historic site. In other words, sharing memories is an embodied and ritualized performance of truth production in the present. As Smith argues in the case of heritage sites, "it [is] the *use* of these sites that made them heritage, not the mere fact of their existence."[7] This is a foundational component of our argument. Shared knowledge of the past is both a form of knowledge and a form of experience. It is produced when narratives are authored and shared, negotiated and challenged, in social spaces. These narratives reside not only in the memory of the individuals who participate but also in the spaces where these narratives are performed.

We are interested in the relationship between mnemonic products (those tangible and intangible things that hold memories) and mnemonic practices (the ways in which people engage mnemonic products).[8] There is an expansive literature describing the roles that mnemonic products play in human memory. Much of this literature argues that influential social, cultural, economic, and political groups use mnemonic products to maintain specific narratives.[9] Things like monuments, holidays, and parades are designed to transmit historical knowledge that can bolster values and morals associated with a particular group identity or to justify existing social circumstances.[10] This literature may answer questions concerning the purpose of mnemonic products, but it leaves unanswered questions concerning their efficacy.

Can mnemonic products actually establish and foster a common interpretation of specific historical narratives? Or are mnemonic products just one source of information among many competitors in our contemporary media environment? On that point, do individuals consult mnemonic products equally to construct a meaningful, useable past? And if so, do they passively receive the messages and values that mnemonic products convey, or do they actively read and interpret them? We argue here that residents rarely simply accede to the narratives conveyed by mnemonic products. Rather, individuals use all available resources to construct narratives that are meaningful to them as individuals and as members of groups.

Our research focused on the Regional Municipality of Niagara. We acknowledge that our research examines the narratives emerging from a colonial imagination. The region's population represents its colonial past. The 2016 census indicated that 78.5 per cent of Niagara residents claimed ancestral ties to Europe while 9.3 per cent declared visible minority status. Indigenous residents represented 3 per cent of the regional population.[11] Our work explored how contemporary Niagara residents engaged with narratives preserved locally in "official" mnemonic products. Through most of the research period, Niagara's stock of mnemonic products did not represent the narratives of ten thousand years of human activity before colonization, nor more recent narratives of racialized and economically marginalized communities. The topics we addressed, then, reflected the interests and biases of a settler society. We applaud those memory entrepreneurs who are now promoting a more inclusive approach to Niagara's narratives, traditions, and identity.

Authority, Affect, Resonance, and Ubiquity

Everyone has stories to tell. These stories inevitably involve a past. However, Christin Köber and Tilmann Habermas argue that individuals' narratives and the memories that support them are rarely stable.[12] Instead, individuals selectively and strategically shape their pasts according to their present needs. The maintenance of group stories – like those of families, organizations, institutions, or nations – suggests a similar process. The frames are clearly analogous: social, cultural, economic, and political groups draw on memories and use them to satisfy contemporary needs. Yet, scholars have struggled with the concept of "shared," "collective," or "popular" memory and how it may function. Indeed, Rauf Garagozov describes the debate this way:

> Perhaps the only thesis which is more or less easily acknowledged by different researchers is the assertion that collective memory is a widely shared knowledge of past social events that are collectively constructed through communicative social interactions which can have significant impact on our behaviour, feelings, and thoughts.[13]

We concur with this characterization, but we stress that such remembering is rooted in experiences that are more complex than a common formal education or shared access to mnemonic products.

Halbwachs suggests that individual memory is linked to wider social structures; that is, individuals remember in specific social and cultural contexts.[14] Provocative research by Alexandru Cuc and colleagues demonstrates this claim.[15] The researchers asked small groups to engage in conversation. Immediately afterwards, individuals were asked to recall and describe the conversation. The researchers found that each individual remembered contributions from other participants in almost equal measure. However, they also found that the addition of an assertive individual shaped the recollections of the other participants; the dominant individual's contributions figured most prominently in everyone else's recollections of the conversation. This echoes Michel Foucault's claim that all historical narratives are products of a will to power.[16] Edward Said makes a similar point, noting that "the art of memory for the modern world is both for historians as well as ordinary citizens and institutions very much something to be used, misused, and exploited, rather than something that sits inertly there for each person to possess and contain."[17] To understand these processes it is necessary to identify the roles played by those individuals and groups – whom Elizabeth Jelin refers to as "memory entrepreneurs" – who dominate historical conversations.[18]

Memory entrepreneurs are social, cultural, economic, or political elites who author historical narratives and assert their symbolic importance with specific audiences in mind. Such elites may occupy a wide range of roles, such as revered family members,[19] professional historians,[20] curators,[21] nation-states, and other government agents.[22] The status of these "experts" – and the narratives they promote – often appear unassailable to their audiences. For example, a powerful memory entrepreneur, like a family matriarch or a federal state, has authoritative weight. Their positions, unique knowledge, and control of content and symbols makes their versions of the past seem believable, trustworthy, and irrefutable.[23] Families, institutions, and nation-states can use such narratives to forge links between diverse individuals by providing a common past, a common understanding of the present, a common path towards the future, and, ultimately, a collective identity. Kenneth Foote and Maoz Azaryahu, for example, argue that,

> In this capacity, public memory is part of the symbolic foundation of collective identity, where the question, "who we are," is answered, at least partially, by answering the question, "where do we come from," and what we share and do together as a community.[24]

Put another way, Eviatar Zerubavel writes, "acquiring a group's memories and thereby identifying with its collective past is part of the process of acquiring any social identity, and familiarizing members with that past is a major part of communities' efforts to assimilate them."[25] Such knowledge can provide populations with common values and morals,[26] heroes and villains, as well as allies and enemies.[27] Thus Charles Withers and Elizabeth Crooke demonstrate how memory entrepreneurs have prescribed memoryscapes to get citizens of Scotland and Ireland to "know themselves."[28] In like manner, Sibylle Puntscher and colleagues argue that mnemonic products have "a significant impact on the current norms, behavior, and beliefs of the community with economic, political and social consequences."[29] Once in place, these beliefs can be hard to shift. For example, Eleftherios Klerides and Michalinos Zembylas demonstrate through their discussion of history textbooks in Cyprus that people who have internalized a set of narratives can become "immune" to alternative versions of the past, especially those that challenge their own.[30]

One possible avenue for understanding the processes of connection and "immunity" lies in the literature on "affect."[31] Kaitlin Murphy, drawing upon Roland Barthes, likens affect to the *punctum*, the thing that touches or moves a person.[32] Kathleen Steward offers this evocative definition:

> They're things that happen. They happen in impulses, sensations, expectations, daydreams, encounters, and habits of relating, in strategies and their failures, in forms of persuasion, contagion, and compulsion, in modes of attention, attachment, and agency, and in publics and social worlds of all kinds that catch people up in something that feels like *something*.[33]

Recognizing affect, therefore, "leads to a focus on embodiment, to attempts to understand how people are moved, and what attracts them, to an emphasis on repetitions, pains and pleasures, feelings and memories" or, in other words, "how ... social formations" – like the collective expressions encouraged by memory entrepreneurs – "grab people."[34]

Mnemonic products possess three characteristics that facilitate their affective potential. The first, as described above, is the weight of authority (be it official, scholarly, technological, or aesthetic). Simply stated, the mnemonic narratives championed by acknowledged experts carry more influence than those from alternative sources.[35] They are also more likely to move people in expected ways; they may inspire pride, belonging, loathing, or some other emotion. Such narratives are more likely to inspire acquiescence, confidence, or belief. In the context of our work, we posit that affect, the ability to move people, is a mnemonic product's ultimate goal. Mnemonic products and practices can move people in expected ways and towards expected emotions, which in turn leads to expected thoughts and actions.

The second characteristic of mnemonic products is resilience established through intertextual repetition. A narrator's reiteration of stories enhances their visibility, reinforces their perceived veracity, and increases the odds that individuals will internalize them. Once internalized, stories may elicit consistent responses from audiences, either favourable or unfavourable, as individuals affirm their own feelings and relationships to their core elements. For example, the Government of Canada honours national heroes and important events by issuing coins and stamps, supporting school curricula, financing museums and archives, designating historic sites, and creating holidays. The goal is to elicit shared, positive responses from multiple constituencies across the country. In Niagara, a confluence of authoritative, intertextual narratives ensures that stories involving the War of 1812 and the Underground Railway cannot easily be impugned. As Tim Cresswell writes, "a story is told over and over until it sticks – until it becomes so much common sense."[36]

Despite Cresswell's assertion, repetition or ubiquity is not enough to ensure a story's affect. This leads to the third characteristic of mnemonic products: resonance. Mnemonic narratives must tell a story that "awakens emotion or affection."[37] They must resonate with individuals in ways that Sara Ahmed describes as "sticky."[38] This type of resonance connotes the ways in which individuals attach specific feelings and then move to expected ideas or thoughts when exposed to affective impulses. For example, and as we will discuss in part 2 of this book, the symbolism of the citizen-soldier resonates with viewers because it moves them to consider values like sacrifice and honour without much conscious reflection. As Rumi Sakamoto writes in her discussion of a controversial museum, such affective impulses may be more important than the historical narratives with which they are associated.[39] We argue, then, that museum displays, historic sites, arches, cenotaphs, wreaths, figures on horses, steles, and super-sized human figures are able to move people in expected ways because their narratives, symbolism, and values derive from authority, are intertextual, and resonate with their audiences. Resonance or stickiness is not merely a matter of individuals simultaneously reacting to mnemonic prompts. Tuuli Lähdesmäki contends in her discussion of historic sites that "affect" can be "sought, regulated or managed."[40] Thus, similar responses towards things like cenotaphs can become socially determined or taken-for-granted engagements. Zerubavel writes,

> The act of taking something for granted ... "has its origin *beyond the individual*, and it is this sociocultural basis that forms the interpretive background to our individual minds." What may seem at first glance to be a strictly personal act ultimately explainable in terms of individuals' personal tendencies thus turns out to actually be a product of essentially *impersonal*, non-idiosyncratic patterned assumptions that are not unique to particular individuals.[41]

Some mnemonic products may become so saturated with affective impulse that they circulate through social and cultural bodies. C. Thi Nguyen puts it this way: "Not only might [a public monument] move me to renew my commitment to a value, but I can also count on other people to encounter it, and I can reasonably hope that many of them will be similarly moved."[42] Ahmed contends that these processes serve as the focus of "affective economies."[43] Thus, as movement leads to feeling and then to cognition among individuals, "histories of association" become "feelings-in-common" or taken for granted.[44] Where things become taken for granted, they may also become difficult to refute and, perhaps more importantly, may become invisible.[45]

At the outset of this introduction, we claimed that shared historical knowledge is fostered through both mnemonic products and mnemonic practices within specific social, cultural, economic, or political contexts. Engagement with such products and practices necessarily varies across individuals, time, and space. Some individuals may internalize specific narratives, others may disavow them, while yet others may be indifferent to them. In writing of identities in the modern world, Manuel Castells suggests that people create identities out of the resources available to them:

> from history, from geography, from biology, from productive and repro-
> ductive institutions, from collective memory and from personal fanta-
> sies, from power apparatuses and religious revelations. But individuals,
> social groups, and societies process all these materials, and rearrange their
> meaning, according to social determinations and cultural projects that are
> rooted in their social structure and in their space/time framework.[46]

Anthony Giddens likens this selection process to making sense of a newspaper:

> A newspaper ... presents a collage of information, as does, on a wider
> scale, the whole bevy of newspapers which may be on sale in a particular
> area or country. Yet each reader imposes his own order on this diversity,
> by selecting which newspaper to read – if any – and by making an active
> selection of its contents.[47]

These choices are not without constraint. Because individuals live in economic, cultural, and political contexts, they are located at the nexus of power relations that constrain their available options, their ability to choose one over another, and their access to alternatives. Constraint may also, however, provide satisfaction and pleasure.[48] Even though individuals' ability to respond to a mnemonic narrative may be limited, adherence to that narrative may provide positive feedback, like connection to a wider community.[49] This kind of bond allows individuals to know themselves and their group identities and to draw boundaries between narrative insiders and outsiders.

We reiterate that individuals need not respond to mnemonic products in an identical fashion. Different individuals may be moved by different narratives, in different ways, and at different times. Laurajane Smith and Gary Campbell describe these complex relationships as a "register of engagement."[50] Their register indicates the range and intensity of individuals' emotional responses to particular narratives. Hot and deep reactions are obviously the most visible, but cool and shallow responses matter as well. For example, symbols do not need to evoke a strong emotional impact every time they are engaged to retain their worth.[51] Michael Billig writes,

> One might predict that, as a nation-state becomes established in its sovereignty, and if it faces little internal challenge, then the symbols of nationhood, which might once have been consciously displayed, do not disappear from sight, but instead become absorbed into the environment of the established homeland.[52]

Apparent invisibility, then, should not suggest that these symbols are without effect. As we demonstrate in the following chapters, individuals' seeming indifference to mnemonic products belies their deep emotional and cognitive investment in them. Responses can likewise shift in an instant. As social, cultural, economic, or political tides change, narratives may fall in or out of public favour. As this occurs, new mnemonic products and practices may be created, invigorated, or become lightning rods for changing public sentiment. Debates in Canada, for example, have highlighted the contentious use of the names and likenesses of heroes-cum-pariahs in the public sphere, among them Samuel de Champlain, Sir John A. Macdonald, Hector-Louis Langevin, and Edward Cornwallis. Such iconoclasm is not limited to Canada. Following the Black Lives Matter protests of 2020, for example, people all over the world damaged, defaced, or demanded the removal of monuments with links to a racist past.[53]

We sought to understand how the public engage with mnemonic products once they are unveiled. Examining heritage activities in the United States, John Bodnar, for example, argues that commemoration is never a simple imposition of elite perspectives on submissive publics.[54] Memorials, he writes, are subject to processes of interpretation that may force compromise between memory entrepreneurs and their audiences. However, while documents may shed light on elites' intentions for promoting heritage activities, there is no comparable data with which to explore audiences' interpretation of them. Here we were inspired by the work of Roy Rosenzweig and David Thelen, who conducted a national telephone survey to explore the popular uses of history in the United States.[55] Parallel studies in Australia by Paul Ashton and Paula Hamilton, and in Canada by Margaret Conrad and colleagues, were equally instructive.[56] All three projects sought to discover how representative samples of the public conceive of and engage with the past through their daily lives. Taken together, their findings indicated that participants were aware of, and actively engaged with, mnemonic products. More pointedly, they found that individuals actively construct personal interpretations of the past by drawing upon multiple sources and mnemonic narratives. Thus, the work of any one group of memory entrepreneurs remains active among their fellow citizens only insofar as it remains relevant and consistent within a wider mix of intellectual resources.

Our research differs from these previous studies in one important way. While they consulted participants to understand their conceptions of "the past," we did not. We consulted participants to understand their conceptions of local culture and identity. We then gauged the extent to which participants drew upon available mnemonic resources in their answers. By not prompting participants to engage the past directly, we remained open to the fact that residents may not employ those mnemonic products offered by "official" memory entrepreneurs. When we discovered that many residents actually did not, we sought to identify the resources they drew upon instead. Given this focus, we might usefully rewrite Giddens's newspaper metaphor to make this point:

> A mnemonic landscape presents a collage of information, as does, on a wider scale the whole bevy of mnemonic products that may be available in a particular area of the country. Yet each viewer imposes their own order on this diversity, by selecting which narratives to engage – if any – and by making an active selection of its contents.[57]

This is the crux of our research: we sought to uncover socially constructed knowledge of the past and its influence on local community identities.

The Niagara Region: A Case Study in Three Parts

This book is divided into three sections. The first section provides an overview of the heritage narratives told in Niagara. Using an inventory of local monuments and plaques and Eviatar Zerubavel's commemogram as an analytical tool, we describe and explain the topography of Niagara's heritage narratives.[58] In doing so, we are less interested in the details surrounding any one monument, plaque, or historic site than we are in identifying the group responsible for it, the narrative it articulates, and the date of its unveiling. We then surveyed residents of St. Catharines and surrounding cities to determine how the stories told by these sites resonated with them. The evidence initially suggested a clear disconnect between the efforts of memory entrepreneurs and our participants' engagement with official mnemonic narratives. We also consider how shallow and cool responses may yield insight into residents' memory making.

The second section focuses on one monument that stands at City Hall in the region's largest city, St. Catharines. Chapter 2 describes local responses to the North-West Resistance of 1885 and the city's only casualty, Private Alexander Watson, whose death inspired the monument. We describe the cultural and political genealogy of the monument in the context of contemporary attitudes towards Canadian citizen-soldiers. Chapter 3 explores how residents engaged the monument during the century that followed its unveiling in 1886, and particularly how public interest in it changed through time as other narratives and other commemorations gained greater attention. Chapter 4 examines how contemporary residents of St. Catharines viewed the monument. In 2005, we surveyed residents to gauge their views of the monument, only to discover that they had very little knowledge of it. In 2020, global anti-racism protests challenged mnemonic products that celebrated troublesome pasts. In St. Catharines, the Watson Monument became the focus of similar protests. We mined social media posts to understand local attitudes towards history, commemoration, and the monument itself.

Part 3 focuses on private-sector marketing and public-sector place branding in Niagara. The heritage value of the local wine industry was often noted by survey participants in 2005. We sought to understand how this relatively new industry had integrated itself into residents' stock of mnemonic narratives. Chapter 5 describes joint private-public efforts to rebrand Niagara as "wine country." While the intended audience for these marketing materials lived in metropolitan Toronto and New York State, Niagara residents also engage them. Chapter 6 explores Niagara residents' responses to the wine industry's rebranding. Our surveys, which took place over four years in five separate communities across Niagara, indicated that residents had embraced the wine industry as a new economic engine and a significant heritage narrative.

PART ONE

Resident Responses to "Official" Mnemonic Products

Assessing Public Engagement with Historical Narratives in Niagara

Long years have passed since the foemen were vanquish'd,
Summers have come and have vanished in distance.
We who now dwell in our peace-blessed Dominion
Owe all our praise to the men who have saved it.
Raise ye monument, crown it with flowers,
Swell yet the shout, let the meadows re-echo
In praise of those men who with patriot spirit
Confronted and vanquished the foes that assailed them.

                                    – E.W. Miller, "On the Erection of a Monument
                                        on the Battlefield of Lundy's Lane" (1895)[1]

Isaac Brock, a career soldier, died a combat death. During the War of 1812 the talented general faced an American invasion force as it crossed the Niagara River into Upper Canada and took command of Queenston Heights. Brock was loathe to see them dig in. Summoning his troops – composed of British regulars, Haudenosaunee warriors, and colonial militia – he led the initial charge to halt the Americans' progress and drive them back. And there he fell, shot twice, the second bullet piercing his chest.[2]

Though Brock died that day, his story did not. Tales were told, poems were written, books and paintings sketched out his life and death, and eventually a monument was raised to honour "the hero of Upper Canada" on the field he sought to recapture. After the monument was destroyed by an Irish patriot in 1840, a larger monument topped with a statue of the general was erected in its place. Brock's name now adorns a city, two towns, and a university, as well as buildings, streets, schools, and businesses across Canada. His likeness is also featured on Canadian stamps and coins. It is fair to say that Brock stands among the pantheon of Canadian heroes. In 2004, for example, a popular vote ranked him twenty-eight on the list of greatest Canadians.[3] He is certainly well represented among the mnemonic products of Niagara. Nevertheless, we wondered whether Brock's story and other well-marked narratives would dwell at the top of residents' minds when we asked them about Niagara, its identity, and their knowledge of mnemonic narratives.

Narratives Told and Remembered in Stone

We know places through the narratives that represent them.[4] Some provide structure to specific local experiences while others provide the scaffolding for national myths. Whatever the scale, narratives provide symbolic resources that allow individuals to make claims about themselves both as individuals and as members of groups. As we noted in the introduction, these narratives are often represented through mnemonic products and practices. Sponsors intend for officially sanctioned mnemonic practices and products to influence audiences to internalize both the stories and the values and assumptions they represent. At the same time, however, the intended audiences for such traditions may choose not to engage with them. In this chapter, we seek to understand how local, provincial, and national mnemonic traditions have contributed to a group identity among Niagara residents.

We recognize there are many forms of mnemonic products (including museums and archives, monuments and historic sites, school curricula, and family collections of ephemera, etc.),[5] but when we began this research we decided to concentrate on memorials, monuments, and historic sites. Our goal was to determine whether people engaged with them and to explore what they may have taken from these experiences. These representations of the past fascinate scholars because they are explicitly ideological devices embedded in material objects. Memory entrepreneurs have used statues and public art to hold memories since antiquity. However, Sergiusz Michalski contends that, between 1870 and the 1990s, Western nations used them aggressively to foster particular social and political orders.[6] He writes, for example, of the "statuemania" of the Third French Republic: "The ruling circles were determined to create with and around public monuments a system of institutionalized democratic pageantry suited to the particular aspirations of the urban middle class and of the intelligentsia in government or municipal service."[7] In strikingly similar manner, Erika Doss explores the "memorial mania" of the early twenty-first century.[8] In this case, however, Doss suggests that monuments have become a medium for groups struggling to convey public identity and belonging. Though the nature of commemoration is different, the principles are the same. Groups with the resources to create material forms of their preferred narratives seek to impress their versions of the past on the public.

Such public presence portrays specific narratives through emblematic moments laden with text and symbolism. For example, classical columns, the monarch on her throne, the general on his horse, or the refugee forging ahead can dramatically augment accompanying text. As we argue in the chapters that follow, the written text may even become redundant.[9] Sponsors simply ask viewers to draw out a monument's lesson by sight, sight augmented by knowledge gained from a supporting intertextual web of like materials and practices. Few Canadian adults, for example, need instruction to understand a war memorial as a marker of sacrifice and valour. These public messages become a ubiquitous part of the public realm, their seeming permanence giving them a timeless quality. As James E. Young notes, "In suggesting themselves as indigenous, even geological outcrops in a national landscape, monuments tend to naturalize the values, ideals, and laws of the land itself."[10]

Public engagement with these outcroppings does not always occur as desired by their sponsors. For example, Kirk Savage as well as Hamza Muziani and Brenda Yeoh illustrate how monuments can be contested even during their planning and construction stages.[11] Such opposition, for example, contributed to the reconsideration and eventual relocation of a proposed monument in Ottawa to the victims of communism.[12] Similarly, several authors have documented how some groups facing political upheavals in Central and South America have used street art and parades to counter the narrative claims of the state.[13] Tensions do not end with an unveiling. Once established, mnemonic products may become the site of violent opposition. Monuments have been attacked,[14] defaced,[15] and bombed.[16]

Time, too, can be the Achilles heel of material commemorations. The literature notes that mnemonic narratives resonate most closely with the generation of individuals who first articulate them.[17] After that, sponsors must repeat or rework a commemoration to ensure its continued resonance or stickiness. Failure to do so may cause a narrative and its representations to fade into ideational or material backgrounds.[18] As Brian Osborne observes of monuments, "Put simply, they last too long! ... They are frozen in space while time moves on around them, their rigid materiality ensuring their estrangement from the

ever-changing values of the society in which they are located."[19] Essay-
ist Robert Musil captures this effect in the following passage:

> You can walk down the same street for months, know every address, every
> show window, every policeman along the way, and you won't even miss
> a dime that someone dropped on the sidewalk; but you are very surprised
> when one day, staring up at a pretty chambermaid on the first floor of a
> building, you notice a not-at-all tiny metal plaque on which, engraved in
> indelible letters, you read that from eighteen hundred and such and such
> to eighteen hundred and a little more the unforgettable So-and-so lived
> and created here.[20]

Even those responsible for monuments or plaques may lose interest in
maintaining them. Alternatively, individuals and groups may reconsider hallowed narratives as time passes and values shift. The massive
removal of monuments springing from the Black Lives Matter movement provides a particularly striking example of this turn against narratives written in stone.[21] We return to this discussion in chapter 3.

The Niagara Region

The Niagara region of Ontario provides an excellent case study to
assess the dissemination of mnemonic narratives. Indigenous Peoples
were the first to occupy the area. Archaeological research has traced
settlement patterns, cultural sequences, trade routes, and geopolitical
struggles spanning millennia.[22] The first Indigenous Peoples arrived in
the Great Lakes basin after the last Ice Age, roughly eleven thousand
years ago.[23] Evidence of their occupancy is revealed through their burial
grounds, middens, and other collections of material artefacts. At one
site in present-day Fort Erie, carbon-dated remains indicate the area
was in use from roughly 2000–900 BCE, with more recent burials dating from the 1300s CE and 1600s CE. During the earliest periods, the
site may have served as an annual camp for migratory communities.
Annual rounds of activities followed game such as fish, fowl, and deer
that once were plentiful. Native species of cherry, plum, apple, grape,
currants, and berries were also available. Archaeological sites dating
from roughly 1000–800 BCE reveal enduring settlements, particularly
in the form of longhouses. Proof of agriculture emerges after 600 CE,
when seeds and agricultural knowledge diffused from other regions.
Remains of maize, squash, beans, sunflowers, and tobacco have been
identified at Fort Erie.[24] Outside of Fort Erie, other notable sites in the
Niagara Peninsula have been uncovered in St. David's and Grimsby.[25]

Chert or flint tools are among the most numerous artefacts found at these sites. The peninsula is rich in limestone. As trading developed among the peoples of the Great Lakes, inhabitants of the peninsula may have found that flints were a valuable commodity that could be exchanged for goods from other regions.[26] As Ronald F. Williamson and Ronald I. MacDonald note, the prehistoric residents of the Fort Erie site "left no lasting monument in any form which would still be easily recognizable. Yet the flint which was a cornerstone of their existence must be considered as permanent a monument as any that could be constructed today."[27]

When Europeans first arrived, Niagara was home to the Neutral. This nation was linked to the Haudensaunee by language, but they resisted the wider geopolitics dividing the Haudensaunee from other confederacies along the St. Lawrence and in the Great Lakes basin. Individual nations had allied themselves with either France or England as they pushed west from the Atlantic Seaboard. Both colonizing powers had recognized the military strength of the Indigenous nations; both sought their aid through trade and strategic alliances.[28] The Neutral chose not to forge independent relationships with either, which prompted the label given them – *les Neutres* – by the French.[29] Descriptions of their agricultural settlements come from French explorers and missionaries who visited them between 1615 and 1650. They flourished along the north shore of Lake Erie, across the Niagara Peninsula, and around the western end of Lake Ontario.[30] A French missionary who spent three months with them in 1626 believed they had twenty-eight substantial townsites as well as smaller villages; eight years later, another missionary estimated their population at over thirty thousand.[31] It was likely their word – *onguiaahra* or *niagagarega* – that gave the river and the peninsula its name.[32] The Neutral met a tragic end, however. Through the 1640s, their communities were devastated by a series of crop failures and waves of European smallpox before they were totally eradicated by Seneca and Mohawk warriors in 1651.[33] Subsequently, both the Anishinaabe and Haudenosaunee peoples occupied the Niagara Peninsula, particularly the Chippewa (Anishinaabe) and Mississauga, but they did not establish permanent settlements.

The vanguard of European colonization – explorers, traders, and missionaries – bypassed the peninsula while travelling further north or west. The Niagara River, however, remained a strategic transportation route. The French established a post at the river's mouth on its eastern bank in 1678. This post became Fort Niagara. British forces took possession of the fort in 1759 and then took control of the entire river, in two ways: they built block houses on the portage routes around the falls, and they constructed Fort Erie (1764) on the western bank of the riverhead at Lake Erie.[34] During the American War of Independence, the Niagara corridor was a bulwark of the Loyalist militia. After the war, it provided safe passage to British territory for allies of the British cause: the Loyalists themselves and six Haudenosaunee nations (Cayuga, Mohawk, Oneida, Onondaga, Seneca, and Tuscarora). The British government facilitated this migration by purchasing title to the peninsula from the Anishinaabe. The first non-Indigenous settlers in the Niagara Peninsula included the fort's Loyalist garrison, Butler's Rangers. Their first villages became Virgil and Niagara-on-the-Lake.[35]

The Haudenosaunee were also granted lands in the peninsula. The American government had dispossessed the Haudenosaunee of their lands south of Lake Ontario due to their alliance with Britain. In compensation, Britain provided them with a substantial tract defined by the Grand River west of Lake Ontario. This tract was also purchased from the Mississauga by the British government.[36] Predictably, white settlers began to encroach on the tract by the 1810s. The original grant was severely diminished in size and its borders remain contested to the present day.[37] These reserve lands remain home to the Six Nations of the Grand River and to the Mississaugas of the Credit First Nation. Although the mouth of the Grand River sits in the peninsula, the reserve is sixty kilometres west of the Niagara region's modern political boundary.[38]

We cannot know if the Loyalist settlers, finding clearings in the forest, understood they had found the abandoned sites of Neutral villages.[39] In establishing a colonial order, however, the Loyalists and later generations of migrants who settled in Niagara would, both wittingly and unwittingly, erase much of its remaining Indigenous imprint. The Loyalists were largely British, German, Dutch, and Black.[40] The new settlers developed their own farms by removing the forests and building their own villages along the region's many creeks. Later migrants included Irish refugees fleeing the Potato Famine and Freedom Seekers escaping slavery by following the Underground Railroad. Traces of the Indigenous era remain, though, in names such as Chippewa, Mississauga, and Niagara itself, while today's major highways still follow the trade routes carved by the Neutrals and their predecessors.[41]

Throughout the nineteenth century, Niagara remained a strategic location during national and international conflicts due to its situation on the border with the United States. During the War of 1812, for example, British, Haudenosaunee, and Canadian troops resisted American invaders over three campaigns fought across the region. After the failed Rebellion of 1837, William Lyon Mackenzie's rebels fled the Upper

Canadian militia by travelling through Niagara to New York. Planning another offensive the following year, the rebels sought support among Niagara's farmers. Instead, they were beaten back by the local militia.[42] Irish patriots also made violent skirmishes across the border. One saboteur destroyed the original Brock's Monument in 1840 and was later accused of an attempt on the Welland Canal.[43] More boldly, the Fenian Brotherhood sought to advance Irish independence by invading Upper Canada from Buffalo, New York. They landed near Fort Erie in 1866 but soon retreated after engaging a determined militia.[44] On at least two other occasions, saboteurs linked with either the Irish independence struggle or the German military sought to destroy locks on the Welland Canal.[45]

Niagara's economic importance grew thanks to a strong agricultural base.[46] The area north of the Niagara Escarpment provides some of the finest agricultural land in Ontario and is still home to orchards, vineyards, and wineries. The area south of the escarpment produces vegetables, grains, and poultry. These strong bases benefited entrepreneurs like William H. Merritt. He promoted the first Welland Canal in the 1820s to link the upper Great Lakes to Lake Ontario and the St. Lawrence River. The canal provided an all-Canadian corridor that ensured a measure of economic independence for a colony seeking to compete with the United States. When hydroelectric power generators were completed at Niagara Falls in 1892 and 1905 (the second under the direction of Nikola Tesla), the region experienced industrial growth to rival Niagara's agricultural production. Among the many manufacturers to locate in Niagara was automotive giant General Motors, which built three plants in St. Catharines. Investments in manufacturing also benefited other canal towns, particularly Welland and Port Colborne.[47]

Niagara changed considerably during the course of our research. Agriculture maintained its strong presence, but the same could not be said for manufacturing. It had declined precipitously after 1990 and faltered again with the global financial crisis of 2008. St. Catharines, Welland, and Port Colborne were hit hard by plant closures and job losses. In St. Catharines, General Motors closed two of its three plants. The second closure occurred in 2010, and the defunct factory's twenty-two hectares of land – a twenty-minute walk from city hall – remained strewn with rubble ten years later. In like manner, Welland lost two of its major employers when John Deere and the Steel Company of Canada both closed factories. These closures rippled through the economy as firms supplying related goods and services downsized or faced closures of their own. Urban managers at the regional and municipal levels responded by adopting the "creative economy" mantra in an effort to reinvigorate economic growth through businesses focused on knowledge production, cultural products, and enhanced lifestyle options. Tourism, especially that anchored by the international appeal of Niagara Falls and a burgeoning wine industry, became a key economic driver.[48] In addition, the promised extension of a commuter rail service centred in Toronto contributed to a residential construction boom as families across the Greater Toronto Area and later Hamilton sought affordable housing. These transitions did little to help Niagara's working-class population. In the late 2010s, demographic data indicated that Niagara had among the highest unemployment rates in Ontario and an affordable housing crisis of its own.[49]

Niagara's Mnemonic Products

Despite recent hardship, residents and governments enthusiastically fete Niagara's successes. Hundreds of mnemonic products rendered in stone, bronze, or steel dot its landscapes. They mark moments of national significance, but they also mark moments that are noteworthy only to select constituencies. In either case, their sponsors attempt to reify specific values and foster a common sociocultural identity.[50] But do they work? Do they really foster identification with their form or content? To answer these questions, we explore how Niagara residents engaged the narratives marked by local plaques and monuments.

Eviatar Zerubavel's concept of the "commemogram" supplies a useful analytical tool for this research.[51] He contends that every community attaches varying degrees of significance to different periods of its past. For example, a community can mark any given period as a golden age, a catastrophic decline, or as simply unremarkable. A commemogram measures the relative significance granted to each period in a community's past by cataloguing that community's commemorations by the dates of the events they mark. Periods that are marked by a greater number of commemorations should, presumably, have greater weight in a community's narrative. When each commemoration is plotted on a graph by its relevant date, the result reveals a topography-like distribution of mnemonic significance. As Zerubavel puts it,

> History thus takes the shape of a relief map, on the mnemonic hills and dales of which memorable and forgettable events from the past are respectively featured. Its general shape is thus formed by a handful of historically "eventful" mountains interspersed among wide, seemingly empty valleys in which nothing of historical significance seems to have happened.[52]

Recognizing a pattern of commemoration among memory entrepreneurs, however, is only a first step in exploring the mnemonic priorities of a community. The next step is to explore how residents engage and then invest in local mnemonic products. To do so, a second commemogram must be plotted. If individuals are asked to identify the people, events, places, or things that represent their community, their answers should reveal the significance they grant to various mnemonic narratives in their community's history. Finally, comparison of the two commemograms should reveal the extent to which memory entrepreneurs and residents share the same mnemonic priorities.

To create a commemogram plotting the work of local memory entrepreneurs, we focused on plaques. Plaques are one of the most salient aspects of memory work because they are publicly visible media through which specific mnemonic narratives are articulated. Further, this observation holds regardless of the individuals or agencies that sponsor them or the artefacts to which they are secured (such as stones, monuments, or buildings). An inventory of a community's plaques therefore can provide a reliable data set for a commemogram.

There were two inventories of plaques in Niagara available when we conducted our research. The first inventory was a database produced by local genealogist and plaque enthusiast Wayne Cook.[53] This online resource was truly comprehensive; it did not focus on any single topic, place, or sponsor, and it provided complete information for every plaque (including the names of sponsors and the dates of placement). Cook's work was aided by volunteers who submitted additional information, occasional corrections, and news of new plaques. The second inventory was a clippings file produced by the Special Collections Unit within St. Catharines Public Library. Library staff had systematically combed through local periodicals for years seeking information on historical narratives relevant to the region. Their mandate was as comprehensive as Cook's had been. Hence, our own plaque inventory was developed by combining the two existing data sets and then adding details through field work when necessary. We are confident that the final inventory provides an accurate representation of how memory entrepreneurs had marked each period of Niagara's past.

Each plaque was categorized in three ways. First, we identified the historic date associated with the person, place, event, or thing commemorated by the plaque; we noted only the decade. If a plaque commemorated multiple dates, we selected one specific year in the spirit of its narrative (e.g., we dated all First World War cenotaphs to the 1910s even if they marked subsequent wars). If a plaque commemorated a subject outside of human history, such as the geological formation of the Niagara Escarpment, we did not include it in our analysis. Second, we identified the sponsor of each plaque. This screen allowed us to seek connections between specific types of sponsors and specific narratives. Last, we identified the location of each plaque. This data allowed us to produce a map of Niagara that revealed its memoryscape (see map 1.1).[54]

Our final inventory identified 261 plaques. This number includes those sponsored by private citizens and all levels of government. We were struck by the commemogram that emerged (figure 1.1). Three major peaks were produced by narratives of white settlement (a period spanning all decades prior to 1810, but particularly 1790 to 1810), the War of 1812 (1810s), and the First World War (1910s). Smaller peaks were produced by narratives of the original Welland Canal (1830s), the Underground Railroad (1840–60), and the Second World War (1940s). Other decades were marked by very few commemorations. Notably, few markers highlighted Indigenous society prior to European exploration, and fewer still marked anything at all after 1960.

Local organizations were the most prolific sponsors of local plaques, accounting for nearly half of them (figure 1.1). This result should not

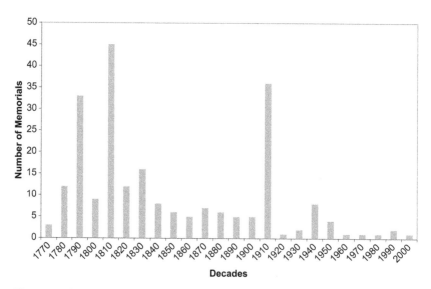

Figure 1.1. Commemogram: plaques in the Niagara region, by decade of event commemorated

be surprising. Residents are more likely than outsiders to grant significance to local narratives. Highlighting this point, local memory entrepreneurs had come together in more than thirty different associations, most of which were funded through voluntary subscription. The narratives they marked represent a diverse range of interests with little overlap. Locals were inclined to honour the everyday (such as notable buildings), the odd (notable feats), and the simply old (notable relics). Thus, we found groups responsible for commemorating the local use of a nineteenth-century tool (the Johansson Bar)[55] and the career of a late nineteenth-century civic politician (Mayor William B. Burgoyne).[56] By contrast, government efforts demonstrated much more concentrated memory work (figure 1.2). White settlement and the War of 1812 had clearly been the focus of both federal and provincial sponsorship.

There was little evidence of dissonance among memory entrepreneurs in Niagara. Plaques did not present opposing interpretations of local narratives. Rather, when local, provincial, and federal narratives overlapped, they tended to reinforce one another. This was true, for example, at Salem Chapel with the marking of the Underground Railroad and Harriet Tubman. Similarly, local, provincial, and federal agencies had all sponsored the commemoration of narratives surrounding white settlement and the War of 1812. There was only one noticeable difference between them: the Governments of Canada and Ontario had commemorated very few local narratives dated after

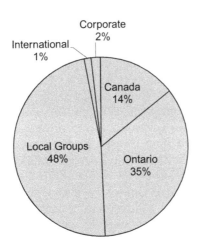

Figure 1.2. Plaques in the Niagara region, by memory entrepreneur

1890, while residents had done so. The two world wars represent a special case. Federal policy dictated that local memorials such as cenotaphs, arches, and buildings would be the responsibility of local organizations and individuals.[57] As a result, the federal government directly financed only two material commemorations of these conflicts in Canada, and both are in Ottawa. It should be noted that the quantity of plaques sponsored by any one agency is not always a clear indication of its mnemonic investment in a particular narrative. The federal government had marked fewer sites in Niagara than other memory entrepreneurs, but its ongoing commitment to three historic sites indicated the significance it leant to the 1812 narrative. Fort George, Fort Erie, and Brock's Monument at Queenston Heights represent massive investments. The ongoing financial support required to maintain them was unrivalled by any other memory entrepreneur producing material commemorations in Niagara. We will return to this theme below.

The plaques in our inventory were dispersed in geographically predictable ways. Among Niagara's many communities, St. Catharines had the greatest share of plaques (map 1.1). The greatest concentration of plaques, however, was along the Niagara River. Most of those associated with the War of 1812 were found along this corridor. These observations reinforce points we have already made. Each Niagara community has its own memory entrepreneurs who had sponsored plaques for worthy local narratives, and the largest community had the most plaques. The Niagara River, however, had also found favour among provincial and federal agencies because the riverside communities were central in narratives describing the creation and defence of Ontario and Canada (figure 1.3).

Resident Engagement with Mnemonic Products

To understand Niagara residents' engagement with local mnemonic narratives, two sets of research were informative. First, as noted in the introduction, we took inspiration from prior surveys of the public conducted in the United States, Australia, and Canada.[58] However, our focus was local rather than national. A second set of research addressed this focus by examining audience responses to mnemonic work. For example, Paul Litt and Alan Gordon both used on-site observation to consider how tourists view and respond to reconstructed historic sites in Ontario.[59] Their insights are valuable, and yet we wanted to know how residents responded to such work. Tourists go home to continue

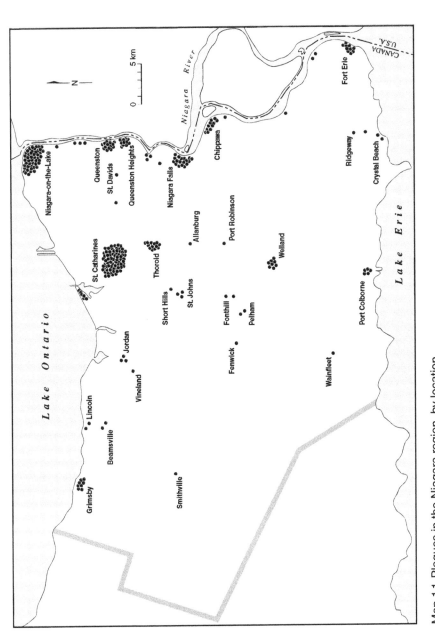

Map 1.1. Plaques in the Niagara region, by location

Cartography: Loris Gasparotto.

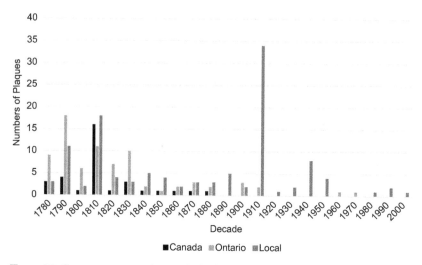

Figure 1.3. Commemogram: plaques in the Niagara region, by memory entrepreneur and decade

their lives in other contexts. Residents, however, live in social, cultural, economic, and political environments that they share with mnemonic products. If individuals habitually experience them in their daily lives, they may add the narratives of mnemonic products to their personal stock of identity markers. Indeed, some scholars have extended this point: if a community is noteworthy for its historical narrative, then contemporary life in that community can become overdetermined by historical interpretations of local identity rather than present-day interpretations.[60]

We sought to test this argument by consulting Niagara residents directly. Data was collected through an intercept survey of pedestrians conducted in centres across the Niagara region. This type of survey is highly useful for identifying thoughts that reside top of mind. If residents absorb the narratives marked by local plaques into their own conceptions of local identity, then we might expect them to mention these narratives when questioned about local identity. We consciously avoided words like "heritage," "history," or "the past" when introducing ourselves and our purpose.[61] Only after participants answered our initial questions did we use these terms to probe their answers. The questionnaire employed a mix of closed and open-ended questions. The closed questions, using Lykert scales, produced a "register of engagement" with mnemonic products and narratives.[62] The open-ended questions generated responses in the participants' own words. Taken together, both sets of questions formed a nuanced picture of participants' thoughts and emotions regarding mnemonic narratives and local identity.

The lead researchers and a team of research assistants (comprised of female and male graduate students) staked out public sidewalks and courtyards. These locations were characterized by a mix of institutional and commercial sites: municipal buildings, public libraries, busy commercial streets, farmers' markets, and public parks. We conducted interviews during months when the weather would be sufficiently warm to foster participation but also ensured that most passers-by were locals rather than tourists. Each day in the field was selected using a rolling week technique to capture the various rhythms of the city, such as market days, court days, or holidays. However, we used substitute days whenever inclement weather occurred on our scheduled day. The time of day was selected to provide the most demographically diverse set of pedestrians: 10:00 a.m. to 2:00 p.m. brought us into contact with individuals working downtown and residents from across the city running errands. Each interview took between ten and fifteen minutes to complete.

All members of the team wore name tags identifying them as employees of the local university. Passersby were asked if they would like to participate in a survey on local culture and identity. We had two screens when recruiting participants. Both were deployed with our demographic questions at the start of the interview. Our primary goal was to understand how residents respond to local mnemonic products, so we only completed the interview with those who were born, or currently lived, in Niagara. We were also most concerned with the casual retention of mnemonic narratives. We did not, therefore, include anyone under the age of eighteen who might still be in school. Local schools feature elements of local history and students are rewarded for their ability to retain such information. At the end of each day, we compared participants' aggregated demographic data against a Statistics Canada profile of the region. As our field work neared its end, we tried to ensure that statistically under-represented groups were invited to participate. We were frankly overwhelmed by the willingness of residents to participate and the generosity of their answers.

We embraced a constructivist approach while drafting our questionnaires.[63] Though a widely varied field of scholarship, constructivism insists that knowledge is socially generated in specific contexts. In this case, our original questionnaire was informed by the sum of our own reading, our own experiences, and our own understandings of the

Niagara region. As some scholars conclude, this foreknowledge may literally create the results.[64] Thus, we understood that our survey may not return "facts" that accurately reflected residents' thoughts as much as a set of individual responses contingent upon the context of our survey (i.e., the time, place, word choice, or researcher performance).[65] To ameliorate this concern we adopted a flexible, open-ended research strategy consisting of a loop of knowledge production:

1. We analysed the results of the survey. We also used participants' responses to interrogate our questionnaire and our approach in terms of the concerns outlined above.[66] Our analysis produced a set of emerging ideas that informed the next survey's questions.[67]
2. We revised the questionnaire and our approach as needed to address our emerging concerns and ideas.
3. We used the new questionnaire to conduct the next survey.

After the initial survey in 2005, we went through this loop four times to draft revised questionnaires for surveys conducted in 2008, 2009, 2012, and 2016. While this process addressed our concerns regarding knowledge construction, it also raised caveats regarding our use of the survey results. The process generated data appropriate for each iteration of the survey. However, the iterative approach does not permit direct comparisons between the five surveys. Differences in the contexts and in the questions of each survey iteration render a comprehensive longitudinal analysis impossible.

2005 Survey

Our first survey was taken in two locations: downtown St. Catharines and Port Dalhousie, a neighbourhood of St. Catharines with a well-defined commercial village. We completed surveys with 162 individuals. All participants were born or resident in the Niagara region and at least eighteen years of age. The majority were residents of the area (89 per cent). Only 39 per cent of the sample was born in Niagara, although another 33 per cent of the sample was born in southern Ontario. This suggests that most participants should have been familiar with provincial school curricula and national media narratives. The majority, 66 per cent, were between twenty-five and sixty years of age. In terms of gender identification, there was an equal number of males and females; no one identified in an alternative category. Participants were asked to self-identify their ethnic background, as they would for a census taker. Those identifying as unhyphenated "Canadians" represented 33 per cent of the sample and "Aboriginal" participants represented another 3 per cent. The majority, or 59 per cent, were of European descent, and the remaining 5 per cent represented participants of African, Asian, and Latin American descent, as well as participants who elected not to self-identify. Statistically, this sample size has a confidence level of 95 per cent with a margin of error of +/– 6.2 per cent.[68]

We assessed participants' engagement with local narratives by asking them to name important contributions to local identity. We offered no prompts and accepted every answer offered by each participant. In total, participants offered 449 different responses. All responses were plotted to a chart using the same categories and rules used for the plaque commemogram. If memorials and their narratives promote specific values and foster common social identifications, then participants should have identified the major mnemonic narratives carried by local plaques. A comparison of the two sets of data reveals only faint parallels but startling differences (figure 1.4). The most astonishing result was that a remarkable 87.8 per cent of responses (n = 394) identified contemporary people, places, and things (i.e., those current in the period 2000–5) as the most important markers of civic identity.

The commemogram in figure 1.4 illustrates that most residents identified with contemporary features of the region rather than its past. Common responses named Niagara Falls, the wine industry, General Motors, tender fruit orchards, and the Shaw Festival (a summer theatre festival dedicated to the work of George Bernard Shaw). Four annual festivals staged in St. Catharines also ranked well ahead of any historical

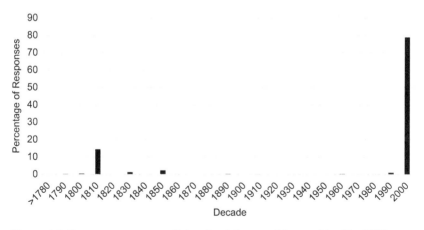

Figure 1.4. Commemogram: participant opinions on Niagara identity, 2005

event in respondents' imaginations: Grape and Wine (which celebrates the grape harvest), Rib-Fest (an outdoor cookout sponsored by a local service club), the Folk Arts Festival (a celebration of multiculturalism), and the Henley Regatta (a national amateur rowing competition). Other responses included Niagara's opportunities for recreation, dining and shopping, and a frank recognition of its shortcomings. Even when identifying things with a notable past, participants focused on their contemporary significance. When respondents spoke of the Welland Canal, for example, they noted its present-day role in Great Lakes shipping rather than its historic impact on the development of Canadian trade and industry. Engaging with memorials and their mnemonic narratives was not a priority. One participant gestured to a plaque on a downtown war memorial, stating, "I haven't read it, but I will someday." Another participant was less evasive when commenting on local heritage: "It's not very significant. People talk about the canals, boats, and festivals and not that stuff." Yet another offered a blunt rationale for such thinking: "It's not significant – maybe only to others. The living are more important than the dead." This comment did not represent all participants, but a clear majority focused on the lived reality of Niagara rather than a past marked by memory entrepreneurs.

2008 Survey

The results of the first survey prompted a revision to our questionnaire. In 2005, we asked participants to name important contributions to local identity. In 2008, we asked a similar question, but in two parts. First, we asked participants whether Niagara had made important contributions to national identity. Second, we asked participants who responded positively to name specific examples of these contributions.

Using the same recruitment strategy we deployed in 2005, we completed interviews with 183 individuals. As in our first survey, all participants were filtered to ensure they were born or resident in the Niagara region and at least eighteen years of age. A clear majority of the participants were residents (93 per cent), but a minority were born in Niagara (31 per cent). All age ranges were well represented, and the average age was forty-six. With respect to gender identification, the sample was 51 per cent female and 49 per cent male; no one identified in an alternative category. In terms of ethnicity, unhyphenated "Canadians" represented 28 per cent of the sample while "Aboriginals" represented another 3 per cent. Most participants claimed European descent (57 per cent), while all those of African, Asian, and Latin American ancestry represented 11 per cent altogether. Statistically, this sample size has a confidence level of 95 per cent with a margin of error of +/– 7.2 per cent.

Answers to the first part of our opening question were positive. A convincing majority of participants (86.8 per cent, $n = 159$) agreed that Niagara contributes something to the national identity. Many participants hesitated, however, before answering the second part. We did not offer prompts and we accepted every answer given. If participants requested a prompt, we asked them to consider "people, places, things, or events." Even with this guidance, some of those who answered positively were either reluctant or unable to name examples. When asked the opening question, one participant replied, "Certainly, but let's not do examples." Another stated, "I would hope so! But I can't think of anything specific."

Ultimately, 151 participants offered 379 examples of Niagara's significant contributions to national identity. Their examples suggest the revised opening question produced a clear effect: when the question was framed in a national context, participants in 2008 were much more likely than those in 2005 to think in historical rather than contemporary terms. This was true even among participants who offered no specific examples; they intuitively aligned national significance with the past. One commented, "Strangely enough, I cannot think of anything … History is harder to come to grips with than the present." Yet another explained an awkward silence by stating, "I'm not a history buff. I don't know."

When specific examples were named, 44.6 per cent of them ($n = 169$) were clearly aligned with "official" mnemonic narratives. Curiously, however, participants rarely linked examples to specific narratives. Many participants named Niagara's historic forts without mentioning the War of 1812 or named historical figures without mentioning their deeds. Other participants contextualized their examples by referencing the source of their knowledge rather than their associated narratives. For example, one noted, "I remember reading about when Brock [Monument] was put up. I also read about [William Hamilton] Merritt and Adam … [Beck]. I read about that on a plaque in the park." Still others were simply unsure of their history and geography, setting the Battle of the Plains of Abraham in 1812, declaring that St. Catharines was a former First Nation reserve, and claiming Pierre Elliott Trudeau and Terry Fox as residents of Niagara. Perhaps there was a common link among these responses. Participants may have been aware that certain people and places have been commemorated and therefore concluded that they must be significant. The reasons for their commemoration, and much of the details, however, remained unclear. We come back to this idea in chapter 3.

We plotted participants' examples on a new timeline and compared the result to our plaque commemogram. There were modest parallels (see figure 1.5). Examples that referenced the past echoed those narratives frequently marked by plaques: the War of 1812 (n = 110), the building of the Welland Canal (n = 40), the Underground Railroad (n = 20), and European settlement (n = 14). It is noteworthy that participants referenced the Underground Railroad. Local memory entrepreneurs at the time were coordinating regional commemorations of railroad sites. These sites were publicized with a common brand name, "The Freedom Trail," and covered by local media.

Despite these references to the past, a slim majority of examples (55.4 per cent, n = 210) referenced contemporary people, places, things, and events. In this way, the 2008 commemogram (figure 1.5) echoed that of 2005. The most common examples referenced high-profile local industries. Tourism led the way. Niagara Falls was the most frequently named contribution to national identity (n = 79), followed by the region's wineries (n = 54), local festivals (n = 15), and Niagara-on-the-Lake, a restored colonial townscape and home to the Shaw Festival (n = 12). Heavy industry was also referenced by those who named the Welland Canal, the hydroelectric plants at Niagara Falls, and General Motors (n = 19). Responses naming the canal were again instructive. As in 2005, participants remarked on its contemporary rather than historical importance, combining its potential as a tourist draw with its significance to the Canadian economy.

In sum, the opening question of the 2005 survey asked participants to name important contributions to local identity. The 2008 survey asked participants to name Niagara's contributions to national identity.

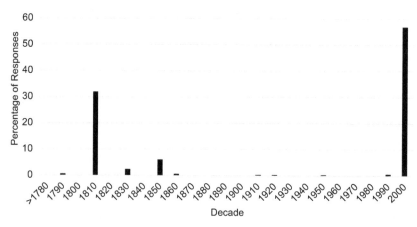

Figure 1.5. Commemogram: participant opinions on Niagara's contribution to Canadian identity, 2008

A majority of participants in both surveys drew their responses from contemporary life. As noted, however, more 2008 participants were inclined to reach for historical identifiers. There are two possible explanations for this difference. First, the 2008 participants may have unconsciously linked national significance to moments of historical import. Second, many participants who named examples imagined how other Canadians or foreign tourists might perceive "Niagara." One participant stated, "Tourism; there's the falls and Niagara-on-the-Lake … Historical tourism." Another said, "Well, Niagara Falls are a wonder of the world. I'm not sure if people come to the area for historical places." Similarly, participants referencing Niagara-on-the-Lake did not mention its War of 1812 sites, but instead mentioned its refurbished colonial-era streetscape. As one person told us, "There is the Grape and Wine [festival] and icewine, and Niagara-on-the-Lake has quaint buildings."[69]

Hence, the revised opening question prompted participants to see Niagara as others might see it. For many, these "others" were tourists drawn by unique narratives and experiences. Be that as it may, most examples offered by participants drew upon contemporary life rather than the past. One individual offered a particularly comprehensive reply: "Aren't we famous for being the doughnut capital of Canada and for fat people? There's also the auto industry, the War of 1812, the falls, and the wine industry."[70] This response suggests that local mnemonic narratives may be significant, but they are no more significant than contemporary representations of Niagara. Another participant rejected even that concession; she believed "Laura Secord and people like that" are historically significant, but "people don't identify with them nationally. There's the Grape and Wine, it attracts a lot of people … And Niagara Falls. You say you're from Niagara Falls and everyone knows where that is."

2009 and 2012 Surveys

Intrigued by our first two surveys, we sought to interrogate one specific set of responses that were common in 2005 and 2008. Participants who named contemporary identifiers for Niagara frequently named the grape and wine industry; participants named only the falls more often. The 2009 survey opened with a new question that forced participants to prioritize local identifiers: What is the first thing that comes to mind when you hear the word "Niagara"? We offered no prompts to guide their answers. We sought only their genuine, first impressions. If participants requested clarification, we again suggested they think of a person, place, event, or thing. Follow-up questions asked participants their views of the wine industry and its links to local mnemonic traditions.

In 2009, we followed the same recruitment strategy we had used in the prior surveys with one change: we conducted the work in downtown St. Catharines but not in Port Dalhousie. In total, 204 participants completed the survey. Once again, all participants were filtered to ensure they were born or resident in the region and at least eighteen years of age. Almost all participants were residents (98 per cent), and a significant majority were also born in Niagara (66 per cent). All age ranges were well represented, and the average age was 45 years old. With respect to gender identification, the sample was equally split, female and male; no one identified in an alternative category. With respect to ethnicity, unhyphenated "Canadians" represented 38 per cent of the sample while "Aboriginals" represented an additional 3 per cent. A majority of participants claimed European ancestry (52 per cent) while all those of African, Asian, and Latin American descent represented an additional 7 per cent. This sample size has a confidence level of 95 per cent with a margin of error of +/– 6.9 per cent.

The responses given to our opening question were strikingly similar to those of the previous two surveys. Participants gave answers that were closely tied to the present. There were, however, two notable differences. First, participants in 2009 did not reach for national mnemonic narratives. Indeed, only three participants (1.4 per cent) referenced *any* mnemonic narratives. This reinforced our belief that introducing the national context in 2008 had affected participants' responses. Second, when participants drew upon the present for that "one thing" that identifies "Niagara," the grape and wine industry was named most often; yes, more often than Niagara Falls (see table 1.1).

Table 1.1. If you had to choose one thing that identifies Niagara, what would it be?

Responses (N = 219)	Percentage that named St. Catharines
Wine industry and vineyards	29.7
Niagara Falls	19.6
Landscape and climate	13.2
Arts and culture	9.6
Agriculture (not grape and wine)	9.6
Negative social aspects	5.0
Tourism	5.0
Positive social aspects	4.6
Historical aspects	1.4
Welland Canal	1.4
Home and family	0.9

Note: St. Catharines was surveyed in 2009.

The grape and wine industry's emergence as a regional identifier was consistent through all three surveys. Having anticipated this result in 2009, a follow-up question asked participants if the grape and wine industry was part of Niagara's heritage. Participants firmly agreed, as 82.6 per cent said yes.

The 2009 results prompted a new question: Did the responses collected in St. Catharines accurately reflect the thoughts of residents across Niagara? In 2012, we deployed the same questionnaire and protocols in four other centres: Grimsby, Virgil, Welland, and Port Colborne. We wondered if proximity to grape-growing areas and wineries would influence participants' responses. We selected Grimsby and Virgil to represent residents proximate to the industry; both towns sit among multiple vineyards and wineries. Welland and Port Colborne, which sit outside the viticultural zone, were selected to represent residents removed from the industry.

We followed the recruitment strategy used in all prior surveys with one change: all participants were filtered to ensure that they were residents of the specific centre where the interviews took place. In all, 133 participants completed the survey. All age ranges were represented in each centre, and the average age across all four centres was 48 years old. The higher average age for this set of participants reflects regional demographics. The smaller towns and rural areas skew older than the larger urban centres. With respect to gender identification, the sample was 54 per cent female, 44 per cent male, and 2 per cent who chose not to self-identify. With respect to ethnic identification, unhyphenated "Canadians" made up 23 per cent of the sample and "Aboriginals" represented an additional 2 per cent. The majority of the sample was of European descent (71 per cent) while all those of Asian and Latin American ancestry, as well as those who chose not to self-identify, represented an additional 4 per cent. The sample size has a confidence level of 95 per cent with a margin of error of +/– 8.5 per cent.

It quickly became apparent that residents of these four centres also draw upon the present rather than the past for identification. The grape and wine industry was frequently named, and proximity did affect the results (see table 1.2). Participants in Grimsby and Virgil were far more likely to name the wine industry as a top-of-mind identifier before the falls, although participants in Virgil named all modes of agriculture before wine itself. Participants in Welland and Port Colborne were far more likely to name Niagara Falls. These results suggest a clear divide separating residents of these two areas. However, the results are complicated by the question regarding the grape and wine industry's place

Table 1.2. If you had to choose one thing that identifies Niagara, what would it be?

Responses	Percentage of responses			
	Grimsby (n = 37)	Virgil (n = 52)	Welland (n = 39)	Port Colborne (n = 43)
Niagara Falls	24.3	3.8	28.2	48.7
Wine and vineyards	43.2	26.9	15.4	7.0
Agriculture (not grapes)	13.6	53.8	12.8	7.0
Landscape and climate	10.8	11.7	10.3	7.0
Welland Canal	–	–	5.1	14.0
Historical aspects	2.7	–	5.1	2.3
Other	5.4	3.8	23.1	14.0

Note: Grimsby, Virgil, Welland, and Port Colborne were surveyed in 2012.

Table 1.3. Is wine part of Niagara's local heritage?

Responses	Percentage of responses			
	Grimsby (n = 37)	Virgil (n = 52)	Welland (n = 39)	Port Colborne (n = 43)
Yes	100.0	75.0	82.1	83.7
No	–	23.1	12.8	14.0
No answer	–	1.9	5.1	2.3

among local mnemonic narratives. A clear majority in all four centres believed that wine was part of Niagara's heritage (see table 1.3). We discuss these results in further detail in chapter 6.

The Bicentennial of the War of 1812: A Case Study in Mnemonic Work

The bicentennial of the War of 1812 provided memory entrepreneurs with a rare opportunity to take centre stage in the public sphere. It also provided us, as researchers, an opportunity to explore residents' responses to such memory work at a local level. Though the War of 1812 extended far beyond the Niagara frontier, it occupies a significant place in the work of local memory entrepreneurs. The war's heroes, sites, and stories are well represented among the plaques in our inventory (see figure 1.1 and map 1.1). The bicentennial celebration was also part of a new national identity strategy. In 1814, colonial elites celebrated the war's end and declared a victory for Britain; the result was cast as proof that loyalty to Britain had its benefits.[71] In 2012, the federal government recast the war as a pivotal moment on the road to Canadian nation-hood. We sought to test the local impact of the anniversary events.

Our surveys to this point indicated that Niagara residents did not look to the past when asked about local identifiers. Contemporary report-ing suggested that this perspective was common across the country. In 2009, the Association for Canadian Studies commissioned a poll in anticipation of the bicentennial. It found that 39 per cent of Canadians did not know enough about the war to identify a winner. Three years later, the Department of National Defence commissioned its own poll and focus group research to investigate public knowledge of the war, with similar results.[72] CBC News related the following details:

> Few people who took part in the survey, and in related focus groups, were aware of the anniversary "and even fewer could identify the War of 1812 by name" ...
>
> The report revealed that even in areas where the war was fought – specifically Ontario's Niagara Region – awareness remained foggy. "In St. Catharines, which is geographically close to where some significant events of the War of 1812 took place, awareness of the 200th anniversary was higher, but still uneven and not widespread," said the document, released Tuesday by the federal government.[73]

The government's efforts would not reap easy rewards.

The federal government was led at the time by Prime Minister Ste-phen Harper, whose Conservative Party held a parliamentary majority. During his ministry, the government sought to reinterpret Canadian history in ways that privileged his party's contemporary priorities, con-servative values, and mnemonic trajectories. This was particularly true of the military. The government, for example, sought to rebrand Canada as a nation of steadfast, combat-ready warriors rather than peacekeep-ing internationalists. The bicentennial of the War of 1812 became a key feature of this project.[74] As Claire Sjolandaer argues,

> The 1812 narrative ... clearly seeks to project Canadian authority interna-tionally by casting Canada as a prudent but potentially belligerent coun-try ... Far from the helpful fixer posture associated with peacekeeping, Canada would not seek to mediate a resolution to conflicts, but would rather choose sides wisely and be prepared to fight.[75]

In this narrative frame, the War of 1812 was a crucial event on the path to Confederation in 1867. Indeed, the war became the "Fight for Canada" rather than a defence of a British colony.[76] The Conservative government funded this project handsomely. The federal budget for 2010 committed $28 million to the bicentennial effort, spread over three years.[77] This money supported several relevant historic sites and a new monument located on Parliament Hill. The Canadian Mint issued anniversary coins, Canada Post issued stamps, and museums offered timely exhibits.[78] Outside Ottawa, the government had the support of ideologically aligned media outlets. This was especially true in Niagara. Postmedia, a corporation known for its conservative perspective and editorial support of the Conservative Party, owned all the local daily newspapers. These papers featured positive coverage of the War of 1812 and its commemoration throughout the bicentennial years.[79]

The federal government had clear objectives for the bicentennial celebrations. A Heritage Canada report identifies the following goals:

Immediate Outcomes

Opportunities are created for Canadians to participate in commemoration activities and events.

Commemorations have a pan-Canadian reach.

Intermediate Outcomes

Canadians participate in commemoration activities and events.

Final Outcomes

Canadians have an increased awareness of their history as it relates to the War of 1812.[80]

Ideally, the government's bicentennial efforts would not simply support its reinterpretation of Canadian history. They would also revive public interest in the Canadian military and reanimate associated national historic sites by placing them at the centre of a heroic national narrative. As such, the bicentennial was designed to serve the purposes of the government, local memory entrepreneurs, and local tourism operators.

Local memory entrepreneurs did their best to capitalize on the anniversary and its accompanying funding opportunities. The War of 1812 was already well marked in Niagara. Various organizations nonetheless decided the existing stock of commemorations was insufficient. Thus, various sites were refurbished, new monuments were planned, and elements with no prior link to the war were renamed "to honour" major figures. For example, a new monument was unveiled in 2013 commemorating the Battle of Beaver Dams, which was fought at what is now the town of Thorold. It was the first monument ever raised at the former battlefield.[81] A year later, the City of Niagara Falls erected a steel arch over Lundy's Lane to commemorate the eponymous battle.

The battlefield is now a high-traffic commercial strip.[82] Planning also began for two projects – at Queenston Heights and DeCew House – that were incomplete when the anniversary ended in 2014. Both commemorated the participation of Haudenosaunee warriors in the defence of Canada.[83] DeCew House, now essentially a ruin, was also made more presentable. Provincial Highway 405, which serves a border crossing into the United States at Queenston, was renamed for Sir Isaac Brock.[84] And similarly, a parkway leading to the main campus of Brock University in St. Catharines was renamed Sir Isaac Brock Way.[85]

There were events too. Over the anniversary's full three years, the major historic sites hosted the usual educational fare such as walking tours, public lectures, research conferences, and workshops. These were augmented by themed anniversary events such as battlefield re-enactments and military-style balls. One event has since become an annual occasion: the re-enactment of Laura Secord's trek from the town of Queenston to DeCew House (in present-day St. Catharines) to advise British lieutenant James FitzGibbon on American troop movements. A more macabre commemoration marked the burning of Niagara-on-the-Lake by American forces.[86] A final event in 2015 concluded the anniversary with a public procession and a church service in Niagara-on-the-Lake that marked the two-hundredth anniversary of the Treaty of Ghent, which formally ended the war.[87]

The memory entrepreneurs behind these events had local media support. Furthermore, our surveys indicated that local radio stations and newspapers retained their influence among Niagara residents throughout the period of our research up to 2020. Niagara's one talk radio station covered local public affairs, but its program content was difficult to document during the anniversary. Hence, to assess the accessibility and tone of local media support for the bicentennial, we examined the region's newspapers.[88]

Local newspapers targeted residents through their print and online editions. During the period under review, the measurement of Canadian print media audiences was streamlined when two non-profit research organizations – the Newspaper Audience Databank (NADbank) and the Print Measurement Bureau – merged to form Vividata. In 2013, NADbank reported that the three Niagara dailies had a combined daily average circulation of 67,353 copies. The following year, Vividata reported that the three dailies had an estimated combined weekly readership of 221,000 unique individuals; the 2011 federal census estimated that the regional population aged eighteen and over was 347,760.[89] These numbers suggest that local papers may have been challenged by digital news sources from outside Niagara, but they remained a widespread source of information capable of reflecting and constructing cultural issues.[90] As Enric Xicoy, Cristina Perales-García, and Rafael Xambó assert, newspapers are "political actors, and as such they intervene in the conflicts they cover, amplifying them, modifying them, and shaping them."[91] Three factors influence the traction of news stories.[92] First, the reader's reception of any one story will be shaped by its association with credible authors or sponsors, be they trusted news outlets, official sources, or celebrity spokespeople. Second, audience reception will also be shaped by the story's alignment with wider cultural themes and familiar narratives. Third, the story's resonance will be reinforced and/ or amplified if it is received through multiple channels, becoming an intertextual presence in the public sphere. Reporting around Remembrance Day serves as a good example. Any one report by any one news outlet sits at a confluence of messages concerning respect, honour, sacrifice, and citizenship, and thereby amplifies them all.[93] These factors are also germane to local newspaper coverage of the bicentennial.

We performed a content analysis of Niagara newspaper stories written between 2007 and 2017 that featured the War of 1812. Internet searches returned eighty-two stories that fit these parameters. We included all of them in our sample. The bulk of the stories appeared in the years immediately surrounding the bicentennial, particularly the years 2011–14. We then utilized Hyper-Research, a content-analysis software package, to code and count newspaper content. The results indicate that local newspapers supported the federal government's agenda for the 1812 commemorations. The war was touted as an epic event in local and national foundation myths, as a source of local and national heroes, and as a component of local and national identity.

The War of 1812 predates Canadian Confederation. The combatant nations were Great Britain, Six Nations of the Grand River, and the United States, as the latter sought to acquire the British North American colonies by force. Nonetheless, newspaper coverage of the bicentennial constructed a particularly "Canadian" narrative of the war. Stories identified Canada (367 mentions) more than four times as often as the United States (88 mentions), and more than ten times as often as Great Britain (34 mentions). Six Nations were also either featured or referenced in articles that described their perspectives and participation in major battles (53 mentions). Stories tended to focus on the significance of the War of 1812 in Niagara (135 mentions) or on its impact in the development of the future Canada (25 mentions). The announcement of federal funding for the anniversary was described by the St. Catharines *Standard* in precisely these terms:

"The War of 1812 was a defining event in Canada's history," [James] Moore said. "Without the War of 1812, Canada as we know it would not exist" …

The announcement highlighted the Stephen Harper government's effort to prioritize national history, the cabinet minister said. It also echoed the federal government's Speech from the Throne that in June referred to the bicentennial as a "key milestone" that ensured the "independent destiny" of Canada.[94]

Moore was then the federal minister of Canadian heritage. The less-than-critical presentation of his claims both echoed and endorsed the notion that the war was a pivotal moment in Canadian nation building.

Narratives of a glorious past often recount the efforts of heroes and describe the sites of their deeds. These elements featured heavily in media coverage of the bicentennial. Most stories reminded readers of well-known individuals. Sir Isaac Brock's lamentable end at the Battle of Queenston Heights was well documented (214 mentions). Similarly, Laura Secord's trek was given prominence in feature articles and opinion pieces (143 mentions). Other notable figures included in the press's honour roll of heroes included Shawnee chief Tecumseh (41 mentions), Lieutenant James FitzGibbon (37 mentions), General Sir Roger Hale Sheaffe (25 mentions), and Captain John Norton (12 mentions). No other figure received more than 9 mentions. The focus on "Canadian" heroism was reinforced by truly spartan coverage of American participants in the war. The most frequently named individuals were Colonel Charles Boerstler, a commanding officer at the Battle of Lundy's Lane (four mentions), and Francis Scott Key, the lyricist of "The Star-Spangled Banner" (3 mentions). Curiously, no article mentioned Major General Stephen Van Rensselaer, the officer leading the American troops at the Battle of Queenston Heights. Indeed, a reader unfamiliar with the war might not have known who the "Canadian" troops were repelling.

Just as local newspapers emphasized "Canadian" heroes, they also emphasized engagements on "Canadian" soil, specifically in Niagara. Again, such coverage reinforced existing narratives of the war. Local battlefields and forts were the key sites named. Most of the coverage referenced Queenston (96 mentions), Fort George (64 mentions), Fort Erie (55 mentions), DeCew House (54 mentions), and Lundy's Lane (38 mentions). The number of stories referencing Queenston reflects its role as a battlefield and the location of Secord's home. The continuing presence of heritage structures at these sites also made them central to bicentennial observations. In many ways, the newspaper coverage reinforced existing narratives of the war that would have been known to informed readers.

Finally, local newspapers also covered newly undertaken commemorative works. Reporters generally echoed the hopes of planners who believed that the public would engage specific mnemonic narratives at local sites and events. For example, a story in *Niagara This Week* concerning the new monument planned for DeCew House quoted a local dignitary:

> This symbolic First Nations monument will amplify a well-known Canadian story [Secord's trek] and generate a deeper understanding of the role the First Nations played in building our nation right here in this region ... Understanding more about that story is something that we need to pass on to future generations.[95]

Similarly, one bicentennial planner was offered a chance to sum up the festivities from a local perspective in the *Niagara Falls Review*:

> We feel it was very successful by the things you can measure, like attendance ... But we also think we were able to increase the awareness of the War of 1812, especially locally. They got a chance to understand how important Niagara was to the formation of Canada.[96]

In these examples, and in many other articles, memory entrepreneurs' voices were conveyed through the local press without comment or question.

Media coverage, then, supplied by Niagara's local newspapers provided readers with reasonably informative pieces on the War of 1812. They recounted narratives that, while they were readily available to residents long before the bicentennial, were deemed worthy of repetition, especially for recent immigrants. A *Standard* article noted, "This is our history and it's important to remember this conflict and the peace that followed it. There are also many new Canadians that we want to reach so they understand the importance and impact of this war."[97] Reporters consistently included elements that hailed national identity and local pride. One article noted, "Even if history isn't your interest, being Canadian is something we're all proud of and for Canadians to connect to the places and the stories and the people that helped shape this great nation, you can't ask for anything more profound than that."[98] Such reporting supported the federal government's cultural agenda by conditioning readers' attitudes towards the war and its key figures. Between the work of the government, local memory entrepreneurs, and the media, the bicentennial was marked in ways that were impressive and wide-ranging. Some believed that it could not fail to raise awareness of the war and of Canadian military history more broadly.[99]

2016 Survey

Did the bicentennial events achieve their goals? To gauge their effect, we conducted a survey in 2016, two years after the bicentennial's end. This should have allowed us to separate the enduring impact of the war's commemoration from immediate memories of specific events or news stories.

Using the same protocols followed in our previous surveys, we spoke with 250 participants in St. Catharines. Once again, all participants were filtered to ensure that they were born or resident in Niagara and at least eighteen years of age. A significant majority of the participants were residents (83 per cent), but true to prior surveys only a minority were born in Niagara (31 per cent). All age ranges were well represented, and the average age was forty-two. With respect to gender identification, the sample was 52 per cent female and 48 per cent male; no one identified in an alternative category. With respect to ethnic identification, unhyphenated "Canadians" represented 28 per cent of the sample and "Aboriginals" made up an additional 3 per cent. The majority of the sample was of European ancestry (57 per cent), while all those of African, Asian, and Latin American descent represented an additional 11 per cent. This sample size has a confidence level of 95 per cent with a margin of error of +/– 6.2 per cent.

We retained the opening question used in the 2009 and 2012 surveys: "If you had to choose one thing that identifies Niagara, what would it be?" The results were remarkably similar to our previous survey (see figure 1.6). The vast majority of participants did not reference the War of

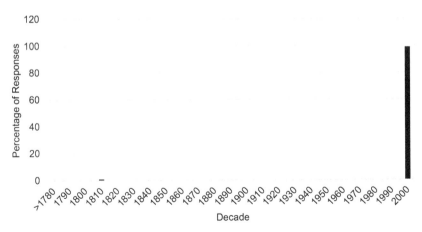

Figure 1.6. Commemogram: participant opinions on Niagara's identity, 2016. "What is the first thing that comes to mind when you hear the word 'Niagara'?"

1812 or any other "official" mnemonic narratives. Rather, 94.8 per cent ($n = 237$) of participants' top-of-mind responses were firmly located in the present. The minority ($n = 13$) included three who named elements of the War of 1812, one who named the Underground Railroad, and nine who referred to "history" or "heritage" without naming any specific person or event. Apparently, the bicentennial events and coverage had no enduring impact on the way most participants thought about Niagara's core identity.

Resonant Narratives, "Feelings in Common," and the Mnemonic Landscape

We sought to understand how Niagara residents engage with mnemonic products. Our measure was the frequency with which specific mnemonic narratives are marked by plaques throughout the region. Our surveys revealed that few participants shared the priorities of memory entrepreneurs. The most commonly plaqued narratives rarely influenced participants' notions of identity. Quite the opposite, most participants drew upon their lived experience rather than their knowledge of the past.

During the 2005 survey, we asked participants how they gain knowledge of local history. Few named historic sites and monuments as a source of knowledge (11.7 per cent, $n = 19$ of 182). This level of engagement was evident in each new iteration of the survey. Participants often volunteered that narratives marked by plaques were important. Nonetheless, they took these narratives for granted or simply forgot them. They were certainly not top of mind. For example, one participant who asserted that commemorations are a significant source of information on local history also had some difficulty with pertinent details:

> PARTICIPANT: Monuments do contribute to my knowledge of Canada. In Niagara-on-the-Lake, there's that big statue of whoever – the general who fought against the Americans. Then there's the Moses of her people ... the Underground Railway. I can see her, ... she's rather severe looking ... I remember from the photos ...
> ANOTHER PEDESTRIAN: [In passing] Harriet Tubman!
> PARTICIPANT: Oh yes, Harriet Tubman.

Material commemorations may mark narratives that are meaningful and significant, but the commemorations themselves may not move the viewer towards any emotion, thought, or action.

Some participants felt apologetic about their lack of engagement with material commemorations. While describing a common experience, one participant explained that such commemorations "are a minor source of local history. Maybe it would be more [significant] if I had time. It is difficult if you are just driving by." Another saw engagement with monuments as an unexpected outcome of a planned outing: "Monuments provide passive knowledge. You go there for some other reasons – say it's a beautiful park – and then say, 'Hey, what's that?'" Other participants contrasted the differing perspectives of residents and tourists. More than one asserted that monuments and plaques are things one "doesn't do" in Niagara as a resident. Another commented, "For the tourists, monuments are good. If you take an interest, you will go see it. I'm not sure if local people visit them or know much about local history." Another echoed these sentiments: "Monuments don't tell me much. I take it for granted. I feel like I know it all. I'm more interested in statues … where I am [travelling]. We live here, but don't do what's here." In these cases, encounters with a material commemoration may be accidental and any affective response to it may be minimal at best.

These participants' responses suggest that material commemorations exist on the fringes of daily experience. Two points can be made here. First, these participants echo Billig's contention that mnemonic products are not inherently moving.[100] In practice, the opposite may be true: mnemonic products may be uncontested by viewers because they offer some welcome familiarity in a banal, daily routine. Second, each mnemonic product may serve as an enthymeme, a simplified and incomplete argument that champions its associated narrative. The gaps in the argument must be filled by the viewer's own knowledge; cursory glances at static displays may register on an unconscious level and provoke vague memories of narratives that are already known.[101] Insofar as the viewer does complete the argument, a mnemonic product does function and may move viewers, even if only slightly. Our surveys left little doubt that some mnemonic narratives remain in circulation despite participants' indifference to historic sites and monuments. Chapters 3 and 4 explore this idea further by examining the career of a monument at St. Catharines City Hall.

Participants seemed willing to grant greater prestige to national narratives over purely local ones, but their comments also suggested an underlying indifference to most mnemonic narratives regardless of their context or scale. One participant, for example, explained that her awareness of local narratives was dominated by the presence of national and provincial historic sites: "I know where the big things are, but really know them? I notice more when I am travelling." This participant's emphasis on the "big things" echoes the modest parallel between the plaque commemogram and the survey commemograms.

If participants' knowledge of local mnemonic narratives was not derived from material commemorations, then presumably it came from other sources. As one participant noted, most material commemorations are not very informative: "We get general impressions from monuments, not detailed knowledge." More diplomatically, another commented that "They're putting more money into plaques [*pause*]. It's always interesting, so I'm interested. So, I'm drawn into monuments, but I'm more driven to go to the library." The 2005 survey asked participants where they learn about local history. The mass media emerged as the most common source of information (48.9 per cent of responses). As noted above, visits to historic sites and museums accounted for only 11.8 per cent of responses, ranking among the participants' social circles (13.4 per cent) and formal schooling (9.2 per cent).

Several participants stated that they simply absorbed information by living in the area, by picking up bits from the people, texts, and places they encountered in daily life (10.5 per cent). We categorized these responses as "passive learning." Once again, Billig is evoked. Casting an eye across the region, one recognizes that mnemonic narratives are omnipresent. Officially, narratives of European settlement, the War of 1812, the Welland Canal, and the Underground Railroad are often rehearsed and have been featured in intertextual arrays by all three levels of government. Beyond historic sites and monuments, they feature in school curricula, event programming, and annual rituals (e.g., museum exhibits, re-enactments, and days of commemoration). More to the point, these narratives also inform local toponomy through the names of streets, neighbourhoods, institutions, and businesses. Physically, these aspects of the landscape form the visible backdrop to everyday experience, particularly for those who regularly cross the canal or who reside near historic sites. Participants referenced these multiple sources of information and noted when they had learned specific narratives from two or more different ones. Chapter 6 explores this idea in more detail by examining residents' engagement with the wine industry.

Ultimately, participants' responses indicated that certain mnemonic narratives were more commonly held than others. A narrative's ability to affect individuals, to move them towards some emotion or action, is a significant feature of mnemonic traditions. That movement must be channelled in a specific direction to foster "feelings in common" among members of a community. Cultural resonance may derive in part from a sponsor's authority and the narrative's repetition, but it also derives from the opportunities a narrative provides for individuals to identify and engage with the values and practices of wider groups. Participants made it clear the wine industry had achieved this effect; they had engaged with the industry's narrative with an enthusiasm they did not share for the War of 1812 or the Welland Canal.

There is an overriding theme here: the effectiveness of a mnemonic landscape lies in the values, needs, and desires of present-day individuals. As a rule, participants knew the official and traditional narratives of Niagara, which are most commonly marked on local plaques, but they did not draw upon them when asked about identity. At best, mnemonic commemorations seemed to serve as secondary sources of information if that information was at all pertinent. As one participant noted, historic sites were good mainly because "It is important to get a sense of where things happen, otherwise they remain abstract." The purpose of monuments, then, may be to serve as spatial markers for specific narratives and their associated values rather than as a primary source of knowledge.

PART TWO

The Private Alexander Watson Monument

Chapter Two

A War Memorial as a Mnemonic Device

The first gap had been made in our ranks and we could faintly imagine how he would be missed in the little home circle down in Eastern Canada. His quaint sayings and cheerful laugh are gone from No. 1 tent, yet, ... in a week he will be almost forgotten.

> – Lewis R. Ord, on the death of A.W. Kippen at Batoche[1]

A correspondent suggests a good way of marking tangibly the indebtedness that all feel to those who fell in defence of their country in the late fighting, by having memorial tablets, suitably inscribed, erected in some public place.

> – *Canadian Militia Gazette*, June 1885.[2]

Step by deliberate step, pallbearers in brilliant red tunics brought the soldier home. Veterans of past conflicts saluted as the coffin passed through the streets of St. Catharines. With them stood civilians and schoolchildren, many holding flags, along with uniformed police officers and firefighters. At the graveside, the traditional rifleman's salute was followed by a deafening silence, filled only with the last post and a piper's lament. These scenes brought a distant conflict very close to home.[3]

This was December 2006. Corporal Albert Storm, a native of Fort Erie, had been killed on active duty in Afghanistan. Storm was a member of the Royal Canadian Regiment based in Petawawa, but the Lincoln and Welland Regiment stationed in St. Catharines opened its armoury and mounted a full military funeral for a fellow soldier. He left behind his wife and two children, his parents, and siblings. His family was surrounded by the troops on hand and the community. Mourners filled the armoury. Ushers had to turn many away.[4]

This was also June 1885. Three blocks away from the armoury, a worn grey monument reminds passers-by of another soldier whose service was similarly honoured. Private Alexander Watson, a native of St. Catharines, fell at the Battle of Batoche during the North-West Resistance of 1885. Like Storm, Watson had served with a unit based in another city: Winnipeg's 90th Battalion, now the Royal Winnipeg Rifles. Nonetheless, the Lincoln and Welland militias, then operating as two separate units, staged an impressive military funeral. Then, too, the local community stood beside Watson's family as they grieved the loss of their only son and brother.

The public rituals surrounding the death of a soldier seem timeless. Canadians are asked to remember the sacrifices of their military personnel each year on 1 July and 11 November. More frequently, however, Canadians are reminded through relevant mnemonic products and the practices that attend them. Snezhana Demitrova argues that public rituals serve an ideological purpose: communities do not simply mourn the dead – they elevate the citizen who exhibits absolute self-denial in the name of sacred duty.[5] In this way, a community rallies to comfort the family and articulate its own collective sense of grief, but simultaneously transforms the deceased into an ideal that can be preserved through time.[6] As timeless as these rituals may seem, their deployment for a common soldier rather than a sovereign or commander was relatively new in 1885. Canadians took inspiration from ancient heraldic protocols and modern democratic innovations to produce commemorations that resonated with their own emerging values as citizens of a new state. These shifting practices prompt certain questions: Why does a community commemorate any one narrative? Who has the authority and resources to mark a narrative? What knowledge is conveyed to future generations? How is that knowledge expressed and preserved? And why were these practices changing in Victorian Canada?

These questions clearly echo those we sought to address in chapter 1: How do people create mnemonic narratives and engage specific mnemonic products? As we began this research, we were interested in how the patterns revealed by the commemograms aligned with public engagement with a specific monument. In this section, we focus on the Watson Monument. Prominently located in front of City Hall in downtown St. Catharines, it is one of the city's oldest monuments. This chapter investigates the commemoration of Alexander Watson in three phrases. First, we describe the North-West Resistance in order to place the soldier's participation in its cultural and historical context. Next, we highlight contemporary funerary practice. Watson was given a public funeral, an honour not typically granted to common soldiers.

Nevertheless, community leaders deemed his death worthy of immediate public commemoration. Last, we investigate the creation of the monument itself. The enthusiasm that flowed from the public funeral service was materially captured in the monument bearing Watson's name.

A Soldier's Death, 1885

The colonization of North America by European settlers is not a heroic narrative. From the Atlantic coast to the Pacific and to the Arctic, Indigenous communities were displaced if not destroyed by agents of the Crown, private companies, and settlers who claimed entire territories as their own. Sometimes the settlers' claims were made by force of numbers, sometimes simply by force.[7] The North-West Resistance of 1885 was one violent eruption in this narrative.

In 1869, the Canadian federal government began a survey of the Red River Valley. It planned to open the Prairies for farm settlements for European immigrants. The local population, predominantly of Métis descent, believed the surveyors would not recognize their existing communities and hunting grounds. To protect their interests, and in defiance of Ottawa, the Métis established a provisional local government. This government, under the leadership of Louis Riel, halted the survey by force of arms and petitioned Ottawa for provincial status within Confederation. It also asserted its authority over the local population. In one instance, it jailed, tried, and executed Thomas Scott, a man who violently rejected its authority. Scott was a member of the Orange Order, a militant Protestant brotherhood. The order had its roots in Ireland, where it championed a politically conservative and Protestant United Kingdom under the rule of the British Crown. Orangemen who migrated to Canada established their own lodges and staked out the same ideological and sectarian terrain. When news of Scott's death reached Ontario, Orangemen were outraged and demanded Riel's arrest for murder. While Ottawa considered the petition for provincial status, it also sent troops after Riel. Riel fled to the United States, and the provisional government henceforth collapsed. Nevertheless, within a year Ottawa granted Riel's request and created the province of Manitoba.[8]

Fifteen years later, similar events unfolded further west. Private investors supported by Ottawa were then building the Canadian Pacific Railway to link British Columbia with Ontario. The span from the Rocky Mountains to the Red River Valley is roughly thirteen hundred kilometres wide and crosses the heart of the Canadian Prairies. These were the ancestral lands of eight Indigenous nations – including the Assiniboine, Cree, and Métis – who maintained multiple settlements and extensive hunting grounds. Once again, the federal government sought to establish European settler communities before it fully acknowledged the territorial claims of the existing residents.

In March 1885, the Métis again became defiant. Riel returned to lead them, formed another local provisional government, and made the village of Batoche its capital. The provisional government then confronted a North-West Mounted Police detachment at nearby Duck Lake. Riel hoped this combination of political posturing and sabre-rattling would again prompt Ottawa to address Métis concerns. His efforts were amplified by sympathetic members of neighbouring First Nations. Acting on their own, Cree warriors raided Hudson's Bay Company stores at Frog Lake, Lac la Biche, and Green Lake, and confronted the police at Fort Battleford and Fort Pitt. The confrontation at Frog Lake was particularly brutal. Nine white settlers were killed and the surviving villagers, both settlers and Métis, were taken hostage in the camp of Mistahi-maskwa (also known as Big Bear). Among the hostages was Theresa Gowanlock, whose parents' farm sat twenty kilometres outside St. Catharines. Among the dead was her husband, John. When news of the raid reached other settler communities, residents believed the Métis and First Nations had gained control of the region and were set to extend their power.[9]

Ottawa responded quickly but offered no negotiation. Instead, the federal government formed the North-West Field Force, mobilizing eight thousand troops drawn from across the country to crush the resistance. A decisive battle at Batoche ended the provisional government. Several of its leaders surrendered or were captured while their warriors dispersed. Riel himself surrendered, was tried for treason, and was found guilty. He was hanged six months later. Cree chiefs Mistahi-maskwa and Pîhtokahanapiwiyin (also known as Poundmaker) also surrendered. They, too, were found guilty of treason and imprisoned for three years, though the charges brought against them were considered spurious. Pîhtokahanapiwiyin was officially exonerated by the federal government in 2019.[10]

Alexander Watson was a volunteer with the field force. In 1881, aged twenty-four, he had sought his fortune by following the new railway west to Winnipeg. In this respect, Watson was following in his father's footsteps. David Watson, a carpenter, had chased work across Ontario before settling in St. Catharines with his young family. Alexander spent his adolescence there, learning his father's trade. He then set out alone for the Prairies. His carpentry skills allowed him to achieve some success in Winnipeg during its railway boom. Outside of work, he was reportedly a popular Sunday school teacher and engaged to the daughter of a prominent Winnipeg family. During his spare time, he found companionship as a volunteer in the 90th Battalion.[11]

The provisional government came into being on 18 March 1885. Watson's regiment was immediately put on notice, and boarded trains for the contested territory twelve days later. Watson first saw action on 24 April. The Métis held a position at Tourond's Coulée, on the land route to Batoche. While attempting to cross Fish Creek, the field force were exposed to sniper fire and were driven back. Watson avoided injury here, but on 12 May he was not so fortunate. He was wounded at the Battle of Batoche during the final charge that took the Métis capital. He died three days later, one of eight soldiers in the field force to be killed.[12]

The resistance and its aftermath prompted conflicting responses from Canadians. Many believed the Métis' and First Nations' land claims were legitimate. Prime Minister Sir John A. Macdonald's Conservatives were the governing party in 1870 and 1885. Macdonald was also the minister responsible for Indigenous relations in 1885. He was also an Orangeman. The Liberal Opposition blamed the Conservatives for both crises because Macdonald had obstinately refused to address Indigenous land claims even as he pushed forward with the railway and European settlement. Indeed, there were strong indications that the government had provoked the Métis and First Nations into their desperate gamble.[13] In Niagara, where Watson's family remained, these feelings were evident among local Liberals. For example, the *Welland Tribune* argued that the federal government was largely responsible for neglecting both the grievances and the warnings of trouble emanating from the North-West for months before the outbreak occurred.[14] Nevertheless, the Métis' and First Nations' decision to use force was roundly condemned, even among partisan commentators. Again, in Niagara, the *Thorold Post* captured this spirit in a heated editorial condemning the resistance for the deaths on both sides: "Their crimes can never be atoned for, even were all their worthless lives sacrificed."[15] Hence, public opinion turned on the issue of responsibility: Who was most responsible for the conflict, the government or the rebels?

The issue of responsibility was clouded by racism and religious bigotry. Judging by the newspapers of the day, most Canadians agreed with the militia's deployment. The field force volunteers were both anglophone and francophone, Protestant and Catholic. However, Canadians did not agree on a suitable punishment for Riel. Much of English-speaking Protestant Canada viewed Riel – a francophone Catholic – as a traitor who led an armed rebellion against constituted authority. The justice of his cause was irrelevant. This was a view championed by the Orange Order and its journal, *The Sentinel*.[16] It was also the prevailing view in Niagara, which was largely anglophone, Protestant, and had very little visible Indigenous presence. The *Post* referred to Riel as "an inciter of treason, and, if not a red-handed assassin himself ... directly responsible for the deeds of the irresponsible hordes who were inspired by him."[17] This line of thinking led to one conclusion: a court found him guilty, the sentence was death, and justice would be served. In like manner, the *Tribune* concluded that "Riel well deserves hanging."[18] French-speaking Quebec saw things quite differently. The Métis' resistance was not condoned, exactly, but was viewed as pardonable because their cause was fundamentally just. Seen in this light, Riel's execution appeared to be unnecessarily harsh. Riel's jury had recommended mercy, and Conservatives in Quebec asked Macdonald to commute his sentence. In response, the Orange Order recalled Scott's execution and lobbied relentlessly for Riel's death. Lodges throughout Ontario threatened to take their revenge at the polls if Macdonald did not do so. In Merritton, now a neighbourhood of St. Catharines, the following motion was passed:

> Members of LOL 844 do hereby unhesitatingly State that the Arch traitor Luis Reil of the North West Teritory is Guilty of having been tried and convicted of High Treason Be Hanged according to the Law and that We Will oppose to the utmost off our powers any government that will commute the sentance of the said Reil.[19]

This was no idle threat. Irish immigrants were the largest ethnic bloc in Ontario, while Protestants formed the largest faith bloc. The order's violent language also suggested that no French-speaking Catholic was welcome on the Prairies. This suspicion was confirmed five years later when Manitoba abolished its Catholic school system.[20]

The events of 1885, then, did not furnish a simple heroic plot in which good triumphed over evil. From the outset, the federal government's response to Indigenous resistance deepened tensions between the Métis, First Nations, and European settlers; between francophones and anglophones; and between Catholics and Protestants.

A Soldier's Funerals, 1885

When battles end, commemoration begins. This was certainly true of the North-West Resistance. Canadians who supported the field force did so with great enthusiasm. Celebration of the troops began with the 90th Battalion itself, whose members wrote a music hall farce – "a musical and dramatic burlesque in two acts" – while returning to Winnipeg.

The 90th on Active Service was a hit on the Winnipeg stage, and the libretto was published for sale. Though the dead were barely five weeks in the ground, the survivors made little mention of them in the show. Similarly, the unit's newly penned regimental march offered but fleeting recognition: "We fought the rebels at Fish Creek and drove them out of sight / While many of our good men and true fell battling for the right."[21] These two lines emphasize the gallantry of the unit rather than the sacrifice of individuals.

If the soldiers themselves were reluctant to acknowledge the dead publicly, civilians were not. Thanks to the telegraph, news bulletins from the front spread across Canada within forty-eight hours. They were not always clear or accurate. The first list of casualties from Batoche reported that "Watson of the 90th" had fallen. This prompted an awkward episode in Winnipeg as there was a Sergeant Major John Watson in the same battalion as Private Alexander Watson. Anxiety gripped the city. Both the *Winnipeg Free Press* and the *Times* were beset by panicked queries from family and friends of the two men. Over the next two days the papers frantically awaited further dispatches to discover the soldier's identity. The *Free Press* noted that the sergeant major "has a wife and six children in this city, and should the report of his death be confirmed it will be a sad bereavement … [He] was universally liked and was a great favorite [*sic*]."[22] The private may have had fewer friends, but the papers made up for this unintended slight. Once his death was confirmed, he and his dead comrades were lionized in biographical notes illustrated with their portraits, regardless of rank. Similar pieces appeared in the leading Montreal and Toronto papers.[23]

Though outpourings of grief and tributes for fallen soldiers were not new, the soon-to-be-held funerals of Watson and his fallen comrades charted emerging funerary and commemorative practices. Under British military custom, dead soldiers were buried near the field of battle. Only the remains of those deemed worthy of the cost were transported home: those with title and rank. Practicality was the key. The natural process of decomposition and the challenges of transportation encouraged swift interment. Sailors, of course, were buried at sea. Soldiers of the Napoleonic Wars and British imperial wars were destined to be buried thousands of kilometres from home in mass graves. When the vanquished were buried by the victors, their graves were often left unmarked. During the War of 1812, for example, those who fell in Niagara, Hamilton, and Toronto were buried in mass graves that were largely forgotten over time. Since the 1980s, a renewed interest in military commemoration has fostered efforts to document them, rehabilitate them, or repatriate the remains.[24]

This treatment of human remains may seem callous. However, the practice of creating individuated, marked graves for commoners – military or civilian – was relatively new within British funerary practice. Until the 1700s, most graves were marked in ways that were meaningful only to the generation who remembered the dead. In rural areas, the deceased might be buried in a cherished spot or under a particular tree, while graves in churchyards might be marked with wooden crosses. Anything more enduring was an inconvenience. By the late 1600s, the churchyards of urban parishes were full. Gravediggers resorted to ingenious if macabre means to cram ever more bodies into them. Predictably, exceptions were made for the titled and wealthy. Those who could afford vaults and mausoleums could trumpet their family names and accomplishments in brass and stone. Even these arrangements, however, endured only so long as descendants had the desire and the means to maintain them.[25]

In the early 1800s, industrialization placed increasing pressure on England's urban land. Just as green fields were converted to new factories and housing, they were also required to accommodate the dead. New space came in the form of privately owned cemeteries after 1819. Free from church rules or customs, private cemeteries offered their paying customers whatever manner of individuation, sanctity, and memorialization they wished. The contemporary fashion for sentimentalism and romanticism encouraged elaborate displays of grief and aggrandizement. Inevitably, the fashions of the titled and wealthy were imitated and became common practice through less expensive forms and materials. By the mid-1800s, a basic but respectable burial required an individuated grave marked by an enduring stone monument.[26]

Christian theology contributed a significant ideological element to the reality of death: a belief in the eternal life of the individual human soul. Both Catholics and Protestants held that the souls of virtuous Christians would be welcomed in the eternal paradise of heaven, while the unrepentant faced an eternity of punishment in hell. For Catholics, purgatory allowed marginally virtuous souls to be rehabilitated while awaiting their call to heaven. Anxious families could not tell where the recently departed might be dispatched, but they could seek clues in the quality of the death itself. A peaceful end signalled a clear conscience and an easy passage into the afterlife. Perhaps this way was prepared by a lifetime of belief in God, attention to the teachings of Jesus, and evidence of good works. In modern times, good works included service that transcended individual needs and benefited the wider society or the state.[27] Nigel Llewellyn has described the main tropes of the good death that appeared in English visual culture: "the innocence or good scholarship of a dead child, the pathos of a dead soldier, the heroism of a death met in a great cause, the sacrifice of a martyr, the tranquility of an old man dying, [and] the virtue of a good wife dying."[28] The inclusion of military service in this list clearly provided Christian men with the means to establish their virtues.

Migrants brought these beliefs, practices, and fashions to North America. In the growing cities of the Atlantic coast, British settlers kept pace with European trends in mortuary practice. In what became Ontario, the ready availability of land delayed the problem of overcrowding. There was ample space to bury the dead, be it in family plots on farms or in land set aside through the Clergy Reserve system. As towns grew, however, familiar issues arose. Towns expanded out and around burial grounds, older plots became forgotten by later generations, and demand grew for new burial grounds.[29] For example, the first settler graveyard in the Toronto area was created in the 1760s for the troops posted at Fort York. It fell into disuse in the 1860s and was repurposed as a city park in the 1880s, now known as Victoria Memorial Square. Similar events produced Montreal's Dorchester Square, Ottawa's McDonald Park, and Kingston's McBurney Park (still known colloquially as Skeleton Park). Graveyards created by faith groups persisted if local congregations remained active. Non-sectarian cemeteries also endured under the administration of municipalities or private corporations. As in Britain, these latter cemeteries began to appear in Ontario in the 1820s. St. Catharines created a municipal cemetery in 1856, now known as Victoria Lawn. Once opened, the graves in these new cemeteries were marked with the same variety of monuments that appeared throughout Great Britain.[30]

Military interment practices shifted with these trends in civilian life. The American Civil War offers a particularly stark example. In 1861, the American government and the breakaway Confederacy initiated a war to protect their political and economic interests, but they used democratic values to justify the war's human costs. Volunteers were summoned to accept the responsibilities associated with individual liberty and citizenship in a republic.[31] Hence, while both armies initially buried their dead on the battlefield following the European custom, their leaders were prompted by public sentiment to honour them. This meant their remains deserved named, individuated graves in pastoral fields in keeping with the expectations for civilian funerals.[32] And yet, the scale of death was horrific. An estimated 650,000 soldiers died during the four-year war. Neither army had sufficient resources to provide individual graves immediately after each battle. At the war's end, however, the American government sought to identify and re-inter the remains of the Union dead in what became national cemeteries. In the former Confederate states, volunteer organizations began the same task for the Confederate dead. In 1898, the Confederate cemeteries were absorbed into the national system.[33]

The American military cemeteries had an impact beyond the comfort they provided to grieving families. Visitors described the effect of the mass graves as unquestionably moving. American president Abraham Lincoln articulated this sentiment in his dedication of the military cemetery in Gettysburg, Pennsylvania. Referring to the values informing the Union cause, Lincoln stated,

> we cannot dedicate – we cannot consecrate – we cannot hallow, this ground. The brave men, living and dead, who struggled here, have hallowed it, far above our poor power to add or detract ...
>
> It is rather for us to be here dedicated to the great task remaining before us – that, from these honored dead we take increased devotion to that cause for which they here, gave the last full measure of devotion – that we here highly resolve these dead shall not have died in vain.[34]

These words conveyed two arguments. First, Lincoln elevated the dead citizen-soldiers above the common citizen, which rendered their motives and service transcendent and unquestionable; they had died the good death. Second, he claimed their endorsement for the war, which was framed as "the great task" rather than as a government policy. Cast in this light, the dead soldiers' sacrifice became a democratic duty to be emulated by the living. The speech did not offer closure but a call to action. Lincoln had fully embraced the symbolic power of the silent dead.[35]

There was also something new in this line of thinking. The symbols and rituals associated with public funerals were traditionally linked to sovereigns in European culture. In an absolutist regime, the death of a sovereign might well produce a vacuum of power leading to the collapse of political and social order. The ritual acts performed at the death of a sovereign and the coronation of an heir filled this vacuum with established processes that maintained political continuity and national cohesion.[36] In England, Wales, and Northern Ireland, the College of Arms determines all protocols governing title, status, and their display during state funerals. Its most prominent work is seen during royal funerals. The rules governing these events evoke the transcendence of the state beyond any one person, place, or time.[37] The widely reported death of Queen Victoria's husband, Prince Albert, maintained the public's familiarity with this pomp and ceremony even in the colonies. A detailed description of his funeral occupied the entire front page of the Toronto *Globe* in 1861.[38] Only four common-born subjects were granted these funereal honours in Great Britain during the nineteenth century, one statesman and three military commanders.[39]

Britain's North American colonists were not simply aware of these symbols and rituals. They employed them as they built a new British society. The recognition given to Major General Sir Isaac Brock following his death at Queenston Heights certainly mimicked them, leaving a lasting impression in Niagara. His remains were buried at Fort George immediately following the battle. In 1824, he was re-interred at the battlefield beneath a monument built in his honour. When that monument was destroyed in 1840, his remains were moved again until the current monument was begun in 1853. His body now lies in a vault beneath that column. Each new burial was marked with greater ceremony than the last.[40] Canadians also afforded state funerals to notable statesmen. Former governor general Charles Thomson, Lord Sydenham, died in Kingston and was buried there with due ceremony in 1841. Sir Étienne-Paschal Taché, premier of the United Canadas, died during the Confederation debates in 1865. He was buried in Montmagny, roughly eighty kilometres from Quebec City. The Grand Trunk Railway ran special trains for government ministers, sundry notables, and two militia units attending the memorial service. An estimated six thousand people viewed the funeral procession. In like manner, Thomas D'Arcy McGee, a member of Parliament who was assassinated by an Irish patriot, was honoured by the City of Montreal in 1868. Again, thousands lined the streets to observe the funeral procession.[41]

The colonists began to acknowledge the ideological value of the citizen-soldier during this period. The roots of the idea were planted in Upper Canada after the War of 1812. Britain oversaw the defence of its northern colonies until 1871. Though professional imperial troops and their Indigenous allies bore the brunt of the American invasion in 1812, colonial militia units – made up of volunteers – were also mobilized and suffered combat losses. The conservative political establishment in Upper Canada lauded their service to curry political support. They cast the loyal volunteer as a true subject of the Crown while denouncing reformers – who openly admired American democratic innovations – for their apparent disdain for British governance. Moreover, the volunteers were represented as gallant men who were not militaristic or mercenary by nature but answered the call when necessary to defend their ethnic patrimony. In sum, the volunteer was portrayed as the model subject. This portrayal has been described as the "militia myth," in part because it overstated the role of volunteers during the war.[42] Nonetheless, later conflicts in Niagara, the Red River Settlement, and the North-West were completed without British regulars. Those actions seemed to confirm that volunteers ably matched the performance of professionals.[43]

State funerals and public memorials were not planned for the volunteers who served in the War of 1812. Over time, however, the institution of responsible government would gradually transform migrant subjects of the Crown into self-governing citizens of an emerging nation state. Further, the confidence of militia leaders and their supporters grew with each campaign. In the wake of these events, the protector of the realm was no longer the sovereign or commander, but a volunteer risking all in defence of civil society. In Europe, memory entrepreneurs began to raise monuments celebrating the rank and file in the wake of the French Revolution. Both the French and the counter-revolutionary Prussians were inspired to acknowledge commoners' roles in important victories during the Napoleonic Wars. In the United States, this impulse was fully articulated in the Battle Monument in Baltimore, Maryland, which honoured citizen service during the War of 1812.[44] Perhaps a similar impulse informed an early war memorial in Halifax. The Welsford-Parker Arch, erected in 1860, honours two Nova Scotians who died in service with the British Army in Crimea. Both were officers well below the rank of general.[45]

Six years later, residents of Upper Canada had a new reason to honour their militia volunteers. The Fenian Raids were part of an international effort by Irish republicans to end British rule in their homeland. Patriots living in Ireland established the Irish Republican Brotherhood to advance their cause through open rebellion. Compatriots in the United States established their own wing, better known as the Fenian Brotherhood. The Fenians hoped to gain support among Irish settlers in British North America by attacking British authority in the colonies. They made five attempts between 1866 and 1871. In late May 1866, an estimated fifteen hundred Fenians crossed the Niagara River near Fort Erie and marched towards the town of Ridgeway. There they met Canadian militia regiments drawn from Hamilton and Toronto. Despite certain tactical advantages, the Fenians retreated to the American side after four days. Ten Canadians died during the fighting and another twenty-two died of disease contracted while in service.[46] Unlike the dead of 1812, the dead of Ridgeway were transported home to their families and given public funerals. Perhaps politicians and militia leaders were embracing the militia myth, and perhaps knowledge of the American Civil War was still fresh in their minds.[47] Regardless, those who fell in battle received public funerals and individuated graves.

In Stamford, now a neighbourhood of Niagara Falls, residents formed a committee to honour a local casualty of Ridgeway. John H. Mewburn had been the son of a prominent local family and a member of the Queen's Own Rifles (Toronto). The committee, the family's Anglican church, and a local company of the Lincoln militia staged a funeral that combined elements of a typical family observance and a military interment.[48] For most families of the day, a typical funeral would be shaped by religious belief, ethnic custom, and available resources. Funeral homes and their amenities were still relatively new. As such, the dead were most often kept in a family home. Kinfolk and friends were invited to commiserate with the grieving family by visiting them and attending the funeral service. Following interment, the grave was marked by a stone reflecting the family's taste and standing in the community.[49] This commiseration was essentially private. By contrast, public funerary custom invited an entire community to participate in a staged event that articulated civic values. Most of this community may have been strangers to the grieving family and to one another. As Abraham Lincoln understood, the deceased may be less an individual to be mourned than a common symbol to be revered or emulated.

Mewburn's funeral reflected these changing practices. His body lay in his grandparents' home in Stamford until the day of the burial. It was then carried to the family church by eight militia men, who were led by a firing party. Following the service, he was buried in the churchyard with military ceremony: three shots fired over the grave. The headstone notes his death at Ridgeway but otherwise is unassuming. A large body of the township's residents and two hundred militiamen led by their commanding officer attended the event. Judging by a report in the *St. Catharines Journal*, Mewburn's funeral became a moment to honour not just the fallen but all the volunteers who faced the enemy. The reporter noted that the family was

> deeply grateful for the affection and sympathy shown them … by all persons and parties who … came forward to testify their sense of the noble spirit evinced by our brave volunteers, in coming forward readily to risk their lives in repelling a ruthless invasion of our beloved country by a band of lawless foreigners.[50]

Clearly, Mewburn had come to represent the ideal volunteer described in the militia myth. Toronto staged similar services for its dead. Afterwards, the Canadian Volunteers Monument was raised in Queen's Park naming nine members of the Queen's Own who fell during the campaign. Mewburn was included. This may have been Canada's first monument dedicated to the rank-and-file citizen-soldier.[51]

The militia's next combat experience came in the North-West. The field force was prompted to navigate a similar path between family expectations and military practice. The Canadian troops were led by a veteran of Britain's imperial wars. Sir Frederick Middleton had served in New Zealand and India and had recently led the Royal Military College in Sandhurst, England. By most accounts, he was an ardent follower of professional military discipline.[52] Notably, the only heroic dead named in the general's description of the campaign were officers.[53] Colonel George T. Denison's combat memoirs are similarly selective when saluting those who died in the North-West, suggesting the death of an officer was a greater loss to the nation.[54]

The first soldier of the field force to die was Private Achilles Blais of the 9th Battalion (Les Voltigeurs de Québec). As the troops assembled in Winnipeg, cold winds and heavy rain made life miserable for those sleeping in tents in muddy fields. Some fell ill and were left behind when their units went to the front. Blais was the only one to die. He was buried in the St. Boniface churchyard on 30 March with no special ceremony. Only the few soldiers remaining in the city attended. His friends, not the militia, erected a headstone. The minister of militia, Adolphe Caron, claimed credit for the funeral.[55]

In the early stages of the campaign, the remains of those who died in combat faced similar treatment. The North-West Mounted Police lost men in the skirmish at Duck Lake. Their bodies remained in the field until Riel told their commander to collect them. They were buried in the region, though as residents this is not surprising. Similarly, Constable David L. Cowan's remains were left near Frog Lake until the militia found them. He was buried by his fellow Mounted Police and given a rifleman's salute.[56] The Voltigeurs lost a second man to illness while they were en route to Calgary. The unit gave him a military funeral in that city, but one of his comrades lamented the distance between his grave and his family.[57] At least six members of the 90th Battalion died at Tourond's Coulée, and there they were buried. A reporter from the *Montreal Witness* described the scene:

> The graves of those who fell had been dug in the prairie sod within sight of the field of honor [sic] and the mighty river, and green fir boughs with

pale anemones, carefully arranged by comrades of the dead, covered them. More enduring than flowers, a dark cairn of boulders taken from the river and carried up with much toil, slowly rose near the graves, and a great white cross of gleaming white poplar marked the burial ground of the patriotic dead.[58]

These graves remain in the Tourond's Coulée National Historic Site and are maintained by Parks Canada.[59]

The imperial protocols were dropped after Batoche, but only for the field force. Middleton left the Métis dead lying in the field until his own casualties were under care. Even then, some of the Métis dead were left in a cart outside the village chapel until the resident priests prepared them for burial.[60] In Ottawa, however, Macdonald's government had seemingly absorbed the powerful propaganda effect of American military interment. As combat casualties mounted within the field force, the Department of Militia decided to honour the dead individually. The low number of casualties made it feasible to return remains to their families.[61] This decision made camp life disquieting as soldiers sought closure with respect to their dead comrades. All bodies were interred at the battleground until those selected for transportation were exhumed. One veteran later wrote of the dead, "As a rule the men shun them until they are carried out. They seemed to be forgotten in the excitement of the moment."[62] The editor of the *Canadian Militia Gazette* simply declared it a bad precedent: "Although the feelings of the relatives of the deceased are worthy of all respect and admiration, it would have been best if the old rule of making the battlefield the soldier's fittest and most glorious burial place had never been departed from."[63]

Among civilians, the minister's decision had startling results. The home community of every dead soldier took possession of his name and transformed him into a national hero.[64] Moose Jaw provided an early indication of how certain Canadians felt about the dead. As a train bearing coffins stopped in the village, the home guard and a host of citizens came to pay their respects. The village was established only three years earlier on lands traditionally held by the Cree and Assiniboine Nations. If these villagers had feared a widespread conflict, stoked by knowledge of Frog Lake, then their respects were no doubt tempered by their relief that the crisis had apparently passed.[65] White settlers throughout Canada probably shared these feelings. Many had family members who had followed the railway west. In Niagara, news of Frog Lake must have crushed the family, friends, and neighbours who had celebrated Theresa Gowanlock's wedding just six months earlier. The first reports to reach Ontario listed her among the dead along with her husband. After she escaped her captors, she returned to her parents' farm in July 1885. Her own account of the experience was published by *The Globe* in June and revised for a book issued in September. Familiarity with the family and her story may well have kept the events in the North-West alive in local lore.[66]

The first fatalities to receive heroic tributes were men of the 90th. The quartermaster stationed in Winnipeg described the events surrounding their common funeral in that city, held on 6 May:

> His Honor [sic] the Lieutenant-Governor and his Ministers, all the military officers in the city, the Municipal Council, and many thousands of people, attended, thus sympathizing with the friends and showing respect to the deceased, who had sacrificed their lives assisting in the maintenance of law and order. All business was suspended, and the people appeared to be in deep mourning.[67]

This public service was vastly different from the private service that buried Blais just five weeks earlier. His battalion made amends when they returned to Winnipeg. Bishop Alexandre-Antonin Taché led a memorial service attended by the entire unit at the St. Boniface Cathedral. Similarly, Cowan's family had his remains exhumed and transported to Ottawa, where he was re-interred near his father.[68]

In this context, Watson had two funerals. The first took place in Winnipeg, where the city council arranged a public, military funeral with the co-operation of Grace Presbyterian Church and the Montreal Garrison Artillery (then awaiting orders in the city). The bodies of Watson and three fellow soldiers arrived on 22 May and laid in state overnight in city council chambers until a service was held at the church the following day. A military procession led by the garrison band then took three coffins to the cemetery of St. John's Anglican Cathedral. One observer estimated that some "fifteen thousand people lined the route, with more viewing from overhead windows."[69]

The 90th Battalion later marked these graves with a common monument. A single stone was carved into a memorial wall sheltered by a modest canopy. The wall bears the names of the unit's two engagements and the names of all those members who died during the campaign. Eventually, seven men were buried at this site, while two others were buried elsewhere: Watson in St. Catharines and James Hutchison in St. Thomas, Ontario. The *Winnipeg Free Press* reported that Watson had been widely known as "a noble Christian and a gallant soldier."[70] He had been in Winnipeg only four years, but there was an obvious desire among locals – not just the unit – to honour the fallen volunteer as one of their own. Following the service, his coffin was returned to the rail yard to continue its journey to St. Catharines.[71]

Again, the occupation of public space became a common feature of events marking the campaign. Winnipeg City Council took especial pride in the 90th Battalion's role. It hosted a massive public reception when the unit returned in July. Their parade route was marked with a victory arch, and local newspapers competed to sing the soldiers' praises. Similarly, the francophone community of St. Boniface feted the Voltigeurs upon their return to Winnipeg, en route to Quebec. These events had their echoes throughout Ontario. The return of one battalion to Toronto was delayed in every town it passed as cheering crowds met the troops.[72]

Two points warrant emphasis. First, the organizers of these events tended to be individuals in positions of authority, either in municipal government, church life, or militia units. These were groups with access to the required resources to claim public space and promote their values through parades, public gatherings, and temporary civic structures such as arches. These events were not spontaneous expressions of public enthusiasm. At the same time, it must be remembered that the public did attend them, and they did demonstrate support for the returning soldiers. As one militia officer remembered,

> What struck me most was the extraordinary enthusiasm of the people …
> [If] we had been returning from a second Waterloo, concluding a long and anxious war, we could not have been received with greater warmth.
>   I repeatedly saw both men and women cheering wildly, with the tears running down their cheeks with excitement. It was a most interesting study.[73]

The same scenes were described in the newspapers of the day.[74]

Second, it is also clear there were rituals associated with the end of a war. The public's response may have been genuinely spontaneous and enthusiastic, but the public also followed British traditions surrounding the return of the troops and the commemoration of the dead. As many scholars have shown, such rituals can sometimes open space for alternative behaviours and give voice to alternative opinions.[75] For the most part, however, participants in English-speaking Canada did not break from their expected roles in public gatherings following the North-West Resistance. Indeed, their overenthusiasm became a point of comment.

Perhaps the Orange Order was influential here too. The triumph of the volunteer militia over an enemy that was perceived to be francophone, Catholic, or Indigenous, boded well for its Protestant, anglophone, and imperialist members.

The same access to public space, the same rituals, and the same enthusiasm were evident in St. Catharines for Watson's second funeral. As with Mewburn's funeral in 1866, Watson's incorporated both traditional and emerging elements. The body was initially delivered to his parents' home for the family to grieve privately. According to the *Hamilton Spectator*, the family consented to a public event only at the "urgent request of a large number of citizens."[76] The city, local churches, and local militia units were eager to stage the funeral on the family's behalf. Indeed, the Lincoln militia that provided an honour guard for Mewburn now sought a similar role in honouring Watson.

The moment his body left his parents' home, it became the object of a public spectacle. It was first taken to the local courthouse to lie in state for public viewing. The closed coffin was decorated with his uniform and the ensign flag, both symbols of his service to the nation. It was also decorated with several floral tributes linking him to various communities, including his church in Winnipeg and the City of St. Catharines. The courthouse was filled to capacity for the visitation. For two hours, every seat was occupied as a steady stream of visitors paid their respects. The *Welland Telegraph* had expected a big turnout, not because Watson had been well liked or came from a prominent family, but because his death itself was significant, "as this is the first victim of the Northwest insurrection to be buried in this district."[77] The actual service was solemn, consisting of prayer, scripture, and hymns.

A military escort took the coffin to the municipal cemetery, Victoria Lawn (figure 2.1). This was a funeral procession modelled on heraldic custom. In St. Catharines, the organizers of Watson's procession carefully orchestrated their own display of shared values, transcendence, and unity. The local police came first, followed by a firing party, the men of 19th Regiment (Lincoln), their band, and the band of the 44th Regiment (Welland). Next came the coffin itself, carried on a gun carriage pulled by four horses, with an escort of six pallbearers representing the infantry, artillery, and cavalry. Behind it were the family and clergymen in carriages. Then came representatives from local organizations: the officers of the 19th and 44th Regiments, the men of the 44th, the men of the 3rd Regiment of Cavalry, a veteran of the War of 1812 (driven in a carriage), the Fire Department, the local lacrosse club, employees of two local firms, followed finally by elected city officials, judges, and

Figure 2.1. "Part of Funeral Procession"

Source: Published in the St. Catharines *Standard*, 18 January 1937, 7. Image in the possession of Niagara Historical Society in 1936, but since lost. Currently no known copy available; image reproduced from microfilm.

prominent citizens (the last three groups all in carriages).[78] The combination of the military, first responders, politicians, working men, and athletes created a highly masculine and visually united show of strength.

Along the four-kilometre route, this unity was reinforced by other groups. Shops were closed, several businesses were draped in black, and flags flew at half staff. Onlookers lined the streets. Again, these symbols and rituals would have been familiar to immigrants who considered themselves British and understood the ceremony surrounding the death of an important figure. However, the protector of the realm who was celebrated in 1885 was not a sovereign, governor, general, or knight. In Watson's case, the protector of the realm was a citizen, a tradesman, and a private in a part-time militia unit. The procession and its reception were not simple demonstrations of brute military strength by the state, but demonstrations of local gratitude and civic strength by the citizens themselves.

The dedication at the cemetery followed common practice: a prayer, three volleys from the firing party, and a rendition of "God Save the Queen."[79] Still, the religious politics surrounding these events must have been interesting. The three largest Protestant faiths in Ontario were represented over the course of the day. The Watson family were members of First Presbyterian. Their pastor, Reverend John H. Ratcliffe, led the funeral service. At his side was the rector of St. Thomas' Anglican, Reverend Oliver J. Booth. Clerics at St. Thomas' had long served as chaplains to the 19th and 44th Regiments, so Booth's presence may have honoured the local militia. More subtly, his presence may have pandered to the Anglicans' belief that they were the established faith in Ontario and rightfully present at state funerals. Finally, the pastor of Welland Avenue Methodist Church, Reverend James G. Foote, spoke at the grave side. Foote was an Orangeman, and soon after served as a deputy grand chaplain in the order.[80] The absence of a visible Catholic leader is not surprising. The local militia's links to the Orange Order, and its brutal treatment of Irish workers on the Welland Canal, prompted most Catholics to shun militia events during this time.[81]

The funeral received detailed coverage in the *St. Catharines Journal*, and it was front-page news in nearby cities such as Welland and Hamilton, and even in distant Winnipeg. The *Journal* reporter described everything surrounding the day in heroic fashion: "The funeral of this gallant young soldier," an "active and zealous member" of the church, was the "largest and most imposing affair of the kind" the city had seen. His remains were enclosed "in a massive polished chestnut casket," and flowers were "a profusion of magnificent ... offerings of rich and rare colors [*sic*]." As it lay in state, the coffin sat on "a raised catafalque draped in deep black, from the sides of which sprung festoons of heavy crape from whose apex was suspended a sculptured snow-white dove." The elaborate funeral procession was also described in detail. The report concluded, "the echo of the falling earth upon the coffin lid told all that was mortal of a brave Canadian soldier, who died in the path of duty, was forever enshrouded in the bosom of mother earth."[82] The reporter's use of superlatives and high diction bid readers to view this common soldier in an uncommon way. It characterized Watson's actions in terms of bravery, faith, honour, and sacrifice. They were heroic qualities to be emulated by those who survived him.

Watson was not the only field force casualty to receive such tributes. The events in Winnipeg and St. Catharines were repeated in Ottawa, Perth, Port Hope, St. Thomas, and Toronto.[83] Most elements of Watson's funeral – lying in state, floral tributes, the military procession, the participation of civic officials, and the patronage of townsfolk – also appeared in these other services. Even the newspaper accounts echoed one another. Historian Desmond Morton notes that this was the first campaign the Canadian militia had completed without the support of British regulars. Those Canadians who embraced the militia myth found much to support their heroic conception of volunteers.[84]

A Soldier's Commemoration, 1886

Within days of Watson's second funeral, residents of St. Catharines formed a committee to preserve his memory for future generations. Among its members were Mayor Henry King, alderman Dr. Edwin Goodman, and Major George C. Carlisle of the 19th. The committee also sought input from Watson's unit, the 90th Battalion. The local community wanted a monument that bore a likeness of the dead man. The committee did not. Instead, the committee decided upon "a generic representation of a Canadian Volunteer" to honour all Watson's comrades-in-arms. By August 1885, $780.00 had been raised to pay for it. A "Watson Memorial Concert" generated $200 of this amount, city council donated $50, and the employees of Riordan Paper Mills donated another $50. Other donations came from individuals near and far, including members of the 90th.[85]

The monument was originally intended for Watson's grave. Stroll through any Victorian cemetery in a Canadian city and you will find imposing monuments marking the legacy of local worthies, while dynastic families lie entombed in stately crypts. In keeping with this practice, the monument was to be the community's gift to the Watson family: a stone befitting their son's new status in death. The town of Perth made a similar gesture on behalf of Lieutenant A.W. Kippen, who also died at Batoche. A monument much like Watson's was erected over the soldier's grave, and it, too, was financed by loyal friends, grateful townsfolk, and the town council. The local newspaper referred to it as "the citizens monument" (figure 2.2).[86]

Watson's family imagined a more intimate burial. In the days following his death, his father acquired a family plot and a simple obelisk that could have borne each of their names.[87] A month before the monument was unveiled at the cemetery, city council called a special meeting to reconsider its location. Instead of the cemetery, the council agreed to place it at city hall. The decision suggests a desire to place the monument in the public eye, just as the funeral itself had been conducted in the public eye. City hall was located downtown on James Street, where it shared a block with the farmers' market, police station, and courthouse. By contrast, the cemetery was three kilometres outside the core and surrounded by farmland. St. Catharines had no public park at the time, and the local militia did not have a permanent armoury. City hall, then, was the most prominent and visible location available for

Figure 2.2. Lieutenant A.W. Kippen's grave monument, Elmwood Cemetery, Perth

Source: Alan Drummond (Perth, Ontario), ca. 2014. Used with permission.

a monument, the most effective location for it to convey its intended message (figure 2.3).[88]

What was the monument's message? We may deduce the committee's thoughts from its placement and design. City hall was certainly visible but the property also carried symbolic weight as the local seat of government. Further, it represented a secular landscape in contrast to the spiritual landscape of the cemetery. If Watson's sacrifice had been viewed strictly in religious terms, it could have been placed at a suitably impressive church downtown. Knox Presbyterian Church faced the city hall lawn, while St. Thomas' Anglican and St. George's Anglican are just short blocks away. The monument's final placement suggests that Watson's service was a contribution to nation building rather than to Protestant evangelism. And that military service lent honour by association to a city that sought a higher profile in the national public sphere.

The decision to produce a monument in stone is also significant. A memorial can take many forms, but stone conveys permanence. Again, absolutist regimes understood that a sovereign was both a mortal human and a symbol. The human could die, but the symbol had to endure to legitimize the continuity of the regime. An effigy of the

Figure 2.3. Watson Monument as depicted in a 1907 booklet, *The Garden City of Canada*. Engraving produced in 1890s for James Monroe, owner/operator of a marble works, for company advertising.

Source: St. Catharines: Standard Printing, 1907.

sovereign rendered in stone could maintain the vivacity of the symbol. With the Watson Monument, there is again the transferral of an absolutist ritual from a sovereign to a citizen-soldier. The ideal that is celebrated is not the transcendence of the great leader but the willing sacrifice of the individual citizen, a local who had served the nation.[89]

A competition was held for the monument's design. The winner was James Munro, owner of a local granite and marble works best known for gravestones. The completed monument stands roughly seven metres tall, carved in limestone drawn from the Niagara Escarpment. As the committee desired, it does not bear Watson's likeness. The statue portrays a common soldier in period uniform standing at ease. That did not deter the *Winnipeg Free Press* from commenting that the visage was a fair portrait of the dead man.[90] A knowledgeable viewer would have known that Watson wore a rifleman's distinctive green tunic and black busby. The soldier portrayed in grey limestone is easily imagined wearing a red tunic and white helmet. Hence, this is not a rifleman's uniform as the 90th Battalion in Winnipeg wore, but an infantryman's as the 19th Battalion in St. Catharines wore. Casting the local hero in the garb of the local militia was a manifestation of reflected glory, glory that Watson had earned on a national stage.

Watson's name appears at the base of the monument and represents the most prominent text. On the front, the original inscription read,

Erected to the memory of
Alexander Watson,
90th (Winnipeg) Battalion Rifles,
Canadian Volunteers,
and his companions in arms who fell in
battle during the Rebellion in the
Northwest Territories, A.D., 1885.

At the rear were the names of the campaign's four major battles. Relief carvings, now lost to erosion, appeared on the pedestal's cap. One featured crossed bayonets and a busby, another an artillery piece.[91] The total effect of this script and imagery is solemn and relatively modest. The viewer is not called to serve the nation, nor even reminded of the Crown or constituted authority. One can imagine a similar monument to Watson the carpenter decorated with the names of the prominent buildings he raised, and reliefs of a hand drill and crossed hammers. The monument is only patriotically stirring insofar as the viewer is aware of the North-West Resistance and has a predisposition to honour the city's contribution to national defence.

This modest representation in stone was reinforced by the landscaping. The statue stood to one side of city hall facing James Street, on a square bed of grass approximately thirty centimetres higher than the surrounding lawn. Notably, the city already possessed two cannons, which were placed on either side of the building's front steps. Council thought to place them on either side of the monument, but they were never moved. Instead, flowerpots decorated each corner of the raised bed.[92] Funerary symbolism may have overwhelmed the patriotic or militaristic symbolism of the space.

The monument was unveiled on Tuesday, 14 September 1886. It was the first monument raised in Ontario for the North-West campaign, but the city did not publicize this fact. It found another way to enhance the pageantry of the day and draw the country's attention: General Frederick Middleton agreed to preside. The unusual midweek scheduling overlapped with a militia training camp held in Niagara-on-the-Lake. All the volunteers were invited to "make a great day of it."[93] They did. The ceremony involved speeches by prominent men, the transferral of the monument's deed to the city, and the unveiling itself. It was a grand event for the small community. One reporter claimed that it was the largest crowd ever assembled in St. Catharines, "fences, trees and every spot that could hold a sight-seer being occupied."[94] A photograph shows people sitting on the roof of city hall. Nonetheless, reporters felt that it was not well planned (figure 2.4). The speakers' platform stood only thirty centimetres high, and, despite the efforts of police, onlookers took "complete possession of the ground right up to the monument."[95] As a result, it was difficult to hear the speakers from any distance. A dismayed reporter noted that "not a single cheer was heard from beginning to end" since the audience missed its cues.[96] Still, the day was not

Figure 2.4. Unveiling the Watson Monument, 1886

Source: Original photograph taken in 1885; now in the John Burtniak Collection, St. Catharines. Used with permission.

lost: it ended with a particularly bloody lacrosse match between the St. Catharines Saints and a Toronto team, won by the locals. The unveiling was front-page news in Winnipeg, but only made the sports page in Toronto.[97]

A Citizen-Soldier Remembered

Ultimately, Watson had two funerals and two gravestones, and his name still appears on three monuments. Winnipeg and Toronto also commemorated the field force soldiers who fell in the North-West. The memory work in both cities had parallels with St. Catharines: committees formed in the enthusiastic aftermath of the campaign, funds were raised through public subscription, and monuments were erected in highly visible and secular public spaces. In Winnipeg, a column dedicated to the volunteers of the 90th Battalion was placed before city hall.[98] The dead were named individually. In Toronto, an allegorical statue of Peace, dedicated to all the units that fought the resistance, was placed at Queen's Park. It, too, identifies each of the fallen. It was the first official commission for sculptor Walter Allward. At its unveiling, he was modest about its execution: "It is not what it might have been, but it was the best I could do. Probably I will do better next time."[99] He later designed the Canadian National Vimy Memorial commemorating the Canadian dead of the First World War.

There was no contemporary memorial raised to the dead of the resistance. Eight Cree men were executed for their roles at Frog Lake and Battleford: Kah-paypamhchukwao, A-pis-chas-koos, Pahpah-me-kee-sick, Manchoose, Kit-awah-ke-ni, Nahpase, Itka, and Waywahnitch. Members of the Sweetgrass Reserve believed the executioners did not return the dead to their families precisely to prevent memorialization. Indeed, their common grave went unmarked for almost a century, until their descendants sought it out.[100] The conditions that prompted the Métis and First Nations to resort to violence – declining economic prospects and lack of land security – remained and were to grow worse. Many of the Métis leaders were dead, injured, or imprisoned. Farms around Batoche had to be rebuilt after the field force looted then razed them. Still, survivors maintained a veterans' organization known as L'Union Nationale Métisse. It endured into the 1920s and accomplished two mnemonic goals: it erected memorials in 1901 to fallen resistance fighters buried at Batoche and St. Laurent de Grandin, and it promoted 24 July as a celebration of Métis culture.[101] From the 1920s to the 1970s, however, succeeding generations tended to repress knowledge of the resistance out of fear of reprisals from white society. This tendency gradually faded as new political movements supporting Indigenous rights took shape. Today in Winnipeg, there are four monuments commemorating Louis Riel, all raised since 1971, and his former home there is now a museum. A street, a park, and an esplanade are named in his honour.[102]

Canada, politically, was still a new country in 1885. Its population, however, was composed of several ethnic groups with long histories of their own. Sir John A. Macdonald spoke of integrating some of these groups into a single, new national identity. Unfortunately, his ministry's policies – focused on immigration, railway expansion, and white settlement on the Prairies – tended to foster division by privileging anglophone Protestants over First Nations, Métis, francophones, and Catholics. The Prairies were imagined as an empty, fertile utopia for northern European migrants where the legal and religious institutions of Ontario, not Quebec, would be extended. It was an economic and cultural agenda that Ottawa used force of arms to protect.

When the resistance began, news from the North-West was limited and filled with speculation. After Frog Lake, it seemed the entire region could be taken by the Métis and allied First Nations. The Macdonaldian conception of Canada's future seemed in peril. When the field force took Batoche, its victory seemed absolute. The relief felt by most white settlers in Manitoba and Ontario is still palpable in their accounts. The task of nation building was back on track.

In St. Catharines, relief was magnified by the realization that one of their own had shared in the glory. The city's own units were not called to the front, but still the city had contributed to the national war effort. Whatever anxiety or grief residents had experienced during the North-West campaign was transferred to Watson's death in combat. The magnitude of that common anxiety and grief was expressed in two ways: a public funeral to honour the man, and a monument to honour his service. Both were arranged by civic officials and the local militia. Both celebrated the city's contribution to the national cause. Furthermore, city hall, the courthouse, and the streets themselves became sites of commemoration in place of the battlefield. The death of an individual citizen-soldier became the cause for pageantry and for the celebration of duty, courage, and gallantry ascribed to an entire community. For the unveiling, city hall hosted an event with the country's top soldier that placed the city in the national public sphere. It was by no means a private family ceremony, or even a purely local occasion. It was a declaration of St. Catharines's place in the nation.

Memorials, however, should probably remind us of something. In this case, Watson died a soldier, was buried a hero, and was memorialized as an ideal. The militia myth gave residents the values and vocabulary they needed to articulate this transformation. Indeed, the militia's performance in the North-West lent the myth greater currency. The glorification of the Union dead after the American Civil War also provided an influential precedent to memorialize Canadian deaths. Ontarians, however, did not draw solely upon the republican tone of American eulogies. They drew upon the heraldic customs of England to transform each commoner into a courtly knight. Residents had not honoured Brock's soldiers in this way in 1812, but Mewburn and his peers were so honoured in 1866. In future, every soldier who died in combat would be marked in similar ways. Albert Storm's service in Afghanistan was so honoured with a public funeral and the erection of a memorial playground in his hometown. Like the mourners of 1885, the mourners of 2006 marked the passing of someone they believed represented the best qualities of their community. What was once a new expression of civic values now seems a timeless ritual.

Chapter Three

The Watson Monument through Time

This snow white tablet ... will we trust awake anon the tales of valor [*sic*] to many a coming generation.

– From the dedication of a soldier's memorial in St. Catharines, 1900

After the Watson Monument was unveiled in 1886, it became the focus for enthusiastic commemorations of military service. Its celebration of the citizen-soldier created a symbolic landscape in which other military engagements, beyond the North-West Resistance, could also be marked. In time, its role in these commemorations was challenged by new memorials dedicated to the dead of other wars. By examining the monument's career as a mnemonic product through time, the monument's contribution to local identity is revealed. It provides an intriguing case study in the retention and loss of narratives, rituals, and heritage at the local level.

It bears repeating that monuments, like all mnemonic products, are means of communication. Their creators tend to be citizens with the economic, cultural, and political resources to command public space for their own causes.[1] And as Iain Hay, Andrew Hughes, and Mark Tutton note, their monuments mark out favoured people and histories and ignore others, denote patterns of authority and power, inculcate views of heritage and values preferred by dominant groups, and promote particular futures.[2] In other words, a material commemoration encourages a particular interpretation of a community's past in order to foster a shared understanding of the present. It will convey its interpretation effectively through time only if its narrative retains authoritative weight, remains ubiquitous through intertextual repetition, and resonates with new audiences. Our surveys indicate, however, that many citizens are genuinely ambivalent towards monuments and historic sites despite their sponsors' authority or their physical visibility. Even if a monument attracts a sympathetic audience, there is no guarantee its message will remain constant through time. Economic, cultural, and political changes can shift audiences' engagement with its central narrative and the symbolism through which it is conveyed. Many monuments can ride out these changes, but some become lightning rods for public fervour or for discontent.

There is nothing necessarily good or bad about this process. Reinterpretation and debate demonstrate engagement with a memorial; they imply that it still matters to someone. Reinterpretation in itself may prolong a memorial's relevance by ensuring that its narrative continues to mesh with contemporary social values. This is true even if a new interpretation of its narrative runs counter to the intentions of its sponsors. In Toronto, for example, the monument erected in Queen's Park to honour the Canadian militiamen who confronted the North-West Resistance has been used by the Métis Nation of Ontario to mark Louis Riel Day.[3] Hence, it is important to consider the ongoing career of a memorial in the context of its use. Its significance cannot be determined solely through archival research, textual decoding, or semiotic analysis focused on the values of its sponsors. The meaning of a memorial becomes manifest in how community members engage with it and animate its narrative.

Our goal is to explore the career of the Watson Monument since its unveiling. Charlene Mires demonstrated how this can be done through her study of Independence Hall in Philadelphia, Pennsylvania. She isolated key moments in the building's history to explore its changing status in the American imagination.[4] Similarly, Robert J. Harding observed how residents of Newfoundland and Labrador memorialized their soldiers' service in the First World War in the years following the Armistice.[5] We do the same for the Watson Monument. Insights are drawn from a wide array of documents and artefacts: contemporary newspaper accounts and letters to the editor, official municipal records, tourism brochures and videos, amateur histories, photographs, and commercial television programs, as well as an interview and field visits to relevant sites located in the city. The result is a cultural history of the monument focusing on its place in public consciousness. As time passed, public perceptions of the monument shifted through three distinct phases:

- 1886 to 1927, when the monument functioned as a service memorial;
- 1927 to 1971, when its function as a service memorial was eclipsed by a cenotaph; and
- 1971 to 2009, when the monument became a "quaint anachronism."

A Service Memorial, 1886–1927

As described in the last chapter, two important decisions shaped the Watson Monument's role in the community. First, it was not dedicated solely to Alexander Watson but to all militia volunteers who fell during the North-West Resistance. Second, it was not placed over Watson's grave but was placed on the lawn at city hall. These two decisions shifted the focus and symbolism of the commemoration. It was no longer a grateful community's private memorial to a valiant son, but an official endorsement of the citizen-soldier. The monument's implicit message honoured civic duty and individual service.

The monument's location also made it highly visible in the community. It was certainly a recognizable civic landmark and drew attention to Watson and his family. The death of Watson's father, David, for example, was treated as front-page news in 1900. David Watson was not a public figure. It was the monument to his son that made his death newsworthy, and it was the monument that was referenced in the first line of the *Standard* story.[6] In like manner, images of city hall often focused on the monument. This view of the building was sufficiently picturesque for a postcard and featured in at least three promotional booklets by 1907.[7] It also caught the eyes of visitors. A Toronto *Globe* article on the city featured a drawing of city hall with the monument again prominent in the foreground. Local groups, too, recorded themselves while posing at its base. Such images include a men's bicycle club (1889) and girls' cadet corps with rifles at ready position (1891). Its sculptor, James Munro, used an engraving of the monument to advertise his firm. Though many families hired him for burial stones and garden ornaments, the monument was still considered his most impressive work more than twenty years after its unveiling.[8]

The lustre of official recognition and public visibility that was conferred upon the monument was in turn bestowed upon later military commemorations. In effect, the monument became a physical and symbolic manifestation of the link between local governance, local defence forces, and the St. Catharines citizenry. Its symbolism was later reinforced by memorials for veterans of Ridgeway, the Second Boer War, the two world wars, and the Korean War.

The Fenian Raids, as noted in the previous chapter, were part of an international effort by Irish patriots to end British rule in their homeland. American Fenians made their bloodiest raid on Canada in May 1866. An estimated fifteen hundred men assembled in Buffalo, New York, and then crossed the Niagara River to land near Fort Erie. The Canadian government, having had advance warning, called out the militia. Some ten thousand men from across the province were posted to the Niagara region, and four units confronted the Fenians near the village of Ridgeway on 2 June. After four days of fighting, the Fenians retreated to the United States.[9]

The Canadian militia took pride in the campaign. In Toronto, veterans raised the Volunteers' Monument in Queen's Park in 1870 to mark the victory and honour the dead. St. Catharines also took pride in its role. Its militia units were deployed across the region as home guards, its city hall was transformed into a field hospital staffed by local women, and the harbour at Port Dalhousie provided a staging ground for the Toronto militia before and after the Battle of Ridgeway.[10] Niagara residents, however, did not raise a contemporary monument. In fact, a monument was not raised at Ridgeway itself until the 1920s. And none has yet been raised to remember John H. Mewburn, the only Niagara resident who died during the battle.[11]

Twenty-five years later, Toronto veterans asked to have the militia's service during the raids officially recognized by the Canadian federal government. The first raid had occurred before Confederation, so while the local militia had protected Canadian territory, they had not technically served the Dominion of Canada. After several years of veteran lobbying, Ottawa granted their wish in 1898. First, it struck a single service medal for militia veterans of both the Fenian Raids and Red River Resistance of 1870 (militia veterans of the 1885 campaign in the North-West had already received medals for their service). Second, it declared the anniversary of Ridgeway, 2 June, an annual day of commemoration known as Decoration Day.[12]

The concept of an annual commemoration originated in the United States. In the immediate aftermath of the American Civil War, African Americans in Charleston, South Carolina, honoured the Union dead by providing them with respectable graves. They also decorated the graves with flowers. As years passed, the act of maintaining and decorating the graves developed into a yearly ritual when African Americans and the families of the dead would mark the war's end. Confederate sympathizers adopted the ritual too, but they marked days that resonated with their own narratives of the war. Over time, several states formally recognized a single day to institutionalize this ritual. These days have since been consolidated in the American national holiday known as Memorial Day.[13]

The Decoration Day ritual inspired Canadian veterans' organizations, particularly those in Winnipeg and Toronto. Beginning in 1886, Winnipeg veterans of the North-West campaign honoured their fallen comrades each year with a church service and a march past the Volunteers'

Monument. This was largely a private service of personal remembrance. They converted this ritual to a decoration service in 1895. With the participation of the 90th Battalion, it then became a public statement about volunteerism and Canadian military readiness.[14] In Toronto, the day was always designed to generate publicity for the militia by valorizing the dead of Ridgeway and the North-West. Veterans like Colonel George T. Denison III, who served in both campaigns, initiated a decoration service in 1890 at the two monuments in Queen's Park. They lobbied schools to let students attend, and boys were invited to participate in the service.[15] These activities received mixed reactions from the press. The Toronto *Globe* faithfully covered each year's planning and speeches, while the *Star* questioned its perfervid oratory.[16] The satirical magazine *Grip* outright mocked its sponsors' intentions:

> The Loyalists of Toronto are rejoicing over the grand opportunity afforded them by Decoration Day last week of instilling into the minds of the public school pupils lessons of hatred and prejudice against the people among whom a very large proportion of them will have to earn their bread. By the way, how can bitter Yankee-phobists like Militia Col. Denison and Jim Hughes reconcile themselves to the adoption of such a purely Yankee institution as Decoration Day[?][17]

The veterans ignored these jibes, at least in print.

When Ottawa added Decoration Day to its official calendar, a veterans' group in St. Catharines invited all Ridgeway veterans to Niagara for a grand reunion. The Niagara District Volunteer Veterans' Association (NDVVA) gathered local men who had served in Crimea, Ridgeway, and the North-West. The invitation was well received. Toronto veterans wished to hold their own Decoration ceremony in Queen's Park on June second, but they welcomed a summer reunion at their old staging ground. The NDVVA readily acquiesced. It postponed its own Decoration ceremony to a Sunday in July and set the reunion for the next day. *The Standard*'s editor believed that combining the two events generated more interest in the Decoration service itself.[18]

The actual observance – the first to be held in St. Catharines – was held in the city's Victoria Lawn Cemetery. The centrepiece was a service staged at the grave of veteran J.W. Grote. Grote had been found dead on the banks of the Welland Canal just weeks before, and he had been given a pauper's burial without service. He was not known to the NDVVA. According to *The Standard*, "when it was learned that the departed had been a soldier of the Queen, the veterans determined to thus honor [sic] his memory."[19] The moment was heavy with symbolism: a forgotten citizen-soldier, found dead in dubious surroundings, without friends or family, whose passing went unremarked even as his comrades were celebrated. The circumstances of Grote's death underscored all veterans' calls to have their service remembered. On this day, *The Standard* reported that hundreds of civilians attended the service and decorated the soldiers' graves. The NDVVA hoped to build upon this strong beginning, but Decoration Day remained largely a private ceremony for veterans and their families in the years that followed.[20]

The NDVVA, the City of St. Catharines, and the 19th Regiment cooperated in planning the Ridgeway reunion. They prepared a full day of events: speeches, a parade, lunch, music, and sports.[21] Veterans arrived from across Ontario. The largest contingent – numbering nearly four hundred – came from Toronto accompanied by the 48th Highlanders band. The *Thorold Post* estimated that eight thousand veterans, militia volunteers, and civilians attended the events despite a severe thunderstorm that morning.[22] The day began with welcoming speeches at Montebello Park, four blocks from city hall. From there, the veterans led a parade through the downtown core that ended at the Watson Monument. Here they met the day's true purpose: the actual remembrance of the dead through the decoration of a monument. The veterans, bands, and local Boys' Brigade formed a square around the statue. Two wreaths were placed there, one each by the veterans' associations of Toronto and St. Catharines. The 19th presented arms and the 48th band played a hymn. *The Standard* reported that, altogether, it was an impressive ceremony, demonstrating that loyal Canadians were willing to remember brave deeds.[23]

Two insights may be drawn from the Ridgeway reunion. First, the monument's central role in a Decoration service demonstrates that public commemorative practices were shifting from battlefields and graveyards – where the honoured dead were buried – to historically empty sites marked only by memorials. The veterans could have held their Decoration service at Watson's grave in Victoria Lawn Cemetery, but they chose the monument site instead. Perhaps the monument's location was significant: it was certainly more accessible in the city's core, close to the meeting grounds at Montebello Park and amid the downtown streets that hosted the parade. More pointedly, Winnipeg and Toronto veterans had already set a precedent by staging their most solemn events at Ridgeway and North-West monuments in those cities. In the context of memorial services, then, the site of a monument was a ready substitute for Ridgeway and Batoche, as well as every graveyard where militia men were buried.

Second, the reunion planners expanded upon the monument's original purpose. The monument was conceived as a tribute to Watson and his peers in the North-West Field Force. This tribute was now extended to all veterans of the Canadian military. In this context, the reunion served as a rededication ceremony for the monument, subtly altering its role as a site of public remembrance to one of public ritual. Reinforcing this notion, the federal government had planned to distribute the new service medals for Ridgeway and Red River veterans during the service at the monument. The medals were not authorized for production, however, until the following year.[24] When they became available, recipients in St. Catharines gathered once again at the monument. A photographer captured the moment for posterity (figure 3.1).

After the North-West Resistance, the Second Boer War was the next major military action in which Canadian troops took part. Through the late 1800s, Great Britain sought to expand its colonial control over the southern tip of Africa by annexing independent states organized by

Figure 3.1. St. Catharines veterans of Ridgeway and Red River receive their medals, 1900

Source: Photo taken in spring 1900; photographer unknown. St. Catharines Museum, Accession no. 9384-N.

Dutch settlers (known as Boers). Various conflicts between the British and Boers led to open warfare in October 1899. When this happened, the British government asked its settler colonies and Canada to demonstrate imperial solidarity by supplying troops for the war. Sir Wilfrid Laurier's government, like the country itself, was divided. Canada nonetheless organized and equipped two volunteer contingents for service. Following a decisive action at Paardeberg in February 1900, the Canadian troops were celebrated in Britain and English-speaking Canada took greater notice of the war.[25] Some of this enthusiasm took the form of mafficking: spontaneous celebrations in public spaces with marches, songs, whistles, bells, flags, and speeches involving hundreds of people. The practice began in England following the successful relief of the town of Mafeking (now Mahikeng) from a Boer siege in May 1900. Subsequently, similar demonstrations took place across Britain and Canada following each major British victory.[26] The war lingered on, however. Victories grew rare, British malfeasance was exposed, and anti-imperial sentiment grew in Quebec. Popular enthusiasm for the war subsequently declined.[27]

Support for the war in St. Catharines followed these trends. The minister of militia did not mobilize existing regiments to support the Canadian expedition. He asked volunteers to join ad hoc regiments for overseas service. As such, while the 19th Regiment was not called, individual members were selected for the volunteer force. The first troops to leave were sent off with a modest farewell. By contrast, two soldiers returning with this first contingent were met by bands, paraded to the opera house for speeches, and feted by militia officers at the city armoury. As national interest in the war peaked, St. Catharines revelled in its own mafficking.[28]

The dead were not brought home. Under British commanders, the remains of Canadian soldiers were interred near the battlefield or camp hospital where they passed. The army arranged the services and resources required for burial. Many graves were marked only with wooden crosses.[29] When popular support for the war surged in Canada, a new patriotic organization began to take shape in Toronto: the Imperial Order Daughters of the Empire (IODE). The IODE was established by middle-class Protestant women who championed a Canadian identity rooted in British heritage and political traditions. They viewed the African campaign as an opportunity to prove Canada's value to the empire. Following the Boer surrender in 1902, the IODE sought to perpetuate the memory of Canada's service and sacrifice. Thus, they worked alongside veterans and other patriotic groups to identify and preserve the Canadian war graves. Their main partner was the Guild of Loyal Women of South Africa. While the IODE provided lists of the dead and financing, the guild completed field research and erected stone monuments. In the 2010s, there were ten known cemeteries in South Africa with Canadian military graves. The IODE's work during this war fostered a lasting interest in military commemoration among its branches, an interest they then encouraged among Girl Guide and Boy Scout organizations.[30]

In St. Catharines, the city did not recognize the war dead equally. The first to fall in Africa was Henry M. Arnold. Like Alexander Watson, Arnold had settled in Winnipeg in the early 1880s and joined the 90th Battalion there. His death at Paardeberg in February 1900 was noted in the press but was not marked publicly in St. Catharines.[31] The first fatality to stir the city was the death of Robert Irwin. News of his demise reached the city in July 1900, the high point of national enthusiasm for the war. Irwin had been a member of the local 19th Regiment, and the same men who staged the Ridgeway reunion two years earlier now planned a memorial service for their late friend. The 19th – along with local veterans of Ridgeway and the North-West – marched from the St. Catharines Armoury to St. Thomas's Anglican Church. The church service was followed by a march to Montebello Park, where the 19th band played a mix of songs mourning the dead and celebrating the empire.[32] In the following months, four more soldiers with local connections died. They were remembered through private rather than public services. In one instance, a local soldier's death in Africa was overshadowed by city council's staging of a public memorial service following the assassination of American president William McKinley in Buffalo, New York.[33]

St. Catharines did not raise a new monument to honour the dead of the African war. Instead, their names were engraved on the Watson Monument. The Ridgeway reunion had demonstrated the monument's value as a general service memorial. Now, it became one. City council felt that it was "both right and desirable to mark in every possible way the sympathy, esteem and admiration ... for the gallant and patriotic volunteers, the soldiers of the Queen from this City and County who have lost their lives while bravely fighting."[34] City staff were directed to add the name of any local soldier who died in service to the plaque. This was done, quietly and without ceremony, for four of the men.[35] At least one other memorial was commissioned privately. Friends and family of Henry M. Arnold honoured him with a marble tablet in St. George's Anglican Church.[36] Otherwise, there was no public event staged to mark the war's end or to acknowledge the soldiers' sacrifice. Nor was the Watson Monument renamed to reflect its new status as a general service memorial rather than one specific to Watson or the North-West.

The First World War prompted a dramatically different response to soldiers' deaths and their commemoration. The number of men from St. Catharines who faced the Fenians, North-West Resistance, or Boers is not known. The 19th Regiment was only mobilized in 1866, and it never faced enemy fire. We do know, however, the names of seven local men who died during one of these three campaigns. By contrast, St. Catharines – like most towns in Canada and Newfoundland and Labrador – saw many young men leave for Europe. The city's entire male population was roughly 6,200, and traditional estimates set its contribution at 1,000 soldiers and 324 casualties. Such estimates are now considered low.[37] Residents were appalled by the loss of life.

In 1918, a new memorial was erected specifically to honour the dead of the First World War. It, too, was placed on the lawn at city hall. The new piece was a wooden cross described most often as the Soldiers' Cross, but it was also called the War Cross, Memorial Cross, and Cross of Sacrifice. It no longer exists; only two photos remain (figure 3.2). It was probably inspired by the grave markers used by the Imperial War Graves Commission (now the Commonwealth War Graves

Figure 3.2. The Soldiers' Cross and Watson Monument, ca. 1919.

Source: Postcard in the personal collection of Dennis Gannon. Used with permission.

Commission). The commission was established during the war by the governments of the British Empire to administer their military cemeteries along the western front. Early on, the commission decided to treat all fatalities individually and equally regardless of nationality, faith, or rank in deference to their common sacrifice. This was a significant break with the mass graves of the past. At first, all known graves were marked with identical wooden crosses. After the war, the crosses were replaced by stones that allowed each soldier's individual faith to be identified. Imperial soldiers of racialized backgrounds, however, were not always granted these courtesies.[38]

The origin of the cross in St. Catharines is not known. It was first mentioned in news reports of that year's Decoration Day service. Later commentary noted that its design and installation were considered temporary.[39] The only images of the cross show it decorated with a wreath in late spring or summer. Given this slim information, perhaps local residents marking Decoration Day created this empty grave to include the European dead in their observances. Its placement at city hall follows the precedent set by the Watson Monument for military commemorations. It could have been placed elsewhere. The city had several public parks and a new armoury by 1918. Nonetheless, the lawn at city hall remained official, secular, and highly visible. This blunt presentation of a battlefield grave in the heart of the city may have provoked strong emotions in those who walked past it.

The proximity of the cross to the Watson Monument invites comparison. By 1914 the monument functioned as a general service memorial. It should have been a simple matter to incorporate the European war in its message. However, the cross was planted in a discrete space away from the statue. Perhaps its sponsors believed the city's new wartime sacrifices were too great and too distinct from past experience to be comparable. This belief was not unique to St. Catharines, but was articulated throughout Canada and Great Britain.[40]

The apparent rupture between the European war and the Watson Monument could not have been more pronounced than on 11 November 1918. Telegraph messages bearing the first news of the Armistice arrived at 4:00 a.m. St. Catharines time. Residents downtown were awakened by church bells, fire alarms, factory steam whistles, and an impromptu 5:00 a.m. parade conducted by steel and auto workers. As the sun rose, bonfires and an effigy of Kaiser Wilhelm II lit the streets while homes, businesses, and vehicles sprouted flags and bunting. Another parade – the third that day – began at 10:00 a.m., when city hall received official confirmation of the ceasefire. Led by the 19th Regiment band, it was described as the greatest ever seen in the city. It started at

Market Square behind city hall, circled the downtown core, and ended at the intersection of St. Paul and Ontario Streets. This was a popular destination for parades and marches because it provided an open space for citizens to gather for speeches, cheers, and song. Churches throughout the city held services.[41]

The Watson Monument was noticeably absent from these spontaneous celebrations. It was the most visible war memorial in the city, but apparently few were thinking of the past. Most thoughts had turned to the sudden vitality of the present. Two events suggest this. First, the city's Victory Loan Committee met its sales objective within a week of the armistice. Success was celebrated with another parade, more speeches, and once again the 19th band was called into service. The parade paused at city hall, where speeches were given from the front step. The monument was not mentioned in the newspaper report of this celebration.[42] Second, the city decided to stage a public Thanksgiving service to mark the war's end. It was planned for a Sunday afternoon in Montebello Park, then moved to the Grand Opera House to avoid inclement weather, then cancelled due to the growing threat of the global influenza pandemic that emerged that fall.[43] Regardless, the availability or size of the space surrounding the monument was not an issue for either event. Both were planned when businesses and city offices were closed. The monument was either dismissed or forgotten as a site dedicated to service and sacrifice.

The symbolism of the cross also challenged the monument. Though its purpose was to commemorate the citizen-soldier, the monument featured a single soldier in name and form, and the statue was placed heroically on a pedestal high above the viewer. The cross conceived by the War Graves Commission was meant to honour every soldier individually and equally, despite its overtly Christian symbolism. One looked down at the cross; it evoked humility and the blunt finality of the soldier's sacrifice. On the lawn at city hall, it represented every local man who would not return. This tension between the monument and cross practically anticipated the variety of Canadian memorials that emerged after the war, as artists grappled with traditional heroic forms and modern simplicity to remember the dead.[44] No doubt there were residents who found solace and inspiration in both. And those who found neither. A new pastor at St. Paul Street Methodist Church, who had arrived during the war, expressed this latter sentiment. In 1922, Reverend George K.B. Adams stated that he had "looked about for our war memorial, our manifestation of our appreciation [to the valorous men], and found nothing." This was a harsh rebuke given that his church was only five blocks from city hall. His son had died at Passchendaele in 1917.[45]

As time passed and celebration turned to remembrance, the monument returned to people's thoughts. On 2 June 1919, the anniversary of the Battle of Ridgeway, the Watson Monument and the Soldiers' Cross were the site of a renewed Decoration Day service. The NDVVA still managed this annual observance, but after 1909 they had been joined by the local branch of the IODE.[46] The 19th Regiment and Boy Scouts also assisted. In attendance were 20 veterans of Ridgeway, roughly 150 veterans of the European war, active members of the 19th, and civilian spectators. In their words and actions, the organizers linked together all military service over the previous century to enhance the significance of Ridgeway itself. The War of 1812, the Fenian raids, the North-West Resistance, the Second Boer War, and now the Great War: these were not marked as isolated conflicts, but as episodes in a narrative demonstrating Canada's ongoing relationship with Britain. Reverend Lewis W. Broughall of St. George's Anglican hailed "all that had been laid upon the altar of sacrifice to keep flying the Union Jack and to uphold all the ideals of Right, Democracy, and Justice for which it stands."[47] His sentiments were reinforced by further speeches from the president of the NDVVA, the mayor, an IODE representative, and the commander of the 19th. Wreaths were then laid before the monument and the Soldiers' Cross to honour the dead of Ridgeway and the Great War. The ceremony closed with a rendition of the last post. Participants then paraded to Victoria Lawn Cemetery.[48]

On 11 November 1920, the same groups gathered again at city hall to mark the anniversary of the Armistice. It took several years for the day's customs to crystallize into an annual ritual.[49] Some of its elements, such as the rifle salute and the last post, were established practices in military funerals and Decoration Day services. Other elements were new. Canadians had adopted Lieutenant Colonel John McCrae's "In Flanders Fields" as a secular prayer before the war ended. Moina Michael, inspired by the poem, pushed her fellow Americans to wear poppies as boutonnières in memory of the war dead. Her idea was soon adopted in France, Great Britain, Canada, Australia, and New Zealand. In St. Catharines, local veterans and IODE chapters began selling artificial poppies in 1921 to raise money for veterans' welfare programs.[50] The official date of the commemoration was not fixed by the Canadian federal government until 1931. When it was, the name "Armistice Day" was replaced with "Remembrance Day."[51]

Two significant elements of Remembrance Day came from the empire. The two-minute silence was first observed in Cape Town, South Africa, during the war. In May 1918, Mayor Sir Harry Hands asked his fellow citizens to observe one minute for those who returned, and one for those who did not, each day at noon until peace was declared. Reports of this practice spread throughout the empire and reached Ontario through the Toronto *Globe*.[52] King George V was advised to adopt the idea for the first anniversary of the Armistice. He asked his subjects to remember the dead by observing the two-minute silence at the stroke of 11:00 a.m., wherever they might be.[53] Earlier that year, the British government marked the signing of the Versailles Treaty – the official end of the war – with a military parade in London. To remember the dead, an empty tomb constructed of wood and plaster was erected along the parade route. The heartfelt response of soldiers and spectators to its austere presence prompted calls for a permanent monument. This became the Cenotaph. It was unveiled on 11 November 1920, the second anniversary of the Armistice, with a memorial service that incorporated the two-minute silence. Thereafter, communities throughout the empire developed their own anniversary services incorporating the silence at local sites of remembrance dominated by permanent memorials.[54]

St. Catharines held its first Armistice Day service on 11 November 1920. As noted above, a 1918 ceremony to mark the war's end was cancelled due to the influenza pandemic. The following year, no one organized a public service or designated a public site to mark the anniversary. Citizens kept the two-minute silence, but as in Britain this was done informally wherever people happened to be.[55] As such, the Decoration Day service in 1919 may have provided the structure for local Armistice Day observations. The same groups were involved. The IODE took charge and was assisted by veterans' groups, local clergy, city officials, the 19th, and the Boy Scouts.

One thing was different, however: the service was held at the Soldiers' Cross.[56] The Watson Monument already named local soldiers who had died in the North-West and South Africa, but adding all those who died in Europe would have been a challenging task. There is no indication, however, that anyone wanted to do so. *The Standard* believed there was a desire for a new memorial. Its editor suggested that "a beautiful work of art in cut stone or bronze ... would be an inspiration today and forever, and would do much to keep the public mind on the high level it attained during the war."[57] The mayor wanted the city to erect, "on or near this spot, ... a monument simple and dignified, which would commemorate in some permanent form, more than this wooden cross, the gratitude and pride of our citizens for those noble sons who in their death have honored [sic] her."[58] The Watson monument was clearly not in either man's vision.

The Royal Canadian Legion tried to address this absence. Prior to 1918, veterans of Ridgeway had maintained the NDVVA as a club open

Figure 3.3. Great War Veterans Association Club with War Shrine

Source: Photo taken ca. 1918; photographer unknown. Copy in the John Burtniak Collection, St. Catharines. Used with permission.

to any man who had served in Canadian or imperial units. It was they who had used the Watson monument as a memorial to military service. Veterans of the European war decided to form their own association; this became a local Legion branch when they affiliated with the national body in 1925. They acquired a clubhouse three blocks from city hall and constructed their own "War Shrine" to the dead of the Great War (figure 3.3). A board, roughly two metres high by three wide and facing the street, listed the names of St. Catharines men who died during the war. In front of it stood a rustic cross made of logs and supported by boulders. White letters placed in the lawn reminded viewers that "They gave their lives that you may live." The entire space was circled with evergreens. This shrine was unveiled for Decoration Day in 1918, by Governor General Victor Cavendish, the Duke of Devonshire.[59]

From 1920 to 1926, Armistice Day observances were planned by three groups: the European veterans, the IODE, and the downtown Protestant churches. Participants started at the War Shrine, then marched to the Soldiers' Cross at city hall. Music was played, silence was kept among the assembled crowds, and wreaths were placed at both memorials. Those who wished to pay their respects through a non-denominational

Protestant service were then invited to St. Thomas Anglican Church. Though the Watson Monument occupied the same lawn as the cross and stood along the route between the Legion and the church, it was never mentioned in newspaper accounts of these services. The focus of commemoration for Armistice Day was strictly the First World War.[60]

A Monument Eclipsed, 1927–71

During this time period, the Watson Monument was fully displaced as an important symbolic and visual marker for St. Catharines. First, in 1927 the city unveiled a new, permanent memorial to commemorate the dead of the First World War. This was a cenotaph modelled on the original in London, England. Second, those who sought to promote St. Catharines as a place to visit or invest began to extol its pastoral environs over its military history.

If the monument was losing its central position in military commemorations, it must be remembered that its original sponsors were also declining in numbers. The NDVVA had always been dominated by veterans of Ridgeway. There were few local veterans of the Crimea, North-West, or Boer campaigns. By the 1920s, most of the association's members were in their eighties. *The Standard* reported in 1927 that there were only nine still living in the Niagara area who regularly attended events. That May, the paper explained changes in the city's Decoration Day observance:

> It was first instigated by the Veterans of 66 and for years they carried out the task of showing reverence to their comrades by placing flowers on Watson's Monument and on the graves of those who had gone on before. The service at the monument at City Hall was in remembrance of those who gave their all in the struggle of 66, the Great and other wars.
>
> Following the Great War the Old Veterans found their ranks being depleted by the elements of time … They found they were unable to carry on the arrangements and work of decorating the graves so handed the work to the veterans of the Great War. The younger men with the guidance of the older heads have carried out the work and feel it a sacred duty to carry out the services.[61]

Among the men, the task of commemoration had passed from one generation to the next. Local, personal memories of earlier conflicts were disappearing.

It is telling that the young men of the Legion associated the Watson Monument with Ridgeway above all else. They kept the battle in their observance of Decoration Day, but orators at the Legion shrine and at city hall in 1927 spoke only of the recent war in Europe. One wreath was placed at the base of the monument. Six wreaths were placed around the Soldiers' Cross. Calls continued for a new memorial.[62]

Veterans and the IODE worked with the city to create a purpose-built memorial for the dead of the First World War. The IODE leadership in Toronto championed a move towards useful or living memorials that would avoid the static, funereal tone set by most traditional monuments. They hoped to finance legacy projects that would bring citizens together, aid the needy, and inspire the young; they proposed public halls, hospitals, and student bursaries. Chapters in St. Catharines worked hard to bring such projects to Niagara, and they succeeded in financing educational materials for local schools and a veterans' rehabilitation centre.[63] Locally, *The Standard* thought these were admirable ideas, but it ultimately concluded that Canadians needed "real" war memorials. Ancient custom and widespread admiration of the Cenotaph in London weighed against the IODE proposal.[64]

The city's new monument, a cenotaph, was unveiled in August 1927. The city provided land in the downtown core that overlooked the original Welland Canal valley. The monument and a reflecting pool were the only features in an otherwise passive park. The unveiling was scheduled to allow Edward, Prince of Wales, to attend during a royal tour of Canada. Newspaper accounts made much of the prince, but very little of the monument or the veterans present.[65] Following the unveiling, the Soldiers' Cross was removed from city hall. Following the unveiling, the Soldier's Cross was deemed redundant and removed from the lawn at city hall.

The Watson Monument remained at city hall. Although the Cenotaph did not replace the statue physically, it did replace it functionally and symbolically. From that point forward, Armistice Day and later Remembrance Day services took place at the new park. *The Standard* hoped that Decoration Day services would also move, feeling that the "annual tribute to the dead should be all the more impressive and inspiring. It will be at the new and majestic memorial … surrounded by the prodigal beauty of Nature in a park setting, of which we think St. Catharines will be proud."[66] Notably, veterans also assembled at the Cenotaph during a visit from Governor General John Buchan, Lord Tweedsmuir, in 1936.[67] By coincidence, 1936 was the fiftieth anniversary of the Watson Monument. However, the city's first war memorial no longer served as an active site of ritual.

The last major observations of Decoration Day were held over the next three years as another war approached. The local Legion invited veterans from across Ontario and New York State to participate in 1938.

The day involved a massive parade through the downtown core and three commemorative services: one each at the Legion hall, the Cenotaph, and Victoria Lawn Cemetery. Curiously, the Legion took down its War Shrine just days prior to this event. It was replaced by a new Roll of Honour: brass plaques naming the dead of the European war were mounted inside the new city hall building. *The Standard* intimated that this was an overdue act of civic beautification. The Legion kept its Soldiers' Cross, however, and that was the focus for the service at its hall. The Watson Monument was not forgotten. The parade halted there briefly while a wreath was placed at its base.[68]

Clearly, the significance of the North-West campaign had diminished in St. Catharines. Canadian participation in the First (and later the Second) World War overshadowed all prior military engagements. The North-West Resistance and Second Boer War may have demonstrated the competence of the Canadian militia at home, but the First World War demonstrated Canada's ability in the company of nations. These sentiments informed the work of local historian Albert E. Coombs. Among his works on the Niagara Peninsula is a fact book commissioned by the City of St. Catharines in 1939. It contains detailed sections on city hall, Montebello Park, and the Cenotaph, but no mention of Watson or the monument. The second edition added several pages on the Second World War, but again Watson was not included.[69] One might glean the reason for this snub from Coombs's major history of the region. The American War of Independence, the War of 1812, the Fenian Raids, the Second Boer War, and the European conflicts are treated in detail, but the North-West Resistance receives only four lines. It held little interest, he felt, since it was evident that the uprising could not reach serious proportions.[70] He seemed to imply that Niagara's volunteers in the North-West were willful adventurers as much as dutiful soldiers, an implication that undermined any claim to heroics.

The monument played a minimal role in civic life over the following decades. Remembrance Day remained the key focus for public commemoration of Canada's military, and the Cenotaph remained the site of the city's official service. And still the city's stock of war memorials grew. The names of the dead of the Second World War were added to the Roll of Honour at city hall. When the city absorbed its suburbs in 1961, it assumed responsibility for their existing memorials and then commissioned one new piece to honour the dead of the Korean War. After 2010, Remembrance Day services incorporated Canada's peacekeeping missions and the War in Afghanistan. Meanwhile, several local organizations – particularly veterans' groups and churches – erected their own war memorials in public view. Among all these memorials, the Watson Monument was still decorated each November. A wreath was placed at its base as veterans and the Lincoln and Welland Regiment paraded from the armouries to the Cenotaph. By contrast, Decoration Day was virtually unknown to the general public. The Legion still joined with the regiment to mark the day, however.[71] A small Canadian flag was placed on each soldier's grave in Victoria Lawn Cemetery, including Watson's, and a wreath was placed at the Cenotaph.[72]

Just as the Watson Monument lost its mnemonic value, it also lost its status as a civic landmark. The first sign of this came in a curious fashion and while the city was developing its Remembrance Day rituals. In 1921, *The Standard* published a photo souvenir for Old Home Week that contained two photos of the intersection of Church and James. Both were taken at angles avoiding the statue. A new scenic spot had pride of place in the booklet: a rose garden planted in Montebello Park in 1913. For the next twenty years, most booster publications included the rose garden, probably to underscore St. Catharines's status as the so-called Garden City. Local entrepreneurs further developed this theme in the late 1930s when they staged a spring blossom festival to generate tourist traffic through the region's fruit orchards. A more enduring venture was started in 1952 to celebrate the fall grape harvest: the Niagara Grape and Wine Festival. Beyond the rose garden and festivals, two other scenic spots emerged as key attractions: a beachfront amusement park in Port Dalhousie and the Welland Canal. These became central to the city's tourist promotions after 1945. In competition with the grandeur and kitsch of Niagara Falls, or the nostalgia-laden imperialist pomp of Niagara-on-the-Lake, marketers working for the City of St. Catharines did not imagine the Watson Monument as a tourist draw.[73]

By 1936, the City of St. Catharines required more space to house its expanding administrative staff. The existing city hall was a converted Victorian home. It was demolished to make room for a new art deco office building on the same corner lot. The most troubling issue for council was the direction the building would face: commercial James Street, or institutional and residential Church Street? The architect preferred Church and drafted plans accordingly without consulting city council. When council realized what was happening, it was too late to alter the plans, and they were simply approved.[74] The Watson Monument received even less attention. There was no official discussion of its fate. At some point, it was repositioned. Having once faced James Street, it now faces diagonally into the intersection of James and Church. A ring of evergreen shrubs was planted around the base, extending roughly two metres on all sides. This landscaping made it difficult to approach the monument and read its inscriptions, a point noted by locals.[75] After the new building opened, few official photos of city hall included the monument. Most emphasized the modern lines of the building's front entrance.[76] A telling image appears in a local history issued by the county in 1956. Watson's story is not told, and a photo of city hall with the monument in the foreground is simply captioned "City Hall."[77]

Council may not have given the monument much thought, but *The Standard* did. When the new city hall was announced, the paper published a brief piece on Watson and the city's response to his death. Its content and tone suggested that Alexander Watson's story was little known. Fifty years had passed since patriotic citizens had erected the monument. The generation that considered Batoche a pivotal moment in Canadian history was now passing. In fact, the mere mention of Batoche was no longer sufficient to describe Watson's significance in a national or local narrative. *The Standard* reporter felt compelled to recount the events of the resistance to a readership for whom the statue and its back story was a novelty. *The Standard* has published a similar article every decade since.[78]

Lending symbolic weight to its lack of popular significance, cracks appeared in the monument in 1970. Two different firms believed the soft limestone was beyond repair and dangerous to passersby. On their advice, council voted to maintain the base as a monument but to remove the statue.[79] This decision sparked little debate in the community. There was no popular call to restore the statue, as there had been to erect it, and only two letters addressed the decision in *The Standard*. These letters are themselves noteworthy. Both authors felt their contemporaries took patriotism and the sacrifices of war for granted. The pseudonymous "Lest We Forget" wrote, "When our country has been at war patriotism ran high, however in peace time we apparently are prone to forget."[80] While both writers felt it important to preserve the statue to perpetuate the memory of wartime sacrifices, neither referred to the North-West Resistance or the Second Boer War. "Lest We Forget" conjured the First World War. *The Standard*'s editorial on this issue struck a similarly ambiguous note. The paper wanted to preserve the monument because it was "something worthwhile to leave to future generations"; Canada's record at monument building was frankly embarrassing when "other countries, with climactic conditions equal to ours, are able to build structures that still stand centuries later."[81] These three voices were apparently alone, however.

The city reversed its decision only when local architect Norman Macdonald proposed a cost-effective way to save the monument. The stone was cleaned and the cracks duly sealed. During this process, the original text and relief carvings were removed because they had grown faint with time and weather. The texts were replaced with new brass plaques that replicated the original wording. The reliefs were not recreated. Again, there was very little debate regarding these decisions. Macdonald was driven by a desire to test new restoration techniques as much as a desire to save an urban landmark. In the latter respect, he shared *The Standard*'s concern for the city's self-image rather than the monument's legacy. He had no agenda in the commemoration of Watson, the North-West campaign, or the Canadian military.[82]

A "Quaint Anachronism," 1971–2009

In the wake of Canada's centennial celebrations in 1967, new research kindled interest in Louis Riel and his legacy. This took place in the national public sphere, through academic publications, journalism, and state agencies. In St. Catharines, however, renewed interest in Riel had little effect on the Watson Monument. Groups emerged to preserve Watson's memory, but the monument claimed little space in the popular imagination.

Canadians had never shared a single view of Indigenous resistance in the Red River District or the North-West. Their various perspectives were informed by partisanship, economic interests, sectarianism, ethnocentrism, and outright racism. During the 1960s, however, some scholars began to cast light on the role played by Sir John A. Macdonald and his government. If one acknowledged that the Métis and First Nations had legitimate claims and grievances that Macdonald studiously ignored, then their actions could no longer be called rebellious. Riel's leadership could be similarly reinterpreted, not as treason, but as principled civil disobedience; his execution as a criminal reinterpreted as state-sanctioned murder. Seen in this light, Riel became a martyr. Such views had always existed, but the new scholarship gave credence to them; they then filtered into history textbooks and popular political movements. Indigenous groups viewed Riel's cause as their own, Quebec nationalists believed his treatment was symptomatic of anglophone prejudice, and some Prairie protest groups compared his grievances to their own Western alienation from Ottawa.[83]

Revision of the Riel story affected even the government that had prosecuted him. The one hundredth anniversary of the Red River Resistance occurred in 1970. Two branches of the federal government sought to mark the occasion with sympathetic representations of Riel. Parks Canada purchased the Riel homestead to create a museum, while Canada Post issued a commemorative stamp. The Province of Manitoba also marked the date. The province was carved from the North-West

Territories to resolve the crisis; the new political jurisdiction allowed residents to manage their local affairs within Canada's constitutional framework. Because Riel negotiated Manitoba's entry into Confederation, the province recognized him as its founding statesman, and a statue was erected on the grounds of the provincial legislature in 1971. The Royal Canadian Mounted Police, whose forces had engaged Riel's supporters in the North-West, had its own centennial in 1973. Its anniversary publications downplayed any suggestion of villainy on Riel's part. In 1985, these interpretations were maintained for the anniversary of the North-West Resistance. Parks Canada expanded and reimagined the Batoche National Historic Site after it acquired much of the village in 1976. In 1923 the site was marked by a single plaque dedicated to the victory of the field force. By 1985 it was a major national park and living museum dedicated to Métis heritage. Once again, the national media marked the occasion with articles and documentaries that echoed the consensus that had emerged by then among academics and public historians. Seven years later, the Canadian Parliament joined the Manitoba legislature by formally recognizing Riel as a founder of Confederation.[84]

Ultimately, the debates surrounding Riel's legacy percolated into commercial entertainment. Numerous books have appeared on his life, including Chester Brown's celebrated *Louis Riel: A Comic Strip Biography* (2003). Riel has been the subject of a play, an opera, and more than thirty songs by Canadian musicians.[85] He has also been the subject of three television movies or miniseries and a significant personage in three feature documentaries.[86] Two of these deserve comment. The CBC's highly rated *Canada: A People's History* (2000–1) not only cast Riel and his allies as victims, it also cast the North-West Field Force as the aggressors. The battle at Batoche is represented as a siege on an Indigenous settlement; Riel's forces do not retreat from battle but are lucky to escape. The field force soldiers are portrayed as murderers and thieves. *The Retrial of Louis Riel* (2002) gives some indication of how Canadians responded to such depictions. The Dominion Institute for Canadian Studies, in association with the CBC and the *National Post*, produced this program to assess how Riel's nineteenth-century legal charges would have fared before a twenty-first-century jury. The character of Riel was represented by a celebrity lawyer and evidence was presented following standard courtroom procedures. The jury was the entire viewing audience, who were invited to submit their decision by Internet ballot. Almost ten thousand people participated, and 86.7 per cent voted that Riel was not guilty of treason.[87]

Within this climate of opinion, it would not be surprising to discover that St. Catharines, too, had a change of heart. If Riel had become a heroic if misguided figure, then perhaps the field force was seen as the villain. One might have expected the monument's place at city hall to be questioned. It was not. Quite the opposite, it appeared that residents of St. Catharines made no link between Riel and the troops who faced him. The North-West Field Force had no place in local discourse. Meanwhile, the First and Second World Wars dominated Canadian representations of the country's military history. The image of Johnny Canuck in Europe, in olive fatigues and steel helmet, was reproduced in newspapers, magazines, and television programs every year at Remembrance Day. Hence, the displacement of the Watson Monument by the Cenotaph in 1927 was perpetuated through the mass media.

Local writers reflected the national mood during this period. For the most part, this meant Watson was not represented in discussions of the city's heritage. Histories and picture books about St. Catharines typically contained chapters on its wartime experiences but made no reference to Watson apart from photographs of the monument. At best, the North-West Resistance and the Boer War were briefly treated as two minor incidents in the continuing support shown for the British Empire. One illustrated history devoted two pages to the First and Second World Wars, but only two sentences to Watson and no photograph. Guidebooks followed a similar pattern. The Chamber of Commerce produced a manual for tour operators that highlighted local points of interest. Historic sites commemorating the area's social and industrial history were noted, as were local wineries, shopping malls, and parks, but there was no mention of local military history. Similarly, the city's tourism department produced a video brochure featuring military history sites outside the city, such as the 1812 battlefields at Queenston Heights and Fort George, but none in the city itself. A brief shot of the Watson Monument appeared during a montage of city parks, without explanation.[88]

Local militia units that had links to the monument seemed to forget them. The 19th and 44th Regiments both participated in Watson's funeral, commemoration, and later memorial services. They have since merged to form the Lincoln and Welland Regiment, based at the armoury in downtown St. Catharines. Their participation in memorial services can be interpreted in two ways: first, as ritual acts honouring fallen soldiers, and second, as public relations exercises. Recruiting was the responsibility of individual regiments until the Second World War. Good community relations were essential to maintain troop strength, and it was valuable to celebrate heroic acts performed by local soldiers.[89] Watson's death and commemoration therefore met a need between 1867 and 1914. Watson and the dead of the Second Boer War provided the regiment with local tales of valour during a lengthy period when no local unit was called up. Surprisingly, however, regimental historian

R.L. Rogers did not include the monument or local memorial services in his discussion of the unit's ceremonial duties. Either the unit's historian was unfamiliar with its long involvement with the oldest war memorial in the city, or he deemed it unimportant.[90]

A handful of writers did confront the changing tenor of commentary on Riel and connected this commentary to Watson. In 1982, *Standard* reporter Andrew Dreschel felt compelled to ask if time had "tainted" the "hero's renown." The city had been proud of its soldier son in 1886, but "Watson's currency as a hero has fallen somewhat since then. Since historians now recognize the Métis had legitimate grievances, … modern sensibilities may find the statue slightly embarrassing."[91] The city's Local Architectural Conservation Advisory Committee expressed similar misgivings in a guidebook for walking tours of old downtown. Its author, Robert R. Taylor, knew that generations of local residents held strong ties to the military traditions of the United Kingdom, and he explained the monument in this context without direct reference to Riel. Apologetically, he noted, "it is a unique reminder of an aspect of our British heritage, which some may find meaningless or embarrassing, but which we cannot ignore."[92]

A different tack was later taken by *Standard* reporter Gail Robertson. In a piece written for the centenary of the Battle of Batoche, she acknowledged Riel's reputation had been rehabilitated. Nonetheless, Robertson refused to condemn the soldiers of the North-West Field Force. She viewed Watson sympathetically by separating his military service from the government's political goals. If Riel had fought for justice, then Watson, too, "was fighting for a cause in which he believed and one he and many others saw as a duty to the government of Canada," regardless of the government's motives.[93]

While journalists and historians contemplated the monument, it remains difficult to know how other residents felt about it. One local theatre company may provide a clue. The Carousel Players marked the city's own centennial in 1976 with a light-hearted musical revue that drew inspiration from local landmarks and stories. It contained two serious moments: two monologues delivered from the perspective of statues in the city core. One was the statue of William Hamilton Merritt, the entrepreneur behind the Welland Canal. The second was the Watson Monument. In "his" monologue, an actor dressed as "Watson" (in red tunic and white helmet) states his name, rank, and battalion number, and then addresses the audience:

> I wonder if you recognize me. I'm on the corner of Church and James. You know I'm there but nowadays you look past me the way you look past the city hall, or the post office, or the library. I've become accepted as part of the scenery, part of the background. But not one in a hundred knows who

I am … Were the actions of my life so great a crime that a statue had to be
built so that everyone could ignore me publicly?[94]

Reminiscing forty years later, playwright David MacKenzie noted, "I
was born in and grew up in … downtown St. Catharines. The statue
was just always there … Watson's complaint that he is the unknown
soldier stems directly from my own experience of seeing him practically
every day and not knowing who he was." MacKenzie also noted that he
sympathized with the anti-war movement during the 1960s, and Buffy
Sainte-Marie's condemnation of rank-and-file personnel in her song
"The Universal Soldier."[95] MacKenzie's words, in the 1970s and 2010s,
evoke a public that did not engage with military commemorations and
that was, in fact, ideologically disposed to dismiss them.

"Awake Anon the Tales of Valor [Sic]"

"This snow white tablet … will we trust awake anon the tales of valor
[sic] to many a coming generation, when the days of the present one
are long numbered with the dead."[96] The rector of St George's Anglican
Church spoke these words while unveiling a private memorial to Henry
M. Arnold, killed during the Second Boer War in 1900. They express
rather well the sentiments of Victorian St. Catharines towards its war
dead, sentiments then common throughout English-speaking Canada.[97]
Nonetheless, the confidence they evoke was not entirely warranted. In
1946, a letter published in *The Standard* suggested how far the dead
of the North-West and South Africa had passed from local consciousness. One of Watson's comrades, a surviving member of the 90th who
had fought alongside him at Batoche, urged St. Catharines to remember
Watson with an annual service at the monument, but he acknowledged
that "It is a far cry from the village of Batoche, N.W.T., May 12, 1885, to
the City of St. Catharines, Ontario."[98] The paper published no response.

Memorials tend to resonate most intimately with the generations who
raise them. In 1885, grateful Canadians celebrated the returning troops
of the North-West campaign and commemorated their dead. These men
were upheld as models for public emulation. Watson was no exception,
and local memory entrepreneurs made use of his story. His monument
was an active site of ritual until the First World War and attained almost
national prominence as the material focus for the Ridgeway reunion in
1898. Nonetheless, its public profile would diminish. The First World
War, with its mechanized fighting and staggering body counts, rendered Watson's contribution relatively inconsequential. Belgium, not
Batoche, resonated in local consciousness and prompted the erection of

a new civic war memorial. Subsequent wars and military actions have added new chapters to the city's military narrative, while conceptions of the North-West Resistance itself have shifted. The repairs made to the monument in 1971 did not reflect enthusiasm for a local hero. It was an opportunity to test a new industrial solution on a public danger that otherwise would have been removed. By 2009, the efforts of a young private who died thousands of kilometres from home, in a battle now thought regrettable, had seemingly lost its ability to move people.

As noted in chapter 1 memorials can suffer many fates: their audiences may lose interest, their significance may be reinterpreted, or other political and cultural circumstances may intervene. Looking at the Watson Monument, it is clear that all three had occurred. Few locals since Watson's generation had been willing to keep his individual narrative alive. The monument may prompt an affective impulse to remember, but it cannot prompt an awareness of what was never known. Hence, the specific message of the monument – the story of Watson and his comrades-in-arms – had largely disappeared from public consciousness by 2009.

Chapter Four

Residents Engage the Watson Monument

Q: To what extent do historic sites and monuments contribute to your knowledge of history?
A: I don't think there are enough of them. They are helpful, both for locals and for tourists.
A: I think they're important, but I don't go and see them.
A: Never look at them. Maybe if I had kids.
A: There aren't many in St. Catharines.
A: I was in Ottawa recently.

 – responses from five different survey participants, 2008[1]

In 2002, Caroline Grech, a contributor to the St. Catharines *Standard*, interviewed folks passing by the Watson Monument. None of them knew of Private Alexander Watson or the campaign in which he fought. Even a local history buff could not explain the reasons for Watson's commemoration. The article was prompted by a decision of the St. Catharines Historical Society; they planned to honour Watson and revive interest in his story. A ceremony was duly held at the monument on 18 May 2002.[2] The story and the ceremony had little effect. Passing through downtown St. Catharines, crossing through a busy intersection, you would be unlikely to recall Watson's story even if the monument had caught your eye (see figure 4.1). It still stood on the lawn at city hall, but decades had passed since it last held a central role in the city's commemorative services.

Over the last two chapters we described the monument's genesis and its seeming disappearance from public view. In this chapter, we describe its dramatic return to the public eye. The context for this re-emergence was twofold. Canadians' changing perceptions of the North-West Resistance played their part, as did the revelations of the federal

Figure 4.1. Watson Monument, former Lincoln County Courthouse, and city hall, 2021

Photo: Michael Ripmeester.

government's Truth and Reconciliation Commission after its report was published in 2016.[3] More pointedly, Canadians campaigning for social justice linked their growing knowledge of Indigenous history to global movements battling systemic racism. Activists elsewhere had questioned their nations' stock of mnemonic products that championed values and narratives no longer held by contemporary society. Canadian activists had similarly challenged the legacies of Edward Cornwallis, Sir John A. Macdonald, and Sir Hector-Louis Langevin, all of whom played leading roles in the destruction of Indigenous communities. Many of these mnemonic products – like currency, stamps, monuments, and toponymy – had been unremarkable elements of everyday life. Activist scrutiny returned them to public view.

The career of the Watson Monument provides a clear example of this process. Material commemorations are explicitly ideological devices embedded in the landscape. Their sponsors occupy public space to convey their values and assumptions through architecture or enduring works of art. Consequently, their presence provides the temporal and spatial coordinates for individuals to contemplate the past. If a memorial successfully resonates with the public, it can also build group identities by fostering an emotional and cognitive sense of consensus and belonging.[4] As we argued in chapter 1, memorials derive part of their power from the authority vested in their creation, as well as their physical presence and resilience. Their success, however, is not guaranteed. The narrative's significance may fade if the new sponsors cannot maintain its social currency by adapting it to changing values and assumptions.[5] The career of the Watson Monument after the First World War demonstrates the difficulty of this task.

The narratives, arguments, and revisions that shape mnemonic products exist in many forms beyond their material existence. The contemporary media environment provides seemingly unlimited access to information, both institutional and informal.[6] Hence, when interpreting a mnemonic product, individuals may draw upon a vast array of sources well beyond the preferred reading its sponsors have provided. *Any* given reading of a mnemonic product may become ascendent if its visual and ideological cues are intertextual, diffuse through multiple contexts, and readily accessible within the community. Peter Stupples contends that any reading supported by such ideological scaffolding then becomes normative.[7] It functions as a "cognitive default" interposed somewhere between viewers' gaze and the mnemonic product they engage.[8]

Given these insights, we should not dismiss seemingly irrelevant or forgotten monuments. If we accept that things like monuments naturalize the narratives, values, and assumptions of their sponsors, then we should also accept that these ideological elements do not have to be openly engaged to be acknowledged or understood. We are trained from early childhood to read our environment in specific ways.[9] Schooling, for example, provides many of the visual and interpretive skills needed to comprehend and navigate our world.[10] We develop reflexive responses to certain sites, imagery, and symbols such that they immediately prompt specific narratives or arguments to come to mind. If most visual artists work within the same set of intertextual cultural referents, then preferred readings become self-reinforcing. M. Christine Boyer, exploring how memory is situated in the cities of western Europe, refers to such elements of the landscape as rhetorical topoi: these are things that situate arguments intended to instruct local consciousness.[11] Once naturalized, their associated values and assumptions may become taken for granted and thereby deflect close interrogation.[12]

Sponsors, of course, hope their audiences will not simply be aware of their mnemonic products, let alone take them for granted. They seek to move individuals, to prompt them towards particular feelings that will align them with wider values and with one another. Here we engage research on "somatic markers."[13] William Connolly writes, "A somatic marker operates below the threshold of reflection; it mixes culture and nature into perception, thinking, and judgment; and it folds gut feelings into these mixtures."[14] Symbols consistently cast in a positive or negative light become tied up in "histories of association" and "gut feelings in common."[15] Writing about Japanese citizens' knowledge of kamikaze pilots, Rumi Sakamoto states, "the affective and emotional reactions to the kamikaze image could place such responses in our long-term memory – and in fact, in our physical nervous system – ready-stamped with intense emotion and associated meanings."[16] In this way a mnemonic product may be marked as special. It carries the weight of authority, its specialness is often entangled in supportive intertextual webs, and its narrative resonates or sticks with audiences. This response may not be hot or deep, but it often remains potent.

A Citizen Survey, 2005

Our first survey in 2005 – described in chapter 1 – sought to understand how residents responded to the Watson Monument. What did it mean to residents? Had they read or contemplated its message? Did they recognize Watson as a local hero, or as an aggressor against Indigenous land claims? Were they prompted to learn more about the North-West Resistance? Did they link Watson's service to a collective past? If so, was that collective past a local narrative or a national narrative? Put bluntly, just what did residents make of it? We conducted the survey in two locations. The first location was in the shadow of the monument itself, in downtown St. Catharines. The second location was the business section of Port Dalhousie, a neighbourhood of St. Catharines on the shore of Lake Ontario. One hundred and sixty-two individuals completed the survey.

Among our many questions, we asked participants "what is your interest in history?" They indicated their level of interest by selecting from the following options: "strong," "mild," "indifferent," "no interest," or "hostile." The last option was informed by many conversations we have had with family, friends, and students over the years who take an active dislike to all things historical. Those with a mild or strong interest in history accounted for 71 per cent ($n = 105$) of our sample; the majority indicated their interest was mild. Another 19 per cent were indifferent to history, while the remaining 10 per cent either had no interest or were, in fact, hostile to it. Three participants stated bluntly that the city had no history to learn. We sought correlations between the participants' interest in history and their ages, genders, and places of residence. There was no significant link between these three demographic identifiers and their level of interest. There was a weak correlation, however, between age and level of interest ($r = 0.16$). Older participants were slightly more inclined to express an interest in history.

Participants were asked how they learn about local history. This was an open-ended question. We gave no prompt to direct their answers, and we recorded every unique response provided by each participant. This strategy sought responses that were frank and top of mind. Confirming this belief, individuals with a stronger interest in history tended to name more sources. In total, our 162 participants provided 274 responses to this question.

The most common response was surprising: 29 per cent of participants stated they learned about local history incidentally ($n = 47$). Acquiring such knowledge was, they believed, a quirk of living in the area. "History" was all around them, and they could not help but absorb it. Participants often explained this belief by listing the historic sites they passed in their daily travels, or noting places named for historic persons. The second most common response may be connected to the first: 24 per cent of participants stated that they learned about local history through word of mouth ($n = 39$) – that is, through conversations with family, friends, neighbours, co-workers, and others. Only three participants named both these sources. Together, then, these two informal and incidental sources of knowledge were named by 51 per cent of our sample ($n = 83$). For 24 per cent of participants, one of these sources was their only source of knowledge ($n = 39$).

The next most common response identified the mass media as a source for local history. We sorted these responses by medium. The most common medium was the Internet, identified by 16 per cent of participants ($n = 26$). It was closely followed by newspapers, books, the library, television, and "the news," in that order. Only one person identified radio as a source for local history, and no one identified film and video. We believe participants referred to the library as a proxy for books, newspapers, magazines, videos, and Internet access – that is, as a repository of local history resources, most of which are mass media products. The central branch of the public library in St. Catharines offered occasional talks and exhibits, but participants did not link the library to the city museum or the local historical society. The ranking itself is interesting. The first four platforms can supply well-targeted local content. There was no television station based in Niagara, so television offered little value as a source for local history. Residents were most likely to see their past portrayed on Toronto-based TVO or the American PBS affiliate in Buffalo, New York, than through any other television provider.[17]

When all these platforms were grouped together, the mass media was named by 56 per cent of participants, a greater proportion than those who identified incidental sources.

The third set of responses suggests an active, participatory interest in local history. Participants reported that they acquired some knowledge by travelling, taking tours, visiting cemeteries, visiting historic sites and museums, or attending meetings of the local historical society. Taken together, these responses came from 28 per cent of the participants ($n = 45$). Travelling may suggest the same incidental acquisition of knowledge that participants gain simply by living in the area, but we believe it is qualitatively different. When individuals travel, they are usually open to learning the history of their destination. Indeed, some historical information may be expected and desired. Such knowledge allows travellers to contextualize a new place, understand it, and perhaps appreciate their experiences there more deeply. This can be true in one's hometown as it is elsewhere. With respect to the Watson Monument, the telling figures in this group are those for visiting cemeteries (6 per cent of participants), visiting historic sites and museums (6 per cent), and taking tours (3 per cent). These participants indicated their desire to engage with material commemorations, but they were clearly in the minority (15 per cent of all participants).[18] Thus, most participants did not hold material commemorations to be important sources of knowledge regarding local history. Their top-of-mind responses were remarkably short on references to historic sites, buildings, monuments, and plaques of any form (see, for example, chapter 1, figure 1.4).

That said, when we asked them directly if historic sites and monuments contributed to their local knowledge, their responses were quite different: 38 per cent of participants stated that a significant portion of their local knowledge came from material commemorations, and another 18 per cent stated that some of their knowledge did so. Altogether, then, 56 per cent of participants ($n = 91$) believed that they had learned something from historic sites and monuments. Another 33 per cent of participants ($n = 54$) were non-committal with respect to this source. This group often explained themselves in ways that suggested they had a mild curiosity in plaques and monuments but felt no gripping need to learn from them. As one participant commented, "I'm aware of them, but I only read them when I need to." Another stated, "Well, if I see one, I'll read it, but ... ," and, rather bashfully, did not finish the sentence. Most of this group, however, were simply indifferent to them. The remaining 11 per cent of participants ($n = 17$) clearly indicated that they had no interest in history and simply did not pay attention to material commemorations.

When asked about the Watson Monument directly, only 6 per cent ($n = 10$) of participants were able to identify Alexander Watson or the North-West Resistance as the object of commemoration. This set of participants included 7 women and 3 men, all of "Canadian" or European backgrounds, with 8 over forty years of age. Seven of the 10 were born in Niagara or southern Ontario, while the other 3 were born outside Canada. Interesting results, perhaps, but unfortunately this subset of participants was too small to draw meaningful conclusions from their demographic profiles. The important number, however, was the proportion of participants who could not identify the original impetus for the Watson Monument: 94 per cent ($n = 152$). Again, there was no significant difference between the participants interviewed downtown, within view of the monument, and those in Port Dalhousie. This number suggests the monument did not fulfil the intentions of its sponsors.

To probe more deeply into residents' thoughts on the monument, we ended the survey with three open-ended questions regarding respondents' perceptions of its significance and how it is engaged by the community. Participants might have lacked specific knowledge of Watson and the North-West Resistance, but many of them shared common perceptions of the monument. For some, the monument was an unobtrusive part of the urban fabric. People referred to it as "a piece of furniture," "familiar," an old "knick-knack," and "public art." One participant stated, "It's just something that you pass all the time." This sentiment was echoed by another who admitted, "I've seen it hundreds of times, but never really looked at it." Some associated their lack of knowledge with their experience of life in a modern urban setting. The monument was erected in a Victorian walking city. The increased pace of urban life and a reliance on automobile transport had altered perceptions of the landscape. Nigel Taylor has argued that the mental acuity associated with increased speed of urban movement has diminished the richness of the urban environment for both pedestrians and those in vehicles.[19] Participants themselves noted the space between a vehicle and a destination is something to be endured rather than something to be enjoyed. Some participants drew attention to this sensibility to explain their limited knowledge of Watson. One told us, "I never heard of him. Most people in downtown are doing business. They go in, they want to get out. We need more parking if we are supposed to lollygag around." Another voiced a strikingly similar opinion: "People don't really walk here. When they do, they are busy doing things and don't really notice things like that." A third explained, "I used to work across the street and never noticed it. It's pretty nice once you notice it, but I wouldn't go over and read it." Even the monument's landscaping presented obstacles. As one participant noted, "the flowers and the hedge and the 'do not walk on the grass' look removes it from the passersby ... [It's] not inviting. It looks inaccessible." For many participants, the monument had become peripheral to their view of the city despite its prominent location. For these individuals, then, the monument had become an unremarkable part of their daily lives. It did very little to motivate a second glance.

Most participants, however, felt the monument was significant to the city's heritage. Notwithstanding their lack of familiarity with its narrative, participants agreed that it was of major (40 per cent) or minor (27 per cent) significance (67 per cent combined, $n = 108$). When asked to describe the monument's significance to the city, one person responded, "It does add a little to the community, but no more than would a pleasing piece of art or architecture. It's nice, well kept." For others, the monument provided the city with historic interest. It was not Watson's narrative that enchanted these participants. Rather, it was the fact that the monument was old; its Victorian origins gave it intrinsic value. One participant stated, "The monument is really important. It reminds us of an earlier time. It's good to carry on traditions. It needs to stay where it is. People would miss it. It's a beautiful structure. People like to appreciate historic stuff. Residents are sad when historic stuff goes. It's well preserved."

Linking these two points – the monument's familiarity and its historicity – some participants thought that residents, like the one quoted above, would miss the monument if it was removed. Some wondered anxiously if a planned removal was behind our survey. Others suggested no one would care if it did go. One participant summarized both attitudes: "Too bad that people don't notice it. It's a beautiful monument, but if it disappeared no one would notice. I see the tourists take pictures of it and wonder: Why do they take pictures of that?" In Niagara, as elsewhere, residents may take their own points of interest for granted, a perception referenced by another participant to explain residents' ambivalence towards the monument: "It's like if you live in the area you don't ride the *Maid of the Mist*. It's not significant."[20]

When informed of the monument's origin in the North-West Resistance, most participants expressed surprise. A few openly questioned the relevance of Watson's service to contemporary residents of St. Catharines. One person captured this sense, stating, "It's kind of unexpected. I would have thought that it was someone more important. I wouldn't associate with this guy in a 'he came from my hometown' way." Another was even more dubious: "One guy died a hundred years ago in a battle that nobody knows about. Even if it was an internal struggle, it's not local. There are simply too many other significant local things to pay attention to rather than something that has very little to do with what happened here." A third person apologetically claimed, "He should be important, but I don't know. I hate to say he wasn't, but he didn't change the course of events in St. Catharines. The event didn't happen here." Yet another participant stated, "He was important at the time, but as the years passed ... [*silence*]." And another retorted, "It has very little relation to our times. It's not my history. Maybe the Métis care." Two participants identified themselves as members of First Nations, and one identified as Métis. None of them knew of the monument. One of the few participants who made the connection between Watson and the Métis cause captured this sentiment:

> He's irrelevant. Why is he commemorated here? Maybe to show the attitude of Upper Canadians. He's not a local hero, it didn't happen here. He fought against the Métis, and they were right. So that makes him sort of an anti-hero, doesn't it? In those days he might have been a hero, but he's not any longer. The monument is significant because it shows the attitude from the past as opposed to the present. I've never bothered to look at it, and I've passed it a lot. It's ironic that they have citizenship court here at city hall. He was sort of fighting against inclusivity, wasn't he?

Among all the participants, such critiques were in the minority. When asked to assess Alexander Watson's significance to the city's heritage after hearing the narrative, 61 per cent of participants believed that he was of major (35 per cent) or minor (26 per cent) significance (61 per cent combined, $n = 99$). Notably, this was less than those who believed the monument itself was significant.

Given time to contemplate the monument's narrative, several participants were moved. Some expressed feelings of guilt or shame when they admitted that they had not known of Watson. One participant confided,

> I feel bad saying that he's nothing, but he's unknown. I can't believe that he's unknown. But it is important for the younger generation to know about people like him. The monument is significant. It's good to see that he was remembered. His importance lies in the history of the city. It's great what they did for him. It's just a good thing for a local.

Even those who questioned Watson's status as a local hero were reluctant to call the monument insignificant. One comment neatly captured this attitude: "I'm big on monuments – speaking as an historian – but this is a weird monument. That shouldn't diminish its significance." This line of thinking may explain why participants invested the monument with more significance than Watson. One participant, bowing to the authoritative weight it carries, concluded simply, "The monument is very significant. After all, they put it up."

Visual literacy and interpretive process were evident in the way participants read the monument. Participants fully understood that the monument commemorated military service. Those who did not know its specific narrative associated it with the Crimean War, the Second Boer War, and the First World War. One person told us, "I've never taken the time to read it. I took for granted that it was a World War One monument. He looks like a soldier, the way he's dressed, and he holds a gun. I never really saw the need to investigate." Another participant stated, "He was a liberator of Canada or something." Similarly, another claimed, "I knew it was regimental or something … perhaps an unknown soldier. Most of us don't know what this guy did. That's the sad part of it." As one individual concluded, "We may just assume we know what it's for."

Despite their jaundiced view of Private Watson and his role in the North-West Resistance, war memorials carried a certain formal significance and deep meaning for many participants. As participants answered the questions, it was clear that many understood that such memorials generally honour a combat death. They also understood that such memorials imply that a soldier's death means something greater than a civilian death. This implication is fostered by the portrayal of individual bravery and sacrifices made on behalf of the greater good.[21] Such emotions ran from gratitude to pride. For example, one participant told us, "It represents all soldiers who have given their all. We need to remember them for their sacrifice. We need all monuments. They are very important parts of who we are." This participants' understanding is underscored by the monument's location, which confers upon it an official sanction and hence greater significance. "It's really important," she continued, "It must be given that it's in front of city hall. And it's big, with the flowers and the flags." Her train of thought suggests that official marking, intertextual symbols, and the sticky narrative of the citizen-soldier served to highlight the monument's higher discursive purpose. In fact, some participants took this idea even further. More than remembrance and gratitude, they found memorials depicting a citizen-soldier to be inspiring, a prompt to participate in civic realms. The selflessness of a combat death prompted them to reflect upon their own responsibilities and contributions. Participants listed virtues like honour, dedication, and inspiration to explain the monument's

significance. Many also commented on the Canadian tradition of volunteerism embodied by the militia. Canadians, participants said, were
reliable soldiers. One noted that Watson was like all the other "boys"
Canada sent to war. Another expanded on this idea: "It's important
to one aspect of the community, especially for people involved in
conflicts elsewhere. It kind of foreshadows future conflicts and other
things to come. It's a reminder of the involvement of this community
in a major event in Canadian history." Likewise, for another, "He sets
a good example of good citizenship. I haven't read the plaque, but I
will someday. I may use it more now that I know about it." For these
participants, the monument may not have prompted recollections of
Watson's narrative, but it did foster contemplation of those characteristics associated with the citizen-soldier and a good citizen generally.[22]
Such thinking might explain why 35 per cent of participants believed
Alexander Watson was of major significance to the city even though
they knew nothing of his story. It might also account for their repeated
feelings of guilt, shame, or sadness when they could not identify Watson or the reasons for his death. Nevertheless, a seemingly invisible
monument sitting among the clutter of a mundane streetscape was
able to provoke emotional responses once its narrative registered in
participants' consciousness.

The Watson Monument, 2009–19

In 2007 the City of St. Catherines asked its Recreation and Community
Services Department to take stock of all its public art and memorials.
There were two concerns: some of the existing stock was deteriorat-
ing, and there was limited appetite to finance repairs. Seeking guidance, the department hired a private consultant to assess each piece and
make recommendations.[23] The consultant's findings, delivered in 2009,
noted the Watson Monument was in an advanced state of deterioration
because the repairs done in 1971 were failing. The prognosis for a long-
term remedy was poor. Even if repair was possible, there was little left
of the monument's original detail, a state that rendered "restoration to
the artist's image and intent impossible."[24] Beyond its physical condition, there was also a growing awareness among department staff of the
monument's contentious narrative of Indigenous suppression.

In the department's subsequent report on public art and memorials,
staff noted the monument's contentious narrative and the costs associated with its maintenance. The consultants believed lasting repairs
would be expensive, in the $75,000–$125,000 range. A less expensive
option might cost $11,500 but would only last three to five years. The

department then provided council with three options to deal with the monument: destroy it, move it to a museum as an educational asset, or repair it in place. Council decided that "the future of the Private Watson statue will require a broad consultation ..., given its distinct placement on the front lawn of City Hall." In this spirit, council voted "that staff be directed to consult with the community, as appropriate, including but not limited to local Legions, the Lincoln and Welland Regiment, 10th Battery, 56th Field Regiment, the Niagara Regional Native Centre, any descendants of the Watson family and any living artists whose work has been assessed and where restoration or maintenance is required."[25] The department did as directed and began a new round of consultation.[26]

The St. Catharines *Standard* carried one report that documented the monument's deterioration, council's motion, and the opinions of interested parties. Reporter Monique Beech concluded that "History buffs are standing ready for a fight of their own in St. Catharines." Her two main contacts were a local historian, Paul Lewis, and the president of the Niagara Region Métis Council, Rick Paquette. Lewis understood that views of the North-West Resistance had changed since 1886, but he was adamant the monument retained educational value: "'It['])s okay that the statue incites debates and leads people to ask questions about Watson,' Lewis said ... 'My Goodness we can't afford to toss it out,' said Lewis. "[I]t's our culture.'"[27]

By contrast, Paquette's comment suggested polite indifference. Paquette was interviewed just prior to a ceremony honouring Louis Riel, and he suggested contemporary Métis would not be offended if the monument was repaired: "Just because he was fighting against our leader at the time, that doesn't change how we would perceive him ... He was doing his job and our people were doing their job."[28] Only one other news item appeared. Reporter Wayne Roberts filed a report for APTN News with comments from Paquette, who repeated his earlier views, and with context from ourselves as scholars who had researched the monument.[29]

Beyond these two items, there was scant public discussion of the department's work, either through traditional media or social media. We found, for example, only two published letters to *The Standard* and one tweet, all seeking to keep the monument in place. The tweet came from a self-identified Anishinaabe poster who stated, "Leave the Watson Statue where it is," then linked to the *Standard* article.[30] Six years later, in April 2015, another Twitter user posted a photo of the statue with the comment, "A real #Canadian hero."[31] Silence ensued. In 2018, the city decided to let the monument deteriorate in place.[32]

The Watson Monument in 2020

In May 2020, four police officers in Minneapolis, Minnesota, murdered George Floyd, a Black man suspected of passing counterfeit currency at a variety store. Passersby, who unsuccessfully attempted to intervene, recorded and posted a horrific video of the crime. News of his death, accompanied by the video, ignited immediate and intense protests against systemic racism in the Minneapolis police force. Though Floyd's death was the impetus, other acts of police violence against Black Americans – including the murder of Breonna Taylor and the shooting of Jacob Blake – provided these demonstrations with ample fuel.[33]

In the wake of these events, the American grassroots movement Black Lives Matter organized further protests throughout the United States demanding concrete policy changes to address systemic racism and police brutality against racialized minorities. The movement had no central organization. It was a network of local civil rights organizers who used social media to coordinate messaging and activities. Twitter provided the movement with a powerful communication platform. Its official Twitter account had over one million followers in 2020, and its two hashtags (#blm and #BackLivesMatter) were used by anyone who wished to reach its members and followers.[34] The platform also gave the movement an international profile. Protests in support of the American movement and against police brutality were held in the major cities of Africa, Asia, Oceania, Central America, South America, and Europe.[35]

Wherever protests occurred, participants were aware of local narratives of colonial oppression and systemic racism. Indeed, many of the protests were held in locations where mnemonic devices celebrated individuals associated with these narratives. Some participants called for the removal of these monuments or plaques, and some simply seized the opportunity to make it happen. In the United States, statues marking Christopher Columbus, slave holders, and Confederate leaders of the American Civil War became focal points for immediate action. One prominent target was a statue of former president and slave holder Andrew Jackson that sits within view of the White House.[36] It did not fall, but many others did. A Wikipedia page documented all instances of material commemorations and works of art that were officially removed by authorized bodies during the protests following the death of George Floyd. As of 1 November 2020, it listed over two hundred removals in the United States, and dozens more in eleven other countries. The list did not include material commemorations that had been damaged but not removed, or those that were still under consideration for removal.[37]

Canadians joined this movement. Protests in support of Black Lives Matter took place in cities and towns across the country, from Victoria to St. John's, from Whitehorse to Windsor.[38] Residents of Niagara also organized protests. The first protests took place in Niagara-on-the-Lake on 5 June and Niagara Falls the following day. A St. Catharines protest, held the next week, took full advantage of symbolic public space. Organizers asked participants to meet at the British Methodist Episcopal Church, the one-time parish of Harriet Tubman. They then marched to city hall, where speeches were given within sight of the Watson Monument.[39]

Soon after these protests, two individuals sought the removal of the Watson Monument. Both used Twitter. On 9 June, Jake Breadman – a public history student with links to Niagara – proposed a petition to his four hundred plus followers. His tweet garnered two supportive replies and thirty-one likes. Over the next few days, Breadman posted several additional tweets explaining Watson's relationship to the North-West Resistance, but he took no further action.[40] On 17 June, Gavin Fearon – a St. Catharines resident and social justice activist – initiated an online petition with a series of eight tweets with a link to the petition website. Notably, Fearon broadcast the request beyond his more than five hundred followers by including a hashtag developed by the mayor's office to engage residents of St. Catharines (#ourhomestc). He also directed his final tweet to city officials.[41]

Fearon's tweets had an immediate effect. The mayor and three councillors tweeted their support within hours of its posting. Offering more than support, Councillor Karrie Porter entered a notice of motion to council to remove the monument.[42] Subsequently, the St. Catharines *Standard* and *Niagara This Week* both published articles covering Fearon's petition,[43] while local talk radio station CKTB interviewed Fearon, Porter, and two local history professionals over the following days.[44] The petition also featured on the evening newscast of CHCH-TV in Hamilton, the closest Canadian television station.[45] Outside of the mainstream media, two alternative news sources ran stories on the Internet. *Niagara at Large* was a daily blog produced by a veteran journalist who offered a progressive take on regional news. It published two pieces on the story: one noting support for the petition from the Niagara Region Anti-Racism Association (NRARA) and the second a report of Council's decision.[46] The *Niagara Independent*, a conservative news site, published one story explaining the motion.[47]

Council scheduled the motion for 22 June, but a busy agenda bumped it to 27 July. Then, following a lengthy discussion, the motion passed unanimously. City staff were directed to explore options for the monument's future in consultation with First Nations and Métis groups, the Watson family, the local historical society, veterans' associations, locally elected officials with both the provincial and federal governments, and affected city departments.[48]

Given our interest in the Watson Monument, we wanted to consult residents directly in downtown St. Catharines as we had in 2005. The COVID-19 pandemic that swept the world in 2020 negated that option. Instead, we investigated the communication channels used by residents both to support and to oppose the petition: social media platforms and the feedback channels of local media outlets. To begin, we consulted Alexa Internet, a research division of Amazon.com that measured global website traffic. This measure was reported as a ranking in relation to all other websites. Among Canadian-based Internet users, the top five social media sites up to mid-September 2020 were YouTube, Facebook, Reddit, Instagram, and Twitter.[49] We searched these sites for all combinations of the keywords "Watson," "monument," "statue," and "St. Catharines" posted during the period from 1 June to 1 September 2020. This period incorporated time prior to Breadman's first post and time after the council meeting that determined the statue would be removed. Facebook, Reddit, and Twitter produced results; the other two sites produced nothing relevant to the petition. Next, local media outlets were searched using the same keywords, in both their traditional formats and their online variants. With both social media and traditional media, only public-facing content available outside password-protection and paywalls was collected. We sought active engagement in the public debate regarding removal, not private conversations. We then analysed every contribution we found that met these conditions. However, to protect the authors' privacy, the account names have been de-identified in our analysis below.

Twitter's communication structure in 2020 was non-linear and non-hierarchical; it was free to use and designed for open participation among associates and strangers, linked only by @s and #s. This structure allowed content to spread quickly at any spatial scale. The growth of Black Lives Matter stands as a testament to its efficacy.[50] In St. Catharines, the initial calls to remove the monument were posted on Twitter, while local media publicized Fearon's tweets. Anyone who wished to engage in public debate over the petition and subsequent council motion could have taken to Twitter at any time. A total of 153 tweets were found posted during the period from 1 June to 1 September 2020; the most recent tweet prior to this period was dated 2015, and it was not added to the sample. All relevant Twitter activity stopped immediately following the council decision. Individuals and two organizations accounted for 121 tweets (79.1 per cent). Another 11 tweets (7.2 per cent) were posted by local politicians and the City of St. Catharines itself, while six local media outlets posted another 21 (13.7 per cent). We are confident this strategy captured most of the public tweets posted regarding the removal, but we recognize our keyword searches may not have yielded a complete sample. We also recognize Twitter users may not have represented the general public since only those users who followed specific hashtags or accounts would have seen the initial posts.

Reddit was also free to use in 2020, and it, too, permitted interactions among associates and strangers. However, where Twitter once limited prose content to 280 characters per tweet, Reddit allowed contributors more freedom to express themselves. Reddit hosted two communities or "subreddits" dedicated to issues concerning St. Catharines: "stcatharinesON" was open to the public and had over four thousand members, while "stcatharines" was private. The public community had one thread that addressed the petition. The thread started when a member posted a link to the first *Standard* story regarding the petition, and simply noted, "Yeah, this should probably happen." From 18 June to 23 June, eighteen accounts posted twenty-nine comments in the thread. Of these accounts, eight supported the petition, five were neutral, and three were opposed. Reddit encouraged readers to indicate their preferred posts in a thread. The top two posts, as voted by readers, both supported the petition. Another seven posts were contributed by accounts that were deleted before we searched the site. The content for five of these posts remained visible, but we did not include them in our sample. For what it's worth, four of them were neutral and one supported the petition.[51]

Facebook, too, was free to use in 2020 and permitted interactions among associates and strangers. Like Reddit, it placed no limit on the content contributors could post. Nine accounts were found with postings relevant to the petition during the period from 1 June to 1 September 2020. Three accounts belonged to the most outspoken supporters of the motion on city council, including Porter and the mayor. Another two accounts were operated by local anti-racism organizations: the Niagara Region Anti-Racism Association and the Brock University Student Social Justice Centre. A sixth account was operated by two independent journalists who produced the radio show *One Dish, One Mic*, a public affairs program focusing on Indigenous perspectives. All six of these accounts supported removal.[52] If any of these accounts received negative feedback, they did not remain posted for long. Only one account retained a single negative post publicly visible on its pages: Porter retained a letter criticizing her motion from a local veteran who self-identified as Métis.[53] There is evidence that one other individual account was opposed to the motion: one regular contributor to the Facebook pages of local media created a counter-petition opposing removal. This account was either closed or locked before we conducted our search.

The remaining three accounts were operated by local media outlets: *The Standard*, *Niagara This Week*, and CKTB. News reports provided by these outlets remained neutral. The newspapers ran no editorials on the issue and no relevant letters to the editor appeared on their pages or their websites before December 2020.[54] Their readership and listeners, however, had a great deal to say about the petition and motion. As CKTB announcer Matt Holmes stated, "The feedback we've been getting has been coming fast and furious."[55] Feedback began to stream into their Facebook pages immediately after news of the petition and motion broke. After that, feedback swelled again around the two council meetings that addressed the motion: those of 22 June, when the motion was originally scheduled, and 27 July, when the motion was rescheduled and passed.

Engaged citizens who wanted to address a local public used Facebook far more than Twitter or Reddit. The petitions were first proposed on Twitter, perhaps to flag their solidarity with activists around the world who sought to remove contentious monuments. However, mainstream local media outlets brought the second petition to most residents' attention. Vividata's *Newspaper Topline Readership* report for spring 2020 estimated that *The Standard* had an average estimated audience of 142,000 unique weekly readers (combining both print and digital editions) aged eighteen and over, or 42 per cent of the regional population in that age range. The Numeris *Topline Radio Statistics* report for spring 2020 estimated that CKTB had an average estimated weekday audience of 27,700 listeners aged twelve and over, or 7.4 per cent of the regional population in that age range. Recordings of its hosted content were also archived online. In addition, *Niagara This Week*, a free weekly paper, distributed 165,000 copies to households throughout the region. It also published its content online.[56] The Facebook pages for all three outlets served as digital extensions of their traditional letters columns and phone-in shows.

Table 4.1 provides aggregate data for all social media posts collected. This table indicates the number of comments posted to social media and their relative sentiments.[57] As each new post was tabulated, we cross-referenced the account name with all previously tabulated posts on all three platforms. This gave us a rough sense of the number of unique accounts that made contributions to the debate ($n = 470$). However, the

Table 4.1. Number of unique social media accounts and posts regarding the petition and motion to remove the Watson Monument

Platform	Citation*	Date	In favour		Neutral		Opposed		Total posts
			Distinct accounts	All posts	Distinct accounts	All posts	Distinct accounts	All posts	
Twitter									
Petitioners	TW	June–July 2020	2	53	0	18	0	0	71
Individuals	TW	June–July 2020	18	24	8	8	4	16	48
Organizations	TW	July 2020	1	1	1	1	0	0	2
Politicians/city	TW	June 2020	4	6	2	5	0	0	11
Local media	TW	June 2020	2	3	5	18	0	0	21
Facebook									
Niagara This Week	NT18	18 June 2020	3	3	13	15	9	11	29
The Standard	ST18	18 June 2020	11	13	15	32	7	10	55
The Standard	ST19	19 June 2020	4	6	14	28	3	3	37
The Standard	ST23	23 June 2020	4	18	10	34	14	25	87
The Standard	ST28	28 July 2020	9	13	22	45	16	20	78
The Standard	ST29	29 July 2020	2	3	6	18	0	2	23
610 CKTB	CK17	17 June 2020	23	72	24	46	88	141	259
610 CKTB	CK22	22 June 2020	1	1	0	0	6	7	8
610 CKTB	CK23	23 June 2020	3	6	4	5	25	33	44
610 CKTB	CK30a	30a July 2020	4	6	6	6	19	21	33
610 CKTB	CK30b	30b July 2020	9	18	4	14	29	42	74
Reddit									
r/stcatharines	RS18	18 June 2020	8	15	5	15	3	7	37
Totals	–	–	108	261	139	308	223	338	917

* Hereafter, quotations from these social media platforms are referenced with these codes indicating the original social media account and the date a contribution was posted.

table cannot be read as a measure of public opinion regarding the petitions or motion. While our sidewalk survey created a sample of residents that reflected the demographic profile of the city, our search of social media created only a database of accounts whose authors felt moved to comment on this issue; the demographic characteristics of the authors were essentially unavailable. As such, we used this data to understand how contributors thought about local heritage and mnemonic narratives, the significance of mnemonic devices, and the Watson Monument itself.

Social media posters in 2020 echoed the opinions expressed by survey participants in 2005. Many posters, for example, commented that they were barely aware of the monument, and few knew of its origins in the North-West Resistance. However, there were two important differences in 2020. First, the media coverage given to the petition educated readers and listeners who followed local public affairs. One exchange on Facebook illustrated this point:

> EA: "Did you know the name of the soldier on the monument before the paper printed it? I worked on that block for seven years and couldn't recall and had no idea what the monument represented."
> WD: "nope, I had to look it up." (CK17)[58]

The irony was obvious, as AB noted: "Seems to me that more people are learning about history through the move to remove this statue than ever learned by it sitting there. I can't say I ever knew anything about this guy or his place in disgraceful events in our country's past" (NT18). The second key difference in 2020 was the petition itself. Our survey asked participants to describe the monument's narrative and its significance for an open-ended academic study. The petition asked council to reject the monument's narrative and its significance. Council's answer, of course, posed an existential threat to the monument. Social media posters were greatly divided.

Opposition to the petition took several forms. Most commonly, opponents defended the monument's educational value. They often began with a simple assertion: "history" is a series of past events that cannot be changed or erased. This claim had a "common sense" appeal. Barring the invention of time travel, the actions of past generations could not actually be undone. Reddit member N23 declared, "Well you can[']t erase history! ... Tearing down these statues will not solve anything" (RS18). This claim was often linked directly to the monument itself. It was not viewed merely as a relic of the past, but as a living record and reminder of its times. If history was worth studying, then the lessons

taught by historical monuments should be preserved for future generations. BM1 wrote, "You can't change history. Learn from it and leave the
heritage alone" (ST28). School children were often hailed in this context,
as MZ did: "[We] Cannot erase history ... What are we going to teach
our younger generation" (CK18)? Or as CL1 wrote, "History is history,
so we remove and hide or we leave and teach our children maybe how
to not be or act the way things used to be. We learn from what bad happened and change it" (CK18). Allusions to "history," "heritage," and
the good will of future generations implied the monument retained an
enduring purpose.

Some opponents built upon the first argument. They acknowledged
the North-West Resistance and a narrative of colonial racism but
quickly posited a new assertion: that past generations should not be
judged by the standards of the present. Yes, opponents noted, societal
values had changed dramatically over the previous century. However,
it was best to understand historical actors in their own context. This
point was made by DW:

> That fellow served his country as every other able bodied [sic] man was
> expected to do. He made the ultimate sacrifice ... We can't judge him 150
> years later for none of us understood the situation brewing out west. (ST28)

Those who took this position sought to preserve the monument in place
with new interpretive materials or with additional statuary representing
the Métis, Cree, and Assiniboine Nations. KF supported this suggestion:
"You learn from the past. Raise awareness. Leave history in its place so
that lessons can be learned. Change the plaque to a more descriptive
position" (ST23). Examples were provided of other monuments in other
jurisdictions that marked unsavoury pasts. Several contributors noted
the solemn preservation of Holocaust sites. For example, JG mused,
"why do you think [A]uschwitz is still standing, because the Jewish
[people] don[']t want anyone to forget what was done to them. History
is a reminder of what has happened and what can happen" (CK17).

Extending this second argument, some contributors drew attention
to the federal government's responsibility for the treatment of Indigenous Peoples. This move was intended to deflect accusations pointed at
citizen-soldiers like Watson. These contributors understood the monument was designed to honour all Canadian volunteers who had died in
service for their country. FS noted, "he was a Canadian soldier obeying
orders and fighting bravely. As a private he was not the architect of
the skirmish" (CK30a). Implicitly, contributors like FS acknowledged
the petition's concerns regarding the North-West narrative. They also

understood the reputations of historical figures like Sir John A. Macdonald were under scrutiny for their racist political doctrines.

> RR: it doesn[']t make a whole lot of sense to remove a statue of a soldier doing a soldier[']s duty and who died a soldier[']s death, while statues of the guy who ordered the troops there in the first place are still prominent around the country (ST17).
>
> DMK: Private Watson was not in charge of Canada's Metis/Indigenous policy. He should not be blamed for things he had no control over, and his statue removed. He was a common soldier who did his duty and died for Canada. Why can't we just honour his sacrifice (ST29)?

On this point alone, many thought the monument should remain to honour the continuing service of local soldiers. On 1 July, DMK, for example, tweeted a happy Canada Day message to the city in which he hailed the citizen-soldier. His tweet featured an image of the Watson Monument. The day prior to council's debate on the motion, he tweeted again with a warning from Lieutenant Colonel John McCrae: "To you, from failing hands we throw, the torch; be it yours to hold high. But, if ye break faith with us who die, we shall not sleep. Though poppies grow in Flanders Fields."[59]

Some opponents defiantly rejected the petition's central premise. They believed Watson and the North-West Field Force were rightly deployed against a violent insurrection; changing times and changing values had not superseded the army's just cause. As MR declared, "It[']s not a generic statue f[or] f[uck's] s[ake]. They chose to honour a St Catharines native who gave his life in service to his country. Going back and looking at it thru liberal eyes doesn't change that" (CK23). Several contributors reiterated the North-West narrative from the federal government's perspective in 1885, thus casting Watson as a hero. A28 provided a succinct summary:

> Louis Riel was tried for treason and the murder of a local man, which required the Canadian Militia and NWMP to twice mount campaigns to arrest him. Riel was found guilty and hung [sic]. Today revisionist history types have made him [a] martyred hero despite his crimes (RD18).

Similarly, TN22 wrote,

> The statue is not only there to hono[u]r Watson but to hono[u]r other fallen Canadians in the NWR, not sure why we would want to get rid of it, as a P[erson] O[f] C[olour] I'm so grateful for the Confederation and the

people who fought to protect it, if anything they should repair the statue, you gotta remember that he was fighting the rebels who wanted to destroy this nation, getting rid of it would be like tearing down a statue of Lincoln and replacing it with [Robert] E Lee (RD18).

Opponents who advanced this argument were dismayed by their fellow citizens' lack of knowledge regarding Canadian history or the petitioners' interpretation of it. BI lamented, "It is very unfortunate that today we don't have the understanding of the times. We choose instead to re-write the past by removing it" (ST28). RB1 chided supporters of the petition: "he was a soldier doing his duty, which he was ordered to do, this was during a[n] uprising and you were not alive and have no idea what it was like" (ST23). Similarly, JS1 offered the following view: "It's concerning that today's citizens want to minimize the contributions that past soldiers have made to their country; something that was directed (ordered) by the elected government officials acting on behalf of the citizens of the time" (CK30b).

Opponents questioned the petitioners' motivations. Breadman and Fearon acknowledged that American anti-racism protests inspired their actions. Many opponents, however, did not view the petition as a principled or sincere fight for social justice. Rather, they believed it was a misguided search for wrongs where none existed. Some went further and accused the petitioners of an opportunistic attempt to acquire the benefits of a reflected glory. The invisibility of the monument came into play here, as BM2 wrote: "People aren't worshipping the statue. Hell no one knew who it was until people bitched" (CK18). Similarly, SL chastised the petitioners who were acting "like removing that statue is going to change things. Most people don[']t even know who the man was" (ST18). KM asked the petition's supporters on Facebook, "seriously, did you even know the statue was there? Why now?" (CK18). Comments like these were couched in language that diminished the petition as a mere political fad that would quickly pass. CN wrote, "the statue never bothered anyone before. Someone needed to find a statue somewhere in order that they could jump on the wagon" (CK18). To underscore the seemingly foreign premise of the petition, AI asked rhetorically, "I am tired of people jumping on a bandwagon. Really you could not come up with one Canadian minority that was attacked unfairly by Canadian Police?" (NT18). JFO seemingly replied, in a different context, "Got to be like the US" (ST28). These comments were posted amid national headlines documenting an officer of the Royal Canadian Mounted Police (RCMP) intentionally using a truck to run down an Inuk man in Kinngait, Nunavut; an RCMP assault on Athabasca Chipewyan chief

Allan Adam and his wife for an expired licence plate; and the shooting death of Chantal Moore, a member of Tla-o-qui-aht First Nation, by Edmundston police during a wellness check.[60] A member of the North-West Mounted Police, the predecessor of the RCMP, is named among the dead remembered on the Watson Monument.

Several opponents saw a more sinister logic in the push for removal. If any one segment of the community could prompt the removal of any mnemonic device, the petition would create a slippery slope towards the erasure of the entire historical record. MM1 asked, "So everyone has to cave to these people? What makes them so much more important than history?" (ST17). Many felt supporters of the petition represented a minority within the community, as FJ claimed: "So council is seriously considering jumping through hoops to please the minority perpetual malcontents?" (ST23). Sometimes the petition was associated with an unsavoury political ideology, which ranged widely from communism to fascism. Whichever ideology was named, the real fear was totalitarianism. PP asserted that the petition was "removing the past, right in the footsteps of Nazism and Communism" (ST18). DH1 pointed out the precedent being set: "Why stop there? Using this rational[e], shouldn't we be renaming Port Dalhousie? It is named after George Ramsay, 9th Earl of Dalhousie … He was a proven racist" (TW). Similarly, PG2 offered other candidates for removal: "Tear down the pyramids because Egypt had slaves[;] tear down the [C]oliseum [R]omans fed Christians to lions. Tear down every religious monument because terrible things have been done in [God's] name. I could go on and on" (CK17). These contributors insinuated that their suggestions were inconceivable and yet fearsomely realistic at the same time.

The slippery slope led opponents to extremes. The petition to remove the monument fostered suspicions that further requests would result in the removal of all other mnemonic forms that embodied challenging narratives, thus posing a threat to individual freedom of conscience and speech. WD made this point: "Let's just completely wipe out our history books because it doesn't fit with your narrow point of view. Soon there will be laws on the books that we all have to paint our houses the same colour as everyone else, vehicles too" (CK17). More than one contributor drew comparisons between the council motion and George Orwell's fictional Ministry of Truth in his novel *1984*. JS2 wondered, "When do we start burning books?" (ST18). KC went darker, telling the petition's supporters, "Hope you will march with torches as you topple it down. I can already hear the jackboots" (ST29). Such fearmongering lapsed into common insults and predictable slurs evoking bestial metaphors, left-baiting, ageism, sexism, ableism, and racism.

Opponents, then, had five arguments rooted in principled claims: they defended "history" for its educational value, opposed the moral relativism of the petitioners' judgments, absolved the loyal citizen-soldier, upheld the constituted authority under challenge, and drew attention to the threatening precedent embedded in the editing of public narratives. They had two further arguments rooted in practical considerations. First, some questioned the priorities of petitioners and council members alike who chose to address a symbolic issue amid a global pandemic. DH2 summarized the situation as follows:

> This is what these elected officials do with their time at city hall. Come on now. [S]eriously ... ?? With all that's going on these days (pandemics, provincial state of emergencies, elderly being denied health care and dying at alarming rates, street rall[ie]s and protests, etc ...) this is what comes to your mind? A 135 year old statue that's been sitting there all this time?? Woooooooow (CK17).

Second, many opponents were mindful that any change to the monument would be costly. PG1 put the removal fees in context:

> Has anyone at the city looked at what it would cost to remove this statue and replant it at the cemetery[?] A statue of this size will require a large crane, transport truck and a foundation to mount [it] on. You just can't plunk it down. Considering that it will be a government job I don't think I would be too far off in estimating the cost at something near six figures (ST23).

Whatever the final cost would be, the city would incur it while operating on emergency financing.

In the midst of these posts, BJ – an active contributor to social media discussions of local public affairs – created a counter-petition on iPetitions.com. Following *The Standard* story posted on Facebook on 23 June, he invited fellow opponents to sign it. Curiously, his explanation of the monument profiled a different Private Alexander Watson, a winner of the Military Medal for bravery who served with the 49th Battalion (Alberta) in the First World War. The petition had collected six signatures by September 2020.[61]

Table 4.1 also reveals that a significant proportion of social media commentary was contributed by individuals who took no clear stand on the petition. Their posts most often took one of two approaches to the debate. First, many neutral posters were keen to have timely, accurate information regarding the North-West Resistance, Watson, the monument's condition, and the administrative process governing the monument's removal. These posters either posed questions or answered questions in interactions with other users. They did not, however, signal their thoughts on the petition itself.

Second, many neutral posters were concerned about the cost of removal. For this group, council's deployment of precious resources such as labour and tax dollars during the pandemic was the key issue. ECM admitted,

> I'm conflicted about this. On the one hand, we as a society shouldn't be glorifying genocidal racists. But that's not what this is about. We are in the midst of a pandemic. Massive unemployment. Economic prospects are bleak. And this is what we are focusing – on spending money – on? (RD18).

Some posters who did not believe removal should be tax-funded were amenable to private financing. JJ elaborated his thoughts over three posts:

> The people that want it removed can pay for it/fundraise for it. I don't want any of my tax dollars supporting the removal because people are butthurt over old statues ... Focus on making a better today and move it to a museum if you don't want to look at it.
>
> Fund for pressing issues like homelessness and mental health resources.
>
> B[y] t[he] w[ay] I'm part [N]ative ... It's a statue ... don't like it don't look (CK17).

LS echoed these thoughts: "I think feeding homeless people or spending the money on people in need is more of a priority in our community [than] wasting money to remove a statue that a lot people didn't even know existed" (ST23). IB thought both sides could be happy with one simple change:

> Put a plaque beside it explaining the controversy and leave it alone. In a couple of years it will fall down on its own. There are a lot more important things to deal with at the moment. This is a distraction we don't really need (ST28).

These contributors believed the pandemic set administrative priorities for council that overrode actions that appeared to be purely symbolic.

Finally, a minority of neutral posters shrugged off the entire debate. CLC offered this frank admission:

> I turn 40 this weekend. I never even noticed the statue was there to begin
> with. I don't have a problem with it staying. Didn't notice the statue, don't
> know who he is, don't care who he is, wouldn't waste taxpayers $$. If you
> move it, set up a private fund for those who care (ST18).

VJC got straight to the point: "It really makes no difference to my daily
life if it stays or goes" (CK23). The monument's apparent invisibility
was deployed once again.

Most social media posts supporting removal appeared to be inspired
by contemporary events outside the region. Breadman's tweets indicated he was an active supporter of social justice causes. His opening
tweet stated, "I've seen a lot of discussion on monuments recently and
I want to write down some of my thoughts on the Watson Monument
in downtown St. Catharines and why I think it should be removed or
recontextualized."[62] Fearon described his motivation during his CKTB
interview: "We're living in a particular moment where a number of
colonial monuments are coming down in the United States and around
the world so it made me want to take a look at my own neighbourhood
and it brought me directly to the Watson Monument at City Hall."[63]
Following these comments, posters on all platforms were consciously
responding to their calls for removal.

The foundation for most supporters was a knowledge of the North-
West Resistance and its consequences for the Métis, Cree, and Assiniboine Nations. One touchstone for supporters was the rehabilitation
of Louis Riel's reputation after 1970, when the federal government
admitted his execution was a mistake. A second touchstone was the
revelations of cultural genocide perpetrated by the federal government through its nationwide residential schools program. Anyone who
acknowledged the racism that informed the government's policies
could not view the Watson Monument as anything other than a glorification of the systemic oppression of Indigenous Peoples. Discussing
the petition, CH noted, "this idea has actually been kicked around the
activist Community for a while … Why should we celebrate somebody
who was part of the group responsible for the execution of Louis Riel?"
(CK17). Her comments were frequently echoed:

> BH: Yes, it glorifies the repression and persecution of [I]ndigenous people
> of Canada – [it] belongs in a museum or cemetery as suggested (CK17).
> MC: For anyone who is whining about this – this one[']s for you[:] nobody
> is erasing history. It happened. It can't be undone. But what CAN be
> done is learning from the past, and making sure it doesn't happen

again[.] A big part of that is being honest about people who were once celebrated, and acknowledging their horrible choices[.] A great way to move forward is to make sure that those people aren't honoured and "placed on a pedestal" … [T]hey don't deserve to have statues memorializing them (CK17).

BC: Why is it there? It's not commemorating anything honourable or worthy. Leaving it up is honouring a genocide. Statues are about honouring. It's not there to remind us of an important history lesson. It's honouring someone. Take it down[;] it was a history to remember not commemorate (CK17).

The plaque added in 1901 for the dead of the Second Boer War did not, for supporters, bolster claims that the monument was a service memorial for all Canadian soldiers. Rather, the Boer link only confirmed its tainted imperialist origins.

Media outlets that covered the global reckoning over racist and imperialist monuments noted the heightened historical consciousness of the activists. Local activists understood the traditional narratives of colonial development to be self-serving justifications for the marginalization and/or eradication of Indigenous Peoples and the enslavement of Africans to speed capital accumulation for European investors. CA declared,

[I] support any change for inanimate objects to be in a museum with "Their"/Our History. [I] Support any Gesture of Acknowledgement of injustice in efforts of "Reconciliation." In the same way that we want Justice for any Murdered Family Member or Friend of Our Own. We Don't Change History We Rectify History (CK17).

This sentiment was expressed in various ways:

MD: referring to school children -- we teach them the truth about how we treated our [I]ndigenous people, we teach them the truth about what REALLY happened vs the whitewashed history that is lies (CK18).

EM: We can remember our history, continue to teach and learn from it, but we don't need to celebrate our mistakes (ST28).

NG: I remember learning VERY LITTLE about [B]lack and [N]ative history in school. It was all explorers made to look like they discovered untapped land instead of what really happened (CK17)!

RC: I am not making the dude a villain, but to continue to celebrate him in such a fashion is past. Time to put him in a museum with the full history (CK30b).

Personal feelings of guilt may also have guided these contributors. As AJ opined, "It's [*sic*] history has been brought to light and if we ignore that we are just as bad" (ST21).

Occasionally, supporters and opponents framed their comments with reference to their individual subjectivities. Those who self-identified as people of Indigenous heritage used Facebook to explain just what the petition meant to them. Though few in number, they placed the debate in an identifiably personal context. EC made this clear: "I'm a father of 3 and I'm also a proud Native American, these statues should have never of been made in the first place ... Kids should be raised to respect others['] feelings and if these statues bring pain to people and represent evil why keep them up? Respect others['] feelings" (CK17). Shifting from the personal to the political, APS noted that, "As a Métis person living in Niagara, it would be nice to be able to have a conversation with the city of St. Catharines without the reminder that the people in this area at one time didn't want us to exist" (CKJ17). Looking askance at comments from those who opposed removal, MM2 wrote, "I'm [N]ative, and frankly I find the lack of empathy even more offensive than the statue" (CK30b). This point – calling out the insensitivity if not sheer racism of opponents – was touched upon by Karl Dockstader and Sean Vanderklis, the hosts of *One Dish, One Mic*. Both were actively engaged with contemporary concerns addressing Western oil pipelines, Maritime fishing rights, and unaddressed land claims in Ontario, and they supported the petition on Twitter and on air:

> KARL DOCKSTADER: I think these discussions are important too, so shout out to Gavin Fearon, shout out to non-Indigenous people for taking some of this weight off of our shoulders by leading this discussion.
> SEAN VANDERKLIS: If you want to be an ally, this is the bare minimum you can do.
> KARL DOCKSTADER: Yeah, this is like the low-hanging fruit: find all the racist statues. There's a bunch![64]

A handful of supporters who self-identified online as people of European descent deemed it important to flag their "white privilege" before addressing the issue at hand. For example, EA wrote, "As someone on the 'winning' side of the Northwest Rebellion, I would like to acknowledge the pain of the Indigenous people and Metis[,] and if this memorial gives them pain, let's move it" (CK17). The opposite also happened. Several supporters called out opponents whose profile images may have presented them as Caucasian. One of the more diplomatic comments came from RB2: "Yikes, there's a lot of white privilege in this thread. I find the volume of people taking time out of their day to be offended by marginalized members of our community speaking up for themselves to be sad and disheartening" (CK30). Less-diplomatic supporters simply lapsed into predictable insults and likened the opponents to residents of the southern United States, racists, white supremacists, Trumpists, and fascists.

Critical historical reflection and inclusive cultural values offered two principled reasons to remove the monument. Two practical reasons also came into play. Opponents believed the monument was educational, serving as an important reminder of military service. Supporters rejected this notion. The common observation that no one knew what the monument represented was repeated here. It was treated as evidence that the monument, as a medium of communication, was simply not working. One exchange illustrated this point:

> EJB: I don't think a single person in this comment thread knows what the North-West conflict was about – and Google exists. Look it up and then explain to me why this guy needs a statue at all.
> TB1: [EJB] thanks I didn't know much about this person (CK18).

As we have noted, all monuments convey their narratives through symbolism and texts. At best, they supply a prompt for the viewer to remember past lessons or to seek new knowledge elsewhere. In this case, it was apparent that this monument's specific symbols and texts did not convey knowledge of the North-West Resistance to contemporary residents. They had never learned the narrative and they never felt compelled to investigate. Responding to comments that suggested that "history" would be lost, supporters offered the following retorts:

> SRK: Quick questions for those opposing the statue[']s removal/relocation. Did you even know who Watson was before this week? Do you know what accomplishments he's recognized for? Can you still say we're erasing history by taking down a statue that … no one knows about or that people refuse to learn about? … No? Looks like a lot of St. Catharines people need to read up on History (CK17).
> BM3: We learn from the past, we don't learn from statues (CK22).
> MG: The fact that you think removing a statue removes history, while also acknowledging the existence of books would suggest you are aware there are actual history book[s] for which one could learn about … wait for it … HISTORY … Clearly there is no risk of erasure (CK17).

> DD: I can learn from history in a number of ways a statue doesn't teach me. You know[,] a book or Google that teaches! No one is saying don't teach about it[;] we are saying take the friggin' shrine down! (CK17).

Supporters used their understanding of historical knowledge and how it is communicated to negate the significance of the monument. They portrayed it as one ineffective channel to communicate a narrative that was better told through other media.

The second practical reason was the monument's fragility. Councillor Porter first suggested the monument be moved to Watson's grave at Victoria Lawn Cemetery, its original destination in 1885. Some supporters viewed this as a sensible plan. SW chipped in: "I personally don't mind this change. I have no problem with his statue being moved to the man's grave as was originally the plan. The plan is only to move it, not destroy it" (CK17). TB2 agreed, asking, "wouldn't it be a better way to honour him?" before offering solace to opponents: "You could still visit it at the grave site of Pvt. Watson" (RS18). The other suggested site for the monument was a museum. Some posters knew the monument was not aging well. If council followed the advice of its 2018 motion, the monument would slowly crumble in place. Removing the monument from city hall might, rather ironically, save it.

For one segment of the petition's supporters, moving the statue from one public space to another was no solution at all. The NRARA qualified its support for removal by arguing the council motion did not go far enough. In the committee's opinion, the monument was "a symbol of identification with white and Anglophone supremacy and militarism and therefore has no place anywhere other than the dust bin [sic] of history. It should not be relocated but should be disposed of and destroyed."[65] The NRARA's posts on its own Facebook page received positive feedback from eighteen different individuals.[66] Supporters online suggested a wrecking ball, rock crusher, burial, and/or dumping it in the Welland Canal (echoing the spontaneous removal of the statue of slave trader Edward Colston in Bristol, England).[67] Some volunteered to help.

Finally, some supporters felt the removal of one monument and its contentious narrative might make room for a new monument with a more inclusive, hopeful narrative. Harriet Tubman and her role in the American abolition movement was suggested by eight different individuals. Opponents of removal commented sarcastically that further suggestions would include Karl Marx, Mao Zedong, Che Guevara, convicted terrorist Omar Khadr, and the mayor of St. Catharines at the time, Walter Sendzik. Supporter MM2 chastised these opponents:

"Don't just keep a statue just because. Make it a choice. And if he's not relevant now … CHOOSE to honour someone who IS relevant now. Choose to make history OURS" (NT18). Such comments made it clear that two groups of memory entrepreneurs were battling over which sets of values were represented by the North-West narrative. Both sides sought control over the mnemonic products that dominated a shared public space.

Ultimately, Fearon's petition itself generated the greatest number of posts from those who supported removal. By late July, when council voted on the motion to remove, the petition counted 848 digital signatures. Unfortunately, it is not known how many of the signees were residents of Niagara because the petition did not request addresses from participants.

A Career Concluded?

The Watson Monument has had a long career. It was created in the enthusiastic wake of the North-West Resistance to honour Canada's volunteer militia. Over time, its role as a site of military commemoration was transferred to new memorials that better addressed the lived experiences of subsequent generations. So how did twenty-first-century residents view it? The 2005 survey and the 2020 debate over removal revealed that most residents had forgotten the specific narrative that inspired its sponsors. The passage of time and changing sensibilities had rendered the historical details of the monument almost invisible. James E. Young has described this process as follows: "believing that our memorials will always be there to remind us, we take leave of them and return only at our convenience. Insofar as we entrust memorials to remember on our behalf, we become that much more forgetful."[68] In 2020, it was not the monument but international events that prompted residents to investigate its narrative.

Both the survey and the social media debate indicated that the monument continued to resonate with residents on an abstract level. Most war memorials construct national identity through carefully crafted narrative, symbolism, and placement in the landscape. Identities, however, must also be rooted in intertextual narratives that carry authoritative, cognitive, and emotional weight such that they become widely accepted. This provides individuals with a common understanding of the past – a sense of belonging – that elides social contradictions in the present and thereby creates the ideological conditions for peaceful coexistence.[69] Citizenship in the nation, however, further implies a reciprocal moral obligation to maintain and preserve the state.[70] As

Michael Ignatieff argues, if one's nation faces an existential threat, then citizen participation in its defence is expected. In combining belonging and obligation, celebration of the military has functioned as a potent symbol of group identity, citizenship, patriotism, and heroism.[71]

Narratives celebrating military service are not dependent on the details of any one person, event, or place. They circulate within society and become manifest in multiple artefacts and practices through commonly recognized symbols and gestures. Janet Donohue makes this case in writing about the generic American war memorial, contending that "the soldiers have lost their identity in becoming representatives of a collective and heroic act that stands as a symbol of American patriotism."[72] Andrew Butterfield extends this thought: "By means of this standardization, it often limits the commemoration of the individual to a single aspect: how the dead personified a social ideal that is regarded as a central and permanent value."[73] The Watson Monument functions in this way. The specific impetus for the monument's creation has been forgotten, yet its symbolism and affective potential remain insofar as it conveys an abstract narrative of the citizen-soldier.

Throughout Canada, the abstract narrative of the citizen-soldier is an element of cultural capital. It can serve to reaffirm – both internally and publicly – an individual's allegiance to the nation.[74] State agencies, charitable organizations, and the private sector maintain memorial practices. Federally, both Conservative and Liberal governments have long promoted Canadian military service through the production of commemorative ephemera, recruitment advertising, and Remembrance Day observations. Provincially, schools have been expected to maintain Remembrance Day curricula appropriate for children of all ages. In St. Catharines, the Watson Monument was one of five war memorials maintained by the city that hosted Remembrance Day ceremonies. Veterans' associations lobby to keep veterans in the public eye. The annual poppy campaign of the Royal Canadian Legion is itself a ritual of remembrance, while the War Amps of Canada works throughout the year to aid wounded veterans. The War Amps of Canada logo, not coincidentally, includes a silhouette of the National War Memorial in Ottawa.[75]

The commemoration of Canadian military efforts gained a new sense of urgency between 1991 and 2014, a period framed by the Gulf War and the War in Afghanistan. Our research on the monument began in 2005. Contemporary deployments in Afghanistan gave the reality of service and sacrifice a new immediacy. Eight Canadians died on duty between 2002 and 2005; this number rose to 158 before the Canadian Armed Forces withdrew in 2014. Some 40,000 veterans returned, with over 2,000 having suffered injuries.[76] As noted in chapter 1, the

Conservative government (2006–15) used this campaign to embellish Canadians' sense of military honour and readiness.[77] These efforts were matched by citizens across the country who participated in commemoratives practices such as "Red Fridays" (when civilians were bid to wear red clothing to demonstrate their support for the military).[78] In 2002 residents of southern Ontario also sought to honour the repatriated dead. The bodies of troops were flown into the Canadian Forces Base located near Trenton, Ontario. They were then transported via hearse to the coroner's office in Toronto before being sent to their final resting places. Knowing this, crowds began to gather along the route each time a hearse carrying a soldier's remains was scheduled to pass.[79] Notably, in 2007 the federal government affirmed this practice, officially declaring the stretch of highway between the air-force base and Toronto the "Highway of Heroes."[80]

The federal government stoked these rising sentiments by way of a shift in its recruiting campaigns. Through the 1990s, the Canadian Forces had enticed young Canadians to embrace the adventure of a military career, proclaiming, "there's no life like it." Focus group research in 2006 indicated that group participants wanted, in Commodore Roger MacIsaac's words, "realism and transparency. They wanted to know what we were doing today."[81] Advertising firm Publicis Montréal used this insight to produce a series of dramatic advertisements, in French and English, that appeared on television and theatre screens across the country. A cinéma-vérité style depicted soldiers in tense moments facing combat threats and natural disasters. Recruits were now bid to "Fight Distress, Fight Fear, Fight Chaos – Fight with the Canadian Forces." The shift in tone from earlier campaigns was clear: an emphasis on Canada's past service in international peacekeeping had been replaced by a warrior mentality. Later iterations of the campaign were extended to social media platforms and sporting events.[82] A year after its launch, *Marketing Magazine* surmised, "The ads clearly struck a nerve. The number of applicants rose 40% to 40,000 and Canadian Forces signed up 12,862 full-time and reserve members between April 2006 and March 2007, about 400 more than the target."[83] The magazine named the Canadian Forces among its elite Marketers of the Year.[84] There were some pointed criticisms of the campaign, but the positive responses suggest that it did resonate with many Canadians. The campaign's tone and execution were only conceivable if the citizen-soldier narrative had already been reduced to a recognized social value or ideological ideal that contemporary audiences understood.[85]

We believe it was successful. In St. Catharines, the 2005 survey participants understood this narrative was present in the Watson Monument.

The symbolism in its form was legible to all, regardless of their knowledge of the North-West Resistance. Even with its colonial-era uniform, the statue and its demeanour prompted thoughts of Canadian citizen-soldiers. Participants never felt the need to examine it more closely. Residents had seemingly internalized this ideological message much as the monument' s original sponsors might have hoped.

These feelings were still evident in 2020. However, discussion of the monument's fate revealed a newly complicated relationship with its symbolism, narratives, and values. Residents who engaged the public debate offered thoughtful insights regarding the function of commemorations despite the monument's supposed invisibility. Opponents believed the monument's symbolic message – the valorization of the dutiful citizen-soldier – overrode any regret felt for the conflict it marked. Implicit here is the command to not "break faith" with the fallen. Supporters may have shared this belief in general, but global events in 2020 had shifted the frame through which they viewed this one specific commemoration. For them, the Watson Monument marked a tragic and shameful past that could no longer be valorized in public. The monument's narrative of duty and valour was undermined by its unspoken narrative of racial oppression. When council agreed to explore its removal, the petitioners and their supporters became successful memory entrepreneurs in their own right, with their own values and narratives now shaping the lawn at city hall.

PART THREE

The Niagara Grape and Wine Industry

Chapter Five

Viticulture as a Mnemonic Product

The Niagara peninsula, in the blossom season, is recognized by world travellers to be one of nature's rarest works of art ... The millions of fruit trees for which this section is so widely famed, clothed in their floral splendor [*sic*], give the entire peninsula the appearance of an endless range of tiny mountains ... The journey borders on historic places and every town and village en route is worthy of inspection.

– United Hotels Company newspaper advertisement, 1920[1]

Niagara has witnessed pivotal moments in North American history. As described in earlier chapters, memory entrepreneurs have identified specific narratives they believe were foundational to the region's development. These narratives describe international intrigue, combat, and migration; Indigenous dispossession and settler colonialism; and massive engineering projects that ushered in the modern era. As impressive as these narratives may be, they constitute only one set of resources available to residents when constructing individual and group identities. Residents may not prioritize mnemonic products that preserve foundational narratives when other, more relevant resources are readily at hand.

Chapter 1 described our 2005 survey of resident knowledge of local mnemonic products and the mnemonic practices that support them. We expected residents to identify their hometown with the dramatic tales captured in the region's monuments and historic sites. Instead, a majority of participants identified contemporary people, events, and places as markers of regional identity. Most notably, the wine industry figured prominently among residents as a key marker of "Niagara." Though long established, the wine industry had garnered little attention until the late 1970s.[2] Still, many participants associated the industry with the region's "heritage." How had residents integrated this narrative with the region's culture, traditions, and history?

An answer lies in wine marketing and regional place branding. Memory resources emerge from the diverse efforts of interested parties and come in many guises. Niagara's toponomy, landscape, and narrative history have been shaped by a variety of actors who advanced narratives pertinent to their self-interests: talented volunteers in local historical societies, dedicated public history professionals, engaged journalists and academics, as well as lobby groups and politicians of every stripe. Alongside these groups there have always been those who see historical narratives and mnemonic products as economic assets.[3] Among the most active producers of mnemonic resources in recent times are those working on behalf of Niagara's grape and wine industry. They have positioned themselves as the promoters of Niagara's agricultural heritage and they have cultivated this association through corporate brands and marketing. Local governments have lent their authority to endorse and reinforce the industry's efforts. In this chapter, we explore the branding of Niagara as "wine country" to understand how this brand became a resource available to residents assembling collective views on local identity.

The emergence of a high-quality wine industry happened at an opportune time. As manufacturing declined across North America in the 1990s, municipal governments sought alternative economic engines. The options open to lower-tier governments coalesced around neoliberal policies in the guise of the "creative economy" concept. In this context, the question of "place" and "identity" became increasingly significant. Promoters of the concept argue that municipalities must either be attractive to entrepreneurs and workers in creative industries or resign themselves to becoming ghost towns.[4] Simultaneously, cities and regions are competing for tourists. There are synergies between the creative economy and tourism; "creatives" and tourists both actively seek unique, consumption-based amenities.[5] Municipalities courting favour among either group quickly find themselves vying against one another. The success of any one municipality, then, lies in its ability to activate a unique blend of quality of life, authentic experiences, aesthetics, and green living that can be articulated through a coherent brand.

In the Regional Municipality of Niagara, municipal leaders see the wine industry as one local attribute that combines all four of these elements. Hence, it offers a strong foundation for a brand. Economic development publications bear this out. A 2007 report, *Energizing Niagara's Wine Communities*, made it clear that wine was growing as a Niagara tourist attraction:

NIAGARA, the name conjures many diverse images – a majestic falls, a rugged river, surging rock faces, orchards, agricultural hamlets, quaint lake-side villages, and bustling urban centres. People the world over visit Niagara every year in vast numbers to experience these remarkable riches. Rolling vineyards and the fine wines crafted from their bounty are included in these riches and contribute to Niagara's lure as an international destination ...

The purpose of this Study is to develop a strategy to energize Niagara's Wine Country communities. A variety of economic development opportunities have been identified to stimulate investment. A number of community infrastructure improvements have been identified to enhance Wine Country as a recognizable place and destination.[6]

This quotation is important in three ways. First, the authors suggest the wine industry can become a tourism driver rivalling the internationally renowned falls. Indeed, they imply that the industry is already a recognizable brand unto itself, and therefore an integral component of the Niagara experience. Second, they point to economic investments – financed by taxpayers – that should enhance the industry's international reputation. Finally, the authors contend that such investments should not simply alter the *perception* of Niagara but its *actual material form*. The region's built landscape should embody the wine industry, thus making the place and the "wine country" brand interchangeable.

After the publication of *Energizing Niagara*, stakeholders sought to make good on the economic promise of wine. In 2010, the municipality unveiled its *Niagara Culture Plan*.[7] Its authors positioned grapes and wine to bolster Niagara's allure as a cultural destination for tourists and new residents. In 2013, the Niagara Chamber of Commerce echoed these hopes with its own "blueprint" for the region's economic future. It claimed that

Agriculture is one of the pillars that built Niagara's economy, and it continues to be a significant contributor to the overall GDP of the region. The growth of Niagara's wine industry and the emergence of value-added agricultural production processes has [*sic*] created more opportunities for growth in this sector.[8]

The chamber explicitly linked wine to economic development, but it also did something else: it situated the industry within the realm of agriculture and that sector's long history in the region. The lustre of an authentic heritage narrative was grafted onto a new economic driver.

Branding is a marketing process. In a competitive environment, it can be difficult for any one good or service to stand out from its rivals if they all offer the same utility, quality, and experience. A company can, however, raise its profile through a strong brand. At its core, a brand is simply a name or symbols fixed to a good or service. This branding occurs primarily through names, logos, trademarks, and marketing campaigns. It can also be extended into the material design of the goods and services themselves, including packaging, storefront architecture, interior design, and employee dress codes.[9] These associations serve as a promise to the consumer: that the named good or service will deliver a consistent experience whenever and wherever it is purchased. Strong brands do more; they associate the good or service with a relevant trait that consumers value. Traits can take many forms, but economy, convenience, social acceptance, and fear are common associations. No matter how a brand is articulated, its strength will be rooted in consumers' perception of its promise, sincerity, relevance, and actual performance.[10]

Places, too, can be branded.[11] As with goods and services, places can only develop strong brands through their association with relevant, desirable traits. This means that a place cannot root its brand in either the natural or built landscape alone. The landscape might provide the material inspiration for a brand, but the beliefs and impressions that people bring to that landscape are crucial.[12] For example, decaying farms may be perceived as a rustic vista, just as barely visible earthworks may be perceived as a thriving Viking outpost. If a place is to be commodified successfully, then it is the consumer's beliefs and impressions – what John Urry called the "tourist gaze" – that must be shaped through branding.[13] Any tourist who visits a location for the first time may have invested considerable time and money based on its brand promise alone.

A place brand may be bolstered by unique aspects of the local culture. In economic terms, a local culture is comprised of "intangible assets" that can be commodified. If a location can be associated with a compelling narrative, such as Loch Ness and its elusive monster, then the tangible elements of the location may gain a lasting mystique via association. Visitors then carry this narrative with them. As Erving Goffman puts it, "A bubble or capsule of events thus seems to follow the individual around, but actually of course, what is changing is not the position of the events but their at-handedness; what looks like an envelope of events is really like a moving wave front of relevance."[14] Thus, intangible assets, if carefully handled by tourism marketers, can perform three important functions: they can entice visitors to spend time in a particular place, manage the visitor's experience while there, and possibly shape the visitor's memory after returning home.

Culture, of course, develops locally. Residents must share certain beliefs, attitudes, and practices to maintain peaceful relations and to develop a communal identity. Elements of this culture may change over time, but this change also occurs within the bounds of communal existence; proposed changes are negotiated, enforced, or rejected. These negotiations, as described in the introduction, rarely occur among equals. The development and maintenance of a cultural realm is subject to the vagaries of social status and power relations that cut across spatial and temporal scales. Local authority figures and their "regimes of truth" may decide which elements of the local culture are acceptable and which are not. The use of such elements may also be prescribed. Individuals who ignore or reject such decisions and prescriptions, who pursue alternatives, also risk social approbation or persecution. In short, local cultures and their elements are socially constructed, open to continuous change, and remain open to contestation. Thus, when local officials or entrepreneurs claim any one resource as an "intangible asset" for the sake of place branding, we must assess how that asset benefits the entire community.[15]

Some scholars, planners, and consultants believe that cultural mapping is a useful way to identify these traits. This type of mapping putatively identifies the beliefs, relics, and rituals common to a community, and then isolates the common thread that connects them: "a true brand essence, namely what a specific place *is*."[16] Among those who work in the industry, cultural mapping is said to reveal an "authentic" representation of a community. In the competition for tourist dollars, perceived authenticity is a highly prized trait.[17] This is particularly true if marketers wish to target affluent, internationally mobile travellers who seek temporary relief from mass-produced culture.[18]

The Niagara region offers a peculiar example of place branding thanks to Niagara Falls. The cataracts became a destination for travellers from around the world after the first European observer described them for European readers in 1697. Indeed, it was an important stop on the North American version of "the Grand Tour" by the 1820s.[19] Standing at the precipice of the Horseshoe Falls, they present a scenic and mesmerizing reminder of nature's unrelenting power. Their appeal has been described in terms evoking the sublime and the sensual, nature's beauty and nature's horror.[20] These descriptions of the falls predate any organized effort to brand the place. Quite the opposite, the municipalities that encompass the falls in Canada and the United States adopted the name "Niagara" to benefit from its existing profile. Complaints about the commercialization of the area were voiced as early as the 1850s, but the falls remain the most recognizable and visited feature of the region.[21] In the 2010s, the City of Niagara Falls drew an average of twelve million visitors annually. Among these visitors, roughly 30 per cent came from outside Canada.[22] Their draw led other attractions to position themselves as things to do "after the falls," a phrase repeated in tourist guidebooks.[23] These other attractions tend to skew their appeal to those seeking aesthetic experiences in intellectually stimulating pursuits (such as theatre, music, or history), or in the sensual through the carnivalesque (such as a midway, magic shows, clubbing, or gambling).[24]

In the 2010s, most attractions that enjoyed a high profile beyond the falls were established after the Second World War. The region was situated within a day's drive of 130 million people in Ontario, Quebec, and the United States. It was well served by highways and rail lines, including four automobile bridges to New York State.[25] Thus, any new enterprise that provided a unique consumer experience could access a considerable population if it could associate itself with the Niagara brand. The two most successful were Niagara-on-the-Lake and the wine industry.

Niagara-on-the-Lake sits on Lake Ontario at the mouth of the Niagara River. It is the oldest European settlement in the peninsula, established by Loyalists during the American War of Independence. The town flourished from the end of that war in 1791 to the opening of the Welland Canal in 1829: it served as the first capital of the new colony of Upper Canada; it hosted a garrison of British regulars who watched the American border; and it served as a market town for the region. But it also suffered setbacks. The colonial government moved to York (now Toronto) in 1797, though the town remained for a time the county seat. Two forts were undertaken to house the troops who protected the region from American aggression, Fort George in 1796 and Fort Mississauga in 1813. Neither prevented an American invasion that destroyed much of the town's infrastructure during the War of 1812. Still, residents returned to rebuild their lives and turn the battlefields into tourist sites. More damaging than the Americans was a local shift in economic power to settlements along the Welland Canal. St. Catharines was the greatest beneficiary. As Niagara's largest centre, it became the county seat in 1862.[26]

All was not lost, however. Niagara-on-the-Lake's founding narrative and built landscape provided local tourism developers with a brand concept: the once bustling capital of a promising young country was now cast as a quaint reminder of gentler times that were steeped in British culture. The original colonial town plan remained, along with the Georgian and Victorian architecture that had risen from the ashes of 1812. Cecilia Morgan has documented the many ways boosters exploited these historical assets after 1860 and publicized the town's role in the building of Canada. However, the rebuilt "Old Town" had limited appeal. Its main visitors included church picnics, history buffs, and affluent cottagers from Toronto and upstate New York.[27]

In 1962 residents sought to refresh the town's economy with live theatre. The works of Irish playwright George Bernard Shaw inspired a new repertory company that staged a slate of plays over the summer season. Shaw was a personal favourite of the festival's founder, and Shaw's plays were then enjoying renewed interest internationally thanks to the musical adaptation of *Pygmalion* (1913) – titled *My Fair Lady* (1956) – that enjoyed extended runs on Broadway and London's West End. The schedule of the new "Shaw Festival" in Niagara-on-the-Lake was well synchronized with summer tourism traffic in Niagara Falls. It proved to be a sustainable venture, and in 1973 it moved into a purpose-built theatre. By the 2010s, it was home to the second-largest repertory company in Canada and its attendance surpassed 250,000 patrons annually. This traffic provided the Old Town with an economic boost that saw hotels, restaurants, pubs, and stores cultivate a British veneer. The continuing presence of colonial heritage sites, the nineteenth-century townscape, and middle-brow entertainment combined to produce a gracious, antimodern experience for tourists seeking a quaint Niagara, an alternative to the carnivalesque atmosphere found in the amusements, wax museums, and hucksterism of Niagara Falls.[28]

The brand concept for the wine industry developed along strikingly parallel lines. It takes inspiration from global wine marketing practices and the agricultural history of Niagara. Internationally, wine marketing has long drawn upon place and local heritage to create enduring brands for both regional styles and individual wineries.[29] This is what scholars in marketing and tourism refer to as "winescapes."[30] At a macro level, European vintners began to categorize the effects that local soils, climate, and rainfall levels imparted to the quality of a region's product in the 1700s. This combination of factors later became the basis for the French concept of *terroir*. Subsequently, vintners throughout Europe established legally defined regional appellations to build recognition for their own regional specialties and brands.[31] To acquire an appellation, a wine must be made in the specified region with grapes grown there. As with any brand, the consumer should expect a certain experience whenever they enjoy a bottle from a specific appellation. Producers and aficionados accept that variations will occur within a *terroir*, from winery to winery and from vintage to vintage.

Individual wineries also make use of local associations, mainly to differentiate themselves from competition within the same *terroir*.[32] Association with a specific past – like a long family history or charismatic artefacts – becomes a potential resource that complements and extends the unique characteristics of *terroir*. Thus, if a winery has a pedigree rendered visible by old buildings, its own vineyards, and intriguing stories, then it has both the tangible and intangible assets that are prized by marketers. These assets are identified in the literature as "winescape atmospherics"[33] or the "human dimension of *terroir*."[34] As Joanna Fountain and Daisy Dawson put it, "*the past* becomes a resource, and, through a selection process driven by political and marketing imperatives, the history of vineyards and winemaking processes are interwoven with the broader heritage and cultural traditions in narratives of the region."[35] When deployed well, these assets can foster an aura of authenticity sought by connoisseurs and tourists alike.

The Niagara wine industry sought a similar aura. Lacking well-established traditions, however, it drew upon the region's wider history of white settler agriculture focused on fruit. The fruits native to the Great Lakes region included strawberries, raspberries, blueberries, currants, and grapes. Archaeological evidence indicates that Indigenous populations incorporated these fruits into their diet, but it is unknown if they cultivated them. When French and British garrisons arrived, they adopted the North American species and worked to domesticate them. They also brought European tree fruits, particularly apples, pears, and peaches. All of them were cultivated as seasonal treats rather than as commercial crops. As Loyalist settlements took shape in the 1780s, the new residents were encouraged to grow wheat and other grains to support themselves and the British military. The first agricultural society imported trees from New York in 1794, and one member developed a sizable orchard of pears and tender fruits before it was razed by American troops during the War of 1812. Nonetheless, wheat remained the dominant crop and major source of income for the region until the 1860s.[36]

Fruit crops grew in stature after 1860 for three reasons. First, wheat fields began to suffer from soil exhaustion in established settlements in Niagara and across Ontario in the 1850s, prompting a search for alternative crops. Second, farmers north of the escarpment gradually realized they occupied a prized location. Their fertile soil was created by the retreating waters of a post-glacial lake, while its climate remained warmer than neighbouring regions throughout the year thanks to the humid influence of Lake Ontario and the protective wall of the escarpment. It was actually ideal for apples, tender fruits, grapes, and berries.

Only two other regions in Canada were thought be as fortunate: the Okanagan Valley in British Columbia and the Annapolis Valley in Nova Scotia.[37] Last, Niagara also benefited from ready access to emerging urban markets in southern Ontario and western New York State. Steamboat service linked Niagara ports in Queenston, Niagara-on-the-Lake, Port Dalhousie, and Grimsby to Hamilton and Toronto by 1840. The Great Western Railway, based in Hamilton, offered service between Niagara Falls and Windsor by 1854, then added a branch line to Toronto soon after. With the completion of the Niagara Suspension Bridge across the Niagara River in 1855, the railway also offered service into Buffalo, New York. The region's unique produce could be picked, boxed, and delivered fresh to urban consumers overnight.[38]

The shift to fruit took institutional shape in 1859. There were enough farmers making the transition that a local Fruit Growers' Association formed that year in affiliation with the newly created Ontario Fruit Growers' Association. They were probably aided by the region's first peach nursery, which opened in Grimsby in about 1855.[39] By the 1890s, the region was popularly known as the "fruit belt" among locals and outsiders. No one took credit for the term. In English parlance, "belt" had long been used to describe swathes of land. In the mid-1800s, Americans had adapted it to describe entire regions of the continent with distinct conditions suited to specific crops, such as the southern cotton belt and midwestern wheat belt. The regions surrounding the lower Great Lakes were designated a fruit belt and the term apparently crossed the border. One Toronto *Globe* writer, describing Niagara in 1902, noted, "This has long been known as the great fruit belt of Ontario."[40] The term was most often applied in an economic context when describing Niagara's agricultural specialty or its prospects for an upcoming harvest.[41]

Much of Niagara's fruit lands produced apples until 1900. A shift came when southern Ontario suffered a series of cold winters from 1897 to 1904 that devastated tender fruit orchards in other regions, particularly the counties of Essex and Kent. Farmers there replanted their orchards with apples. Niagara, unaffected, capitalized on its climatic advantage by expanding its acreage in tender fruits and grapes.[42] This transition brought industry. By the 1920s, there were canneries in all the larger towns and villages, and a scattering of wineries and jam makers. Two pioneering brands remain: vintner Brights opened in 1874 in Niagara Falls, while jam maker E.D. Smith opened in 1882 in Winona (now part of Hamilton). Companies manufacturing baskets, barrels, orchard ladders, and pickers' aprons also flourished across the region.[43] Public-sector investment followed the private sector. The provincial growers' association lobbied the Ontario government to invest in horticultural research to advance their knowledge of local conditions. The government did so through the Ontario Agricultural College in Guelph. It opened thirteen research stations across the province between 1894 and 1907. Niagara's station, established in Vineland in 1906, supported the development of hardier fruit cultivars and pesticides. It remained in operation in the 2020s. The federal government operated a second research facility in St. Catharines from 1912 to 1968, when its operations merged into the Vineland station.[44] Altogether, these investments supported the growers' own investments in their farms. From 1900 to 1950, the number of acres in orchards grew, as did the number of acres in grape vines.[45]

A second shift would follow. A perception developed that the agricultural landscape was not just an economic asset but a cultural asset as well. Early signs of this transition came in a curious form. Spring market projections gauged the health of local blossoms to forecast the summer harvest. Published in the business sections of major papers, crop reports generally offered reassuring commentary from select farmers and weather forecasters.[46] These reports gained a new angle after 1900. Toronto-based reporters duly noted expert opinions on the year's prospects, but their gaze wandered from the science and statistics to the blossoms themselves. A 1908 report in the *Daily Star* conjured the scene:

> Looking out from a train or trolley car, or driving through on the old Hamilton to Queenston road, the foliage on the orchards hides the earth. Mile after mile is covered with practically nothing but the green banks of tree tops heaving and rustling in the hot summer wind ... In the open stretches here and there are the strawberry rows and the berry pickers working in between them in quaint kneeling groups; "bare foot boys" and calicogowned girls ... Further along the road may be a cluster of "stemmers" ... waiting for the patient farm horse to heave in sight with a consignment of fruit for them to stem.[47]

Framed in the romantic aesthetic of anti-modernism, the landscape, rural scenes, and rustic characters composed a postcard view of a supposedly bucolic life.[48]

This reporter was not alone in such imaginings. Life in growing, industrializing urban centres after 1890 fostered an urge to return to the land: to gardens and parks, to seasides and lakes, and to camp sites and cottages.[49] Niagara benefited from this impulse. As previously noted, Niagara-on-the-Lake had persevered as a cottage community, and similar enclaves later developed in Grimsby, Jordan, Port Dalhousie, and

Crystal Beach. They drew clientele from Toronto, Hamilton, and Buffalo. An 1846 gazetteer noted that the village of Grimsby was "beautifully situated on the St. Catharines road, ... in the midst of some very fine scenery. A good mill stream flows through the village. During the summer season Grimsby is a favourite resort for pleasure parties from Hamilton."[50]

The Niagara Parks Commission was driven by a similar impulse. It was established in 1885 to manage the landscape along the Canadian side of the Niagara River. After establishing lavish garden parks with views of the falls, the commission also undertook management of two 1812 sites – Old Fort Erie and Queenston Heights – and a rugged park on the river gorge known as Niagara Glen. Then, in 1905, it began work on a parkway from Fort Erie to Niagara-on-the-Lake that eventually linked the falls and 1812 sites while passing through farm fields and woods.[51] Cottagers and day parties also benefited from the regular train and steamboat schedules serving these Niagara centres. Electric tram service was added in the 1890s linking Hamilton to Grimsby, while two later circuits linked St. Catharines and Niagara-on-the-Lake to Queenston, and Queenston to Chippawa. In all cases, the trip offered escape into farmlands, gardens, parks, and beaches.

The appeal of actual farms was heightened by their blossoms. Autumn brought the harvest and markers of the year's success: baking, canning, and fall fairs. Spring, however, brought the end of winter and the promise of renewed life: the thaw and new buds. The fruit belt had blossoms in abundance. When this fact became popularly known outside the region, it fostered a new and unexpected source of tourism. G.H. Hamilton, a botanist with the Niagara Parks Commission, later wrote,

Of spontaneous origin, Blossom Sunday in the Niagara Peninsula has become an occasion of province-wide interest. If the weather on that day is suitable, carloads of visitors from all parts of Ontario and adjacent United States will travel the highways of this area drinking in the inimitable beauty of the orchard trees in bloom.[52]

Before cars were common, however, it was travel companies that first capitalized on this interest. In March 1902 the *Toronto Star* reported that the SS *Lakeside*, serving the Toronto-Port Dalhousie route, had "jumped into immediate popularity last season because of the delightful boat ride and trolley ride through the famous fruit belt."[53] Readers were encouraged to buy their tickets early. Similarly, a puff piece in *The Globe* noted that, "White and Pink Blossoms ... are now conspicuous between Toronto and Niagara Falls on the Grand Trunk double-track route, the orchards being in full bloom," then helpfully added this was a "Delightful week-end trip."[54] This sentiment made its way into the society columns. Mabel Burkholder made it known that May was "Blossom-Time":

> In this glad season of the year when spring is turning the whole face of the country into one immense picture-gallery, she reserves for her masterpiece the fruit district of the Niagara Peninsula. Here are seen her finest touches; here her most daring color [sic] combinations.[55]

Travel companies themselves began to feature "scenic Niagara" and "Blossom Time" in their print advertisements for Niagara destinations. In 1910, two St. Catharines hotels bid visitors to stay with them during "Blossom Time."[56] The nascent Ontario Motor League, a precursor of the Canadian Automobile Association, also promoted itself with "honeyed hours motoring through the garden of the Niagara peninsula in blossom time."[57] This was an early example of what has been called "the aesthetic value of agricultural land as a commodity for tourism."[58]

The travel companies' promotion of Blossom Time grew more coordinated after the First World War. In 1920, the United Hotels Company addressed "motorists" in a series of sponsored articles that appeared in newspapers accompanied by advertisements for upscale hotels stretching from Hamilton to Niagara Falls.[59] Similar pieces ran over the following years with occasional photographs framing the rural vista from the escarpment. These publicity efforts were often bolstered by the newpapers' own reporting of the spring buds, sometimes meriting the front page. An especially florid piece in 1926 proclaimed "'Blossom Time' at Its Height in Niagara Fruit Belt, Where Chill of City Air Gives Place to Semi-Tropic Warmth, Sweet-Scented and Exhilarating."[60] Images of peach blossoms, a happy child, and an orchard dominated the page. Beyond these pieces, the steamboat lines, railways, and tram systems continued to advertise their spring schedules to Niagara, as did new "motor coach" lines.

These promotional efforts followed the rising popularity of the family car. While travel companies delivered their passengers to established tourist sites, other centres invited Sunday drivers to enjoy day trips along country lanes that led to village shops and restaurants. Grimsby took the lead. Its residents were fiercely proud of their local orchards. They readily identified peaches as a civic emblem, a point underscored by the names of its Girl Guide troop (Peach District), girls' baseball team (Peach Buds), curling club (Peach Growers), and travelling men's hockey team (Peach Kings).[61] The impetus for a spring festival may have been provided by the *Toronto World* newspaper. Its 1922 editorial on the blossoms was reprinted in full in the Grimsby *Independent*:

> It's a wonder that Ontario has never adopted the custom ... in having an annual excursion to its famous fruit district when the orchards are in bloom. In the Annapolis Valley of Nova Scotia they have an annual Blossom Sunday ... Hundreds of automobile parties leave Halifax and various other cities and towns for the occasion.[62]

The World suggested that a festival would pay dividends long after the visitors returned home. It encouraged the farmers' cooperatives to view spring tourism as "an annual district advertisement" that could enhance the profile of their regional industry. In hindsight, this was a moment when the term "fruit belt" shifted from its geographical and economic origins to be become a nascent place brand.

With or without local promotions, *The Independent* estimated that 15,000 to 18,000 cars passed through Grimsby the Sunday following *The World*'s editorial. It also noted that additional passenger cars had been added to the regular Sunday train service.[63] Still, Grimsby took *The World*'s advice to heart. By 1929, residents had developed a slate of activities to coincide with Blossom Time that gradually coalesced around a choral festival and a grand parade on a day duly dubbed "Blossom Sunday."[64] Other popular events included dances, concerts, dramas, and an annual Blossom Queen pageant. A map was published in 1937 that placed Grimsby at the centre of a "Blossom Time Tour" route from Hamilton to Buffalo, and it helpfully advertised local restaurants and inns.[65] The following year was a peak for Grimsby. The parade numbered one hundred floats and was covered live on CBC Radio across Canada. The *Globe and Mail* reported that thousands of visitors had begun arriving early, already clogging local roads the Thursday before the big weekend.[66]

Over the next thirty years, other centres throughout the fruit belt established farm-focused events to capture the imagination of tourists. Enthusiasm for Blossom Time diminished with the Second World War but fully rebounded by 1950.[67] Beamsville residents – already hosts of the county fall fair – supported the events staged by their Grimsby neighbours.[68] Niagara-on-the-Lake developed a range of events similar to Grimsby's but added a "blossom blessing" given by local Christian clergy.[69] Niagara Falls took a different approach by emphasizing the formal gardens established by the Niagara Parks Commission along the Niagara Parkway. Through the 1950s, the commission supported an annual art show held by the city's Art Association during Blossom

Time. Then, from 1959 to 2003, it staged an annual parade on the park-way with the falls as a backdrop.[70] Amid these many attractions, St. Catharines offered travellers its urban amenities: restaurants, hotels, and the spectacle of the ships on the Welland Canal.[71] Community groups in two other centres later established similar events with their own niches: the Welland Rose Festival was established in 1961 and the Winona Peach Festival was established by community groups in 1967. Both relied on local community groups for their administration and labour.

Regional coordination came in two bursts. In 1937, the Grimsby Chamber of Commerce invited businessmen from Hamilton and St. Catharines to discuss a shared advertising campaign. This cooperation produced the Blossom Time Tour map and an advertising campaign that spoke directly to tourists through newspapers and radio.[72] In the 1950s, the municipal government assumed the lead. It hired the region's leading public relations firm. Louis J. Cahill opened the Ontario Edito-rial Bureau in 1936 to serve Niagara businesses seeking growth out-side the region. He had great faith in the modern industries developing along the Welland Canal that were powered by Niagara's hydroelec-tric generators. However, he also championed the allure of the natural landscape and the region's historical ambience. When approached by the county, the bureau's staff sought ways to emphasize all three nar-ratives. They first sought to foster tourist traffic by working directly with major travel promoters: the provincial government's Department of Travel and Publicity, the Canadian Pacific Railway, the Canadian National Railway, and the Good Roads Association.[73] The government office gratefully accepted the bureau's materials when developing its own tourist brochures and spring radio campaigns.[74] As family cars became ever more popular, the last organization was literally "wined and dined" at its annual business meetings. Delegates were invited to cap their days at a Lincoln Hospitality Room conveniently booked at the same hotel. The 1960 invitation boasted, "Famous as one of the most productive fruit growing areas in Canada, Lincoln County is featuring a display of Niagara-grown fruit, as well as a wide selection of wines pro-duced in the Peninsula."[75] These events were reportedly well attended.

The county invited the bureau to participate in wider civic plan-ning in 1960 when it created a Historical Council. Local history had gained a high profile four years earlier during the county's centennial year. A number of commemorative events had been staged by schools, churches, and service organizations across the area, and a formal his-tory was written by local researchers and published by the county itself.[76] Soon after, the county's Publicity and Development Committee financed a short film titled *Lincoln County: Century of Harvest* that interwove tales of local agriculture with the dramatic moments of the area's history. It gained wide distribution and earned a National Film Board certificate for its educational merit.[77]

The new Historical Council built on these community-focused events by using the area's rustic charm to attract tourists from outside the region, particularly from urban centres. The council's members were representatives of local historical bodies, the county's own Publicity and Reception Committee, and staff from the Ontario Editorial Bureau. The council was remarkably productive over the next three years. With respect to infrastructure, it began an inventory of historical buildings across the county, consulted noted architect and preservationist Anthony Adamson on municipal heritage strategies, provided input on the restoration plans for the Old Town in Niagara-on-the-Lake, and supported the conversion of a pioneer-era industrial site into a scenic conservation area with restored historic buildings.[78] With respect to publicity, the council provided guidance on a new county crest, established a scenic route winding through farms and villages with helpful signage, maintained its relationships with the travel promoters, and continued its support for the annual Blossom Time events. All these efforts fostered a coherent county place brand: it offered the natural beauty of the landscape and narratives of noble pioneers as a balm to the weary urban dweller's life in the modern city.

The prosperity of the tender fruit farms would not last. Fruit growers had emerged from the Depression and Second World War in relatively good condition because their produce had remained in demand. Ironically, two greater threats emerged in the 1950s: the continental expansion of American fruit marketing and the growth of local suburban development. American fruit growers held a competitive advantage over their Canadian counterparts due to the scale of their operations. American investments in national highway infrastructure and improved refrigeration systems led to their eventual penetration of Canadian markets.[79] Meanwhile, the well-publicized appeal of Niagara's small towns and villages drew urban residents seeking affordable home sites during the post-war baby boom. Concerns arose first in the villages of Winona and Fruitland, where Hamilton residents began building homes within an easy drive of the city's factories and downtown offices. Developers purchased entire farms for conversion to subdivisions, and the unique tender fruit lands subsequently began to shrink. The provincial government undertook a study of the region in 1967 to comprehend the long-term consequences of suburban growth. Its author quickly identified the threat:

> The general economic growth of the area ... will increasingly depend ... on its environmental appeal, on the amenities of the natural and man-made landscape. From this point of view, the effort that is now being expended on the idea of a ... Scenic Drive along the Niagara Escarpment is sound but the whole concept is meaningless if "the scene" – the charismatic landscape of the area – is not preserved.[80]

Despite multiple studies of suburbanization over the following decades, little was done to protect the unique agricultural zone that was the fruit belt.[81]

Blossom Time planners in 1951 could not have known that orchard production peaked that year. Changing consumer tastes and less expensive foreign imports meant Niagara produce could not compete even on the shelves of the region's grocers. From 1962 to 1971, total acreage in tender fruit fell by 5,000 acres even though total tonnage of processed tender fruit had increased from 50,000 to 65,000 tons.[82] By the 1980s, farmers who specialized in peaches, plums, and apricots were considering new crops to plant, or were selling their land for housing. By the early 2000s, the provincial government was paying them to remove fruit trees.[83]

Farmers who considered new crops were encouraged to plant grapes. North American varieties of *Vitus riparia* and *Vitis labrusca* were native to the region. Their flavour profiles were best suited to fruit eaten whole, pressed into juice, or prepared as jellies. They were not well suited to wine, however, because they did not compare well against European varieties of *Vitis vinifera*. Still, local vintners persevered. George Barnes opened what may have been the first commercial winery in Niagara in 1873, near Port Dalhousie, and his company's enduring financial success inspired later entrants such as Brights, Jordan Wines, and Chateau-Gai.[84]

Vintners who sought better grapes for Niagara wine began experimenting with imported varieties in the 1930s. Brights and the Vineland Research Station were leaders in this regard, but thirty years passed before their efforts showed promise. One solution was to cultivate *vinifera* vines such as Chardonnay and Riesling by grafting their branches onto North American root stocks. A second solution was to cultivate hybrids of *vinifera* and North American grapes, which produced new varieties such as Vidal and Baco Noir. Once the new vines were established, new wineries emerged that produced higher-quality products. The first craft winery to open was Inniskillin, founded in Niagara-on-the-Lake in 1975. It was the first commercial winery to be newly licensed in the province since 1929. There were more to come. By 1993 there were seventeen wineries open in Niagara, more than forty by 2003, and more than eighty in 2018.[85]

Icewine emerged as a key innovation in the early 1980s.[86] The sweet dessert wine was developed in Germany in the 1700s, but Niagara's climate is especially well suited to its production. The specialty wine is made from ripened grapes that are allowed to freeze on the vine. Once picked, ideally at minus ten degrees Celsius, the grapes must be pressed while still frozen. These requirements can reduce harvest rates to 5 per cent of a vine's annual production. This puts a significant constraint on the amount bottled. The final product, however, has been described as rich, complex, and voluptuous. Scarcity and perceived luxury have earned Niagara icewines a reputation as a premium drink. They quickly gained an international reputation for their quality. A 1989 Inniskillin vintage won the industry's most prestigious award, the Grand Prix d'Honneur, at Vinexpo France in 1991. The award elevated the profile of all Canadian wines and established icewine as Canada's specialty. Donald Ziraldo, one of Inniskillin's founders, noted the reputational value of the award:

> Every wine-growing region is singularly famous …: Champagne is known for its sparkling wine, Portugal for its Port, Burgundy for its Pinot Noir and Chardonnay, and Bordeaux for its singular chateaux producing the classic blends.[87]

His implication: icewine had become Canada's standard bearer.

The early craft wineries also benefited from a growing interest in wine worldwide. As international brands sought to expand into Canada and elsewhere, their marketing provided new waves of consumer-friendly information on flavour profiles. This new marketing reinforced the reputation of European brands, but it also raised the profile of wine itself as a drink. In Ontario, per capita consumption rose from 6.2 litres per person in 1977 to 8.5 litres in 1982.[88] Additionally, Canada's implementation of the 1988 Free Trade Agreement with the United States opened American markets to Canadian wineries, greatly expanding their potential sales territories. Vintners in Niagara worked hard to compete against European and Californian competitors. To meet the demand for *vinifera* grapes, farmers accelerated the conversion of their fields with financial assistance from the provincial government.[89]

Selling wines pressed from the new hybrids took patience. They had to overcome consumers' preconceptions of past Ontario vintages. One early attempt at experiential marketing developed in Grimsby. The town's Victorian-era railway station, built in 1902, had once been a busy hub for fruit packers, but declining traffic led to its closure in the early 1970s. Local entrepreneurs bought and restored the property to create a tourist attraction inspired by the town's past, called the Village Depot. The building regained its original ambiance while the interior was converted into "a honeycomb of quaint and unique shops offering crafts, antiques, books, pine furniture ... [and] candles" stocked by local vendors.[90] Two vintage dining cars were similarly restored in the parking lot to create a restaurant, while a caboose served as a cafe. Visitors who wished to see more of the region could take a guided coach tour of Niagara's scenic landscape, including orchards. Its featured stops included a fruit farm, art gallery, and winery. Special seasonal excursions highlighted Blossom Time in spring and the vivid colours of the escarpment in autumn. Via Rail was impressed. It created day trip packages for out-of-town travellers that employed its last remaining steam engine on the trip from Toronto. These trips ran from 1977 to 1980.[91]

Once again, a romantic past was conjured for urban tourists seeking an escape to the country. Vintage amenities, handicrafts, and antiques in a village setting were linked to gracious sensory experiences of rural landscapes. The inclusion of wine tourism, however, was new. So was the winery. Andrés Wines had been established in British Columbia in 1961 by a former brewer based in Hamilton. Its early success led to additional plants in Alberta and Nova Scotia before the company acquired an existing plant in Winona in 1969. Consciously or not, Andrés had grafted its new brand onto the much older narrative of Niagara's small towns and fruit farms. It would not be the last to do so.

To be taken seriously by the international wine community, the Niagara craft wineries established a new appellation for the Niagara region in 1988: the Vintners Quality Alliance, better known as the VQA. The VQA was designed to distinguish the higher-quality wines they produced from those of the legacy vintners. Icewine became a prominent feature of the VQA's quality guidelines and promotional work. Eventually, the efforts of these Niagara wine pioneers reached policymakers. Ontario legislated the *Vintners Quality Alliance Act* in 1999, thereby codifying a province-wide legal apparatus for the establishment, monitoring, and enforcement of Ontario appellations.[92] The VQA recognized four appellations by 2020. Niagara was the largest of them in terms of number of wineries and production volume.

To attract consumers to the wineries, and drawing upon precedents in European wine regions, local wine producers re-conceived the Blossom Time Tour as a wine route to take advantage of the scenic potential of the vineyards, villages, and escarpment. Wine routes, unlike short-term festivals and events, are permanent marketing devices that connect vineyards and wineries with commensurate services and attractions, particularly those involving nature, gastronomy, craft markets, and special events.[93] In Niagara, the Wine Council of Ontario – a non-profit trade and marketing association – promoted its own "Wine Route" and strictly controlled its designations and participants in order to maintain its brand integrity.[94] The route, as it wended its way through Niagara, fulfilled its intended purpose with its own printed maps, guide books, and roadside signage. As more wineries opened, the Wine Country Ontario digital planner was created to generate custom maps connecting the user's own choices of wineries, accommodations, and restaurants. It also provided links to local events and tour companies.[95]

In the 1990s, new industrial alliances emerged out of the burgeoning reputation of Ontario wine and the growing investment in vineyards and production.[96] Wineries developed horizontal alliances through the VQA and the Wine Council of Ontario. The wineries also entered vertical alliances with allied companies in the tourism and hospitality sectors. A key step towards mutual inter-industry support was the creation of "Tastes of Niagara: A Quality Food Alliance" in 1993. This partnership brought together farmers, distributors, restaurants, and wineries in local food services to promote a regional cuisine. As Insun Lee and Charles Arcodia note, "the fact that food is expressive of a region and its culture has meant that it can be used as a means of differentiation for a destination in an increasingly competitive global marketplace."[97] In this way, the linkage of the landscape with its produce and cuisine offered a way to reinforce the perceived authenticity of the place brand. In synch with this effort, the wineries also sought partnerships with hotels, restaurants, historic sites, and the Shaw Festival to develop package tours aimed at consumers seeking culture, elegant meals, and refined comforts. Book publishers chipped in with tourist guides, cookbooks, and coffee-table volumes that made the same connections. Many also featured time spent in nature while birding, hiking, biking, or playing golf. The association of wine with Niagara's wider pastoral charms had taken root.[98]

Energizing Niagara, the 2007 marketing report commissioned by the regional government, promised a coherent brand for the Niagara wine industry. Its starting point was the international practice of regional wine branding and the VQA designation. The Niagara appellation gave local wine a distinct profile among foreign and domestic rivals. Additionally, the local history of white settlement, family-run farms, and the fruit belt set in a region of small towns lent the wineries a compelling story. It evoked pastoral values and pioneer gumption persevering amid international intrigues. In 2003, this narrative very much rang true. A study commissioned by the regional government found that the average size of Niagara farms was smaller than those of other regions in Ontario, that almost half had been worked by multiple generations of the same family, and that three-quarters employed family members.[99] This narrative offered the craft wine industry a much older, more romantic, and more authentic pedigree than its recent origins suggested.

The tangible and intangible assets necessary for a strong place brand were, then, readily available. Governments, institutions, and corporations operationalized these assets through three channels: the marketing produced by each individual winery, the materials produced by the Wine Council of Ontario, and the policies of one municipal agency, the Niagara Economic Development Corporation (NECD).[100] Together, these bodies branded Niagara as "wine country" with iconography and materials thick with images of farms, fruit, and rural life.[101] The NECD supported this effort by creating its own brand, "Niagara Original."[102] Its goal was to position the region as a community that is "creative, unique, [and] one-of-a-kind," but also teeming with "natural abundance."[103] The wine industry occupied an integral part of this strategy as it supplied an ideal model of creative frisson that was also environmentally sustainable. Indeed, the first logo for the Ontario Craft Wineries, a trade organization, was a cluster of unique purple thumbprints arranged to look like a cluster of grapes.

This brand strategy began with the landscape. As connoisseurs of Blossom Time knew, life in Niagara lends itself to rural idylls. When viewed from the edge of the escarpment or the heights of Fonthill, the vista is dominated by farms and small towns. Fields of grain, hay, grapes, and tender fruit alternate with wood lots, creeks, and marshes. A.S. Hill reminds us that farmhouses and outbuildings surrounded by acres of open fields mean that farming is highly visible. Such landscapes shape the perceptions of both visitors and residents, whether they are conscious of it or not.[104] Hence, agriculture remains an important facet of Niagara's regional identity. In 2020 there were some twenty annual events celebrating agriculture in Niagara, ranging from traditional fall fairs to single crop festivals for strawberries and pumpkins. The urban residents of St. Catharines, Niagara Falls, Welland, and Fort Erie can share in this vision if they prize ready access to fresh produce, pastoral scenery, and country rambles. The countryside lies within a fifteen-minute drive of any urban address.

Wineries can challenge this rural landscape. Functionally, they are factories. They do not need to enhance their visual surroundings to produce good wine. When cast as tourist destinations rather than factories, however, their tone and appearance take on much greater importance.

Winery buildings often have "welcome centres" and tasting rooms that double as retail shops. The larger facilities may also include restaurants, banquet halls, entertainment venues, and outdoor art installations. Visitors can expect an aesthetically pleasing structure with an enticing ambiance fostered by friendly and knowledgeable staff.[105]

Cultural scholars Caroline Charest and Nicholas Baxter-Moore believe that Niagara's wineries achieved this goal in three ways.[106] First, some wineries reproduced the architectural styles of Europe's renowned wine regions. Their facilities resemble French chateaux and medieval cloisters, with grounds that maintain the selected theme. This style was common among wineries developed by vintners with ancestral ties to Europe. Second, some wineries embraced local landscapes as a central feature. Many have adopted the built heritage of the Niagara region by repurposing the buildings and landscapes of working farms. Such facilities usually incorporate existing farm homes, barns, and other out-buildings. Others have attempted to incorporate the local landscape, particularly the escarpment, into their design and function. These looks are well suited to estate wineries that cultivate and process their own grapes on site. Third, some wineries have erected thoroughly contem-porary architecture. The facilities that feature this aesthetic share little else in common. However, most of these facilities work in relation to the landscape by incorporating the horizontal lines of the surrounding fields, or by offering postmodern readings of iconographic Canadian buildings, such as cottages. All three styles reference the local land-scapes or some notion of "the past," either in form or inspiration. The past can, for example, tie a brand to something older and more socially significant than itself. A seductive brand story embedded in a heroic or romantic past can make this link explicit. In addition, it may also reduce sales resistance. If a link to the past is thought to be genuine, consumers may look upon the brand more favourably and remember it. For example, a new estate winery that adapts an existing farm for its operation may gain some measure of credibility as a heritage farm, even if its lands had never previously grown grapes or produced wine. In effect, the past may be summoned to authenticate the winery's adopted brand and style.

The "wine country" brand packages all three styles in a single promise: Niagara offers pastoral yet culturally refined relief from urban stress, thanks to the creative flair of local vintners, restaurants, and artists. If the industry's place branding has worked, then visitors should arrive expecting a region filled with green vistas, charming architecture, and fine wines. All three styles of winery can deliver that experience. However, to gain an aura of "authenticity," the styling of any one winery should fit with the visitor's conception of its unique appeal, or its position in the market relative to its rivals. If a winery brands itself in the French tradition, then its product and environs should fulfil that promise. Similarly, if a winery brands itself as a hip New World alternative, then its product and style should live up to that promise. Again, what binds the tangible and intangible assets together is the winery's narrative. A local wine journalist, extolling the virtues of winery tours, emphasized this point when he declared, "every bottle of wine has a story to tell."[107] Ideally, a winery will simultaneously integrate its symbolic assets in a coherent fashion, support the Niagara brand, and produce a memorable tourist or consumer experience. Doing so may convert those tourists into repeat customers. Ideally, those tourists also become brand ambassadors who repeat the brand story themselves.[108]

Through the 2010s, some wineries began to align themselves with environmental sustainability movements as increasing numbers of consumers sought "green" products and experiences. Scholars have recognized "green marketing" since 1975. Its use increased significantly in the twenty-first century.[109] Wineries that demonstrated sustainable environmental, economic, and community practices were able to put the Sustainable Wine Ontario logo – a green leaf – on their labels and advertising materials. Further designations for producing organic wines, for biodynamic production, and for LEED buildings (standing for Leadership in Energy and Environmental Design) were also possible.[110] Southbrook Vineyards, for example, boasted all three designations. On its website, the vineyard emphasized its connection to the local: "Organic certification is a pre-requisite for biodynamic certification. Biodynamics (BD) emphasizes the balance and interrelationship of a farm's soil, plants, and animals to grow low-impact, vibrant crops. It treats the whole farm as a single living entity."[111] Sustainable production made good sense from environmental and marketing perspectives, but it also allowed wineries to demonstrate their custodianship of local *terroirs*.

Niagara's place brand seemed to have a strong foundation crafted from the local environment, carefully designed wine infrastructure, and the intangible heritage of local agriculture. On this foundation, the industry and its allies had reimagined Niagara as "wine country." The idea spread through the toponomy of the region. Though the town of Vineland was established in 1894, an increasing number of vineyards and wineries have become valuable reference points for residents, and Niagara now features at least one subdivision in which all the streets are named after wine varietals. In addition, regional welcome signs marked the gateways to "wine country" while the guidebooks and wine routes directed traffic to wineries and restaurants. As wine tourism developed, related industries offering food services, accommodations, and travel – particularly bus and bicycle tours – drew upon the same iconography to promote their own services. Independent restaurants, too, aligned their marketing to draw attention to their use of local produce and their pairings with Niagara wines. Their marketing amplified the visibility and perceived authenticity of the wine industry.[112]

The production of good marketing materials involves several challenges. First, the concept behind any brand can be complex if it incorporates a company's story, product, and appeal. Advertising works best when it can reduce this complexity to a simple, memorable message. Second, the creative teams that produce advertising must work within the formal restrictions imposed by each communication technology. Further, industry codes of conduct limit what is considered socially acceptable in Canada with respect to an advertisement's text and imagery.[113] Third, the proliferation of media outlets on all platforms has produced audiences with different sets of cultural touchstones and triggers. The competition for the public's interest has seen audience attention spans wither, and creative teams have responded with ever-shorter advertisements, including spots produced for YouTube that last only five seconds. Fourth, the costs associated with production and media placement must always be considered in relation to the possible return on investment that each advertisement will produce.[114]

How do creative teams navigate these challenges? Knowing that audience attention must be captured in seconds, advertisements are commonly developed around social vignettes that are immediately recognizable to most viewers. For example, advertisements may open with a family in their kitchen, co-workers at work, or friends enjoying downtime. These scenarios present familiar locations, defined social roles, and associated behaviours and cultural expectations. Their familiarity encourages audience members to identify with the advertisements' characters. A creative team can then insert their clients' brands into such scenarios in ways that link their products to the audience's needs, desires, and cultural expectations.[115]

A winery is no different. It must tell a coherent, compelling story about itself and its wines. Regardless of the platform used, its message should play sympathetically to the values and tastes of its target market. In Niagara, two advertising platforms stand out: pamphlets and websites. Pamphlets offer a highly flexible platform that can be adapted to a wide variety of purposes since they do not need to fit the production requirements of media outlets. Their information simply needs to fit a single sheet of paper or a small booklet to accommodate whatever content a winery wishes to include. When distributed within the Niagara region, they also offer an efficient means to address an audience whose members are already within driving distance of the wineries. By contrast, websites are potentially unlimited in their geographic reach and page count. If desired, marketers can tell a brand story in great detail. The only limitation here is the patience of website visitors.

We sought to do two things: (1) to assess how individual wineries supported the Niagara place brand; and (2) to explore how the region and individual wineries attempted to tie themselves to the area's past.[116] To begin, we examined all the pamphlets available to tourists at a local information kiosk. We collected them in July 2010, three years after *Energizing Niagara* was published. Most tourist kiosks in the Niagara region are stocked and maintained by a single distribution company. As such, most kiosks found in local hotels, restaurants, and attractions display the same selection. The exceptions are those kiosks found in public-sector locations, but even they stock many of the same pamphlets. Hence, we are confident that the pamphlets available in the selected kiosk were highly representative of the material seen by most tourists and residents.

Our sample produced 140 pamphlets in all, advertising not just wineries but destinations and services in all sectors of the local tourist trade (see table 5.1). The largest group of pamphlets touted tourist attractions in Niagara Falls itself (27.1 per cent), and wine and wine-related activities were a close second (22.9 per cent). If those for food and dining were added (5.0 per cent), then wine, food, and dining occupied more shelf space than any other category of pamphlets. By contrast, few pamphlets described historic sites and museums (5.7 per cent).

Table 5.1. Points of interest promoted through pamphlet kiosks in Niagara (2010)

Points of interest (N = 140)	Percentage
Niagara Falls attractions	27.1
Wine industry	22.9
Shopping	16.4
Tours (not wine related)	7.9
Arts and culture	5.7
Heritage sites and museums	5.7
Food and dining	5.0
A specific town or village	4.3
Tourist services	2.9
Miscellaneous	2.1

The imagery and language of the wine industry pamphlets aim to seduce an urban middle class. They conjure a release from frenetic city confines into soothing natural landscapes with sweeping views and rustic details. Readers in Toronto and upstate New York are constantly reminded that Niagara is nearby and ideal for a weekend escape. East-Dell Estates bid readers to experience not just its wines but "The nature of Niagara": it "rules in every aspect of our integrated operation – from the vineyards and winemaking practices, to the restaurant's focus on seasonal cuisine, to the preservation of our property's spectacular Carolinian woodlot and nature trails."[117]

The pamphlets also suggest a particular sense of refined taste. Illustrations portray smart, well-heeled couples in lush surroundings. They appear to be charmed in equal measure by the warmth of winery stores conceived in exposed brick and pine planks and by elegant restaurant settings sparkling with polished crystal glasses and silver set on crisp linens. Such aesthetics bridge all three styles of winery. Regardless of their individual styles, the wineries' pamphlets bid readers to indulge their passion for fine food and drink.

Each pamphlet also invited readers to discover more by visiting a website or social media platform. We visited sixty-one winery websites, starting with those mentioned in the pamphlets and then linking to others (see table 5.2). Digitally as in print, Niagara wineries staked claims to rustic authenticity. The Niagara place brand and appellation were apparent but rarely central to their appeals. This may be easily explained: sixty-one wineries cannot render themselves unique by referencing the same *terroir* or appellation. Instead, they emphasized those elements of their narratives and operations – their specific "winescape

Table 5.2. Themes manifest on winery websites (2010)

Themes (N = 61)	Percentage
Family ownership	52.5
Local agricultural heritage	44.3
Terroir/Niagara appellation	42.6
Farm setting	34.4
Local setting	27.9
Small-scale production	27.9
European tradition	23.0
Immigrant experience	16.4
Environmental or climactic setting	16.4
Local history (other than agriculture)	13.1

Note: Each website contained two or more of the identified themes.

atmospherics" – that set them apart from regional rivals. Ironically, then, a majority of these sites evoked similar claims to family owner-ship and local heritage.

Often, this common narrative was set on a family farm with small-scale production facilities. The farmstead at Ravine Vineyard, for example, has been home to five generations of the same family. Their website boasted of this long Niagara connection:

> [In] 1869, David Jackson Lowrey planted one of the earliest commercial vineyards with 500 vines. Five farming generations of Lowreys would eventually reap the benefits of his decision, growing all kinds of tree fruits and grapes ... [In] 2003 ... Lowrey descendant Norma Jane (Lowrey) Harber and her husband Blair Harber prepared to return the upper farm to grapes.[118]

The site made further references to both "family" and the winery's col-lection of heritage buildings:

> We welcome you to visit to see for yourself. Our wine shop is open daily and hosted by our extended family of staff who are always happy to help you explore our small batch wines, while sharing our family's rich history. Find out why our heritage buildings, family farm and vineyards are the main characters in some amazing stories.[119]

In similar fashion, Henry of Pelham invoked long family histories and original buildings as part of their offerings:

> Henry of Pelham Family Estate Winery offers a visit experience unlike any other winery in Niagara. Featuring the original buildings of our United Empire Loyalist forefathers and a state-of-the-art wine production facil-ity, you get the best of old world charm and new world winemaking. We promise a visit steeped in history, aged in oak (we also boast Canada's largest underground barrel cellar) and poured with enthusiasm. Whether you're here for a tasting, a tour, a picnic or open air casual dining at our Café, we invite you to come as a guest and leave as family.[120]

As a final example, Rief Estate Winery made claims associating its prod-uct with local landscapes and heritage:

> The gardens are designed to complement and reflect the heritage of our original 1870's Coach House while featuring plantings representative of the colour, flavours and aromas of different grape varieties typically used to produce wines in the region.[121]

In very clear terms, then, the authors of these sites asked the past to authenticate the brand. Even wineries lacking a distinguished pedigree in the Niagara wine industry often embedded their operations in the region's agricultural heritage beyond grapes.

While references to the region's agricultural history were common, references to its dramatic history of conflict and industry were rare. A notable exception was Henry of Pelham, which linked its owner's family to the region's founding:

> Owned and operated by the Speck family since 1988, the land was first deeded to our great, great, great grandfather, Nicholas Smith, in 1794. Nicholas fought with Butler's Rangers in the American Revolutionary war. His youngest son Henry of Pelham built the building that houses our wine store and hospitality rooms. Henry built the former carriage house in 1842.[122]

Still, the Speck story emphasizes the family's local pedigree over its role in moments of national significance. More tellingly, it is rare to find labels at any winery that commemorate the region's military past like that of Hillebrand Estates' Cuvée 1812, which was produced in the 1980s with a label featuring Brock's Monument. Palatine Hills Estate Winery, for example, walked a fine line between commemoration and pedigree with a blended wine simply called "1812." The label had no imagery, and the estate's website made the following claim:

> THE STORY: About the time we resolved our little dispute with our neighbours to the south, our forefathers turned their interests and energies to winemaking. For 200 years, we've been using our land for something that produces … Harmony. The Harmony in this bottle is our salute to resolving differences and a promise of its continuation.[123]

Like other brands, the Palatine Hills "story" or brand narrative for this vintage was about the region's history of wine production, not the war. During the bicentennial period, however, the winery did promote the region's commemorative events as extensions of its own appeals to tourists.

Many wineries also invoked cosmopolitan associations. Beyond their obvious pride of place in Niagara, some wineries claimed pedigrees rooted in their owners' ancestral homelands. This claim was particularly common among growers and vintners with roots, either through kinship or training, in the wine regions of Europe. Those with kinship ties often provided family histories that focused on the immigrant experience.[124] These intimate histories were clearly useful on at least two levels. First, consumers may have appreciated stories that put a face to the wines and fostered a sense of personal connection with the wineries. Such a connection may have deepened consumers' appreciation of the wine and the experience of their visit. It may also have echoed consumers' own family histories among those who recalled an immigrant past. Second, references to European traditions augmented the wineries' claims of quality in a market that lionized European wines. Niagara wineries Château des Charmes, Konzelmann, and Pilliteri could foster a local as well as a global identity in a single brand.

Regional branding for the wine industry was also supported through experiential marketing. The most notable events were annual festivals: the Grape and Wine Festival, the Icewine Festival, the Homegrown Festival, and Cuvée Weekend. The Grape and Wine Festival deserves special mention. Every major town in the region once hosted a fall fair, though only three remained by 2020 (i.e., Welland, Wainfleet, and Grassie). These fairs were harvest festivals that marked the end of the growing season, celebrated agricultural traditions, and put family reputations on the line through various competitions. Fairs also brought sundry entertainments to town in the form of music, midways, and gambling. Currently, the oldest remaining local fair is the Niagara Regional Exhibition in Welland, which began in the 1850s.[125] Grape and Wine is more recent. It was established by the Grape Growers Marketing Board and the Canadian Wine Institute in 1952, in cooperation with the City of St. Catharines, to celebrate that city's largest harvest. In the mid-2010s, its industry support came from the Wine Council of Ontario and VQA. It regularly drew over one hundred thousand participants and generated over two million dollars in revenue each year. The organizers advertised it as "Canada's Largest Wine Festival."[126]

In the tradition of rural fall fairs, Grape and Wine held a decidedly carnivalesque appeal among residents of the city. Several elements became part of its annual schedule: a local grape grower was crowned "grape king," a local debutante was crowned "grape queen," and two parades wound their way through the city's core. The first parade – "the Pied Piper" – invited children to participate in the revelry. A week later, "the Grande Parade" provided traditional fare in the form of floats, bands, acrobats, and other entertainers. Both led to the festival midway at Montebello Park in the downtown core. The midway featured children's games, rides, local arts and crafts vendors, and various fast-food vendors. Over time, the fast-food vendors were replaced by local restaurants and wineries offering sample menus paired with local wines. These were housed in tents raised in a circle in the shade of large trees. Picnic tables were then set within the circle, and a stage provided live music. The parades and midway were scheduled over two successive weekends in September for roughly eleven hours each day. A highly popular addition in the 1990s was the "Event in the Tent," essentially a rock concert that brought in nostalgia acts, where beer generally outsold wine. All these events were well supported by local residents.[127]

Discussions surrounding the "wine country" brand prompted organizers to reconceptualize the Grape and Wine Festival in the early 2000s. Wineries saw the festival experience as an intangible asset and sought to heighten its utility as experiential marketing for their industry. In other words, they realigned the festival experience with the "wine country" brand that promised sophisticated pleasures to urban tourists from outside Niagara. Subsequently, a series of newer events were initiated to deliver on those tourists' expectations.[128] The midwinter Cuvée Weekend originated with the Ontario Wine Awards, established in 1988. The event was designed to be a gala celebration of the province's wine industry for industry professionals. Its signature event was the Cuvée Grand Tasting, a formal event that doubled as a charitable fundraiser for Brock University's graduate program in oenology. The annual Icewine Festival, established in 1996, focused on that premium-priced product. Though residents could attend both events, neither was rooted in community or harvest traditions; they were marketing exercises for the benefit of the wine industry. In the mid-2010s, festival organizers added two more: the Spring Sparkles Festival for sparkling wines and a Homegrown Festival, held in June. The four public-facing festivals (Grape and Wine, Icewine, Spring Sparkles, and Homegrown) created a full-year calendar of wine events celebrating and selling local vintages.[129]

The Grape and Wine Festival itself was reimagined during this brand realignment. The first step was to distance the festival from the harvest traditions of a fall fair: the word "grape" was removed from the festival's name and advertising materials in 2002.[130] Subsequently, other elements of the festival also changed. The Grande Parade became a juried event in 2007, eliminating several past participants that no longer met the festival's standards. The main activities at Montebello Park had once been open to all with free admission. In 2009, fencing was erected around the tents and stage and entrance fees were introduced for some of the events. The Event in the Tent lost its festival affiliation and sponsors, then folded the following year. Similarly, a bicycle race sponsored by a local cycle shop lost its affiliation and was discontinued after seventeen years. In sum, the local harvest festival became a tightly controlled experiential marketing effort staged by the wine industry for tourists and wine aficionados. Locals had become colourful extras – perhaps too colourful – at their own fall fair.[131]

There was pushback. Altering the festival events had altered the festival experience. We surveyed St. Catharines residents in 2009 to further explore the trends that emerged in our 2005 survey. Participants told us in no uncertain terms that they felt excluded by these changes. This exclusion was defined along class lines. One participant noted her family's experience in the following terms:

> We used to go to wine events. But we went to one Cuvée, it cost us hundreds of dollars. The same thing happened at the Grape and Wine Festival. A bottle of ten-dollar wine cost us twenty-seven dollars! The same is true of the Jazz Fest they hold at one of the wineries. These events are not of the people; the average person doesn't have the disposable income to pay that much. If they want to continue to include locals, they need more things that are not so expensive.

While we conducted our survey, several letters to the editor of *The Standard* echoed these sentiments in strikingly similar terms. One letter writer advised festival organizers to

> tone down the "hoity-toity," upscale, cosmopolitan flair that has dominated the festival over the past few years now, to the point of eliminating parade participants that didn't fit that image and giving us a parade that was only 55 minutes long.[132]

Behind the scenes, there were also tensions within the industry. Despite the success of local wineries, grape growers were then facing two significant challenges. The first stemmed from provincial farm marketing regulations. Under provincial laws, wineries could sell wines that contained up to 70 per cent imported grapes under the designation "International-Canadian Blends." The designation was crafted to draw Ontario consumers to less expensive but still seemingly Canadian wines. The original policy was instituted in 1972 to help wineries while grape growers shifted their fields from *labrusca* vines to *vinifera* vines. After that shift was successfully implemented, the wineries lobbied to have the policy extended because local growers could not match the international market price for grapes. When the policy was extended, the result was catastrophic for growers. In 2009, for example, an estimated ten thousand tons of quality grape juice went unsold.[133] Niagara residents who sympathized with local farmers were dismayed. It appeared that the vintners had betrayed both their suppliers and their own promise to maintain local agricultural traditions.

The second challenge facing grape growers was the rising cost of land. Owning a vineyard and producing wine had become fashionable, a fact underscored by the local investments of celebrities such as athletes Wayne Gretzky and Mike Weir, actor Dan Aykroyd, and the rock band the Tragically Hip. The glamour of the wineries brought new investors to the region, and this trend served to increase land prices. One local realtor explained,

> Your local farmer can't afford to expand. The costs are too great. The majority of all land purchases are from out-of-town-buyers that have made money elsewhere, not by farming ... Most have a dream, a passion for wine, and also realize Niagara is still in its infancy stage when it comes to that industry.[134]

For established farmers, it was more profitable to sell their lands than to develop their own operations.

Taken together, the changes made to the Grape and Wine Festival and the pressures faced by grape growers fostered an image of Niagara vintners that ran counter to their marketing efforts. The wineries were keen to portray themselves as family farms firmly rooted in local culture and traditions, a portrayal that melded well with the presumed escapist desires of their urban target market. At the same time, the wineries' business decisions seemed to undermine the local authenticity of the festival and the sustainability of existing grape growers.

Local media coverage of the wine industry generally remained boosterish, emphasizing both its economic and cultural benefits. The opening of a winery was treated as positive economic news, connoisseurs reviewed local labels for weekly columns, and the wine festivals were covered as major cultural events. Indeed, articles on the Grape and Wine Festival revelled in its tradition of community involvement. This excerpt from a St. Catharines *Standard* article, which quotes a local resident, provides a ready example:

> When my daughter was two, we took her to the parade and to my delight she was so excited that I thought that she would burst ... That would have been enough to have been my best memory of the festival, but no, it was only second best. My favourite memory was years later when we took her son to his first parade.[135]

Positive stories like this, told by the wineries through their marketing and their boosters in the media, placed the industry in both intimate family narratives and local heritage narratives. To be fair, local papers did not ignore controversial issues, such as the problems faced by grape growers and disagreements among players within the wine industry itself. There was also concern as American, Chinese, and French conglomerates acquired land and wineries. Such ventures muddied the perception that most wineries were small, family operations.[136] Still, the overall tone of these articles described a thriving and glamorous industry that was important to the region's economic future.

Wine as a Mnemonic Product

Through their on-site architecture and landscaping, marketing, and events, the wineries and their allies spoke in a singular voice. This voice was directed to an external audience of wine consumers, tourists, and investors. It inspired spin-off development in food experiences, entertainment, and local tours that in turn advanced the region's allure for foreign tourists. As wine journalist Tony Aspler declared, "After seeing Niagara Falls, everyone, it seems wants to discover Ontario wines."[137] Local residents, too, received the message.

Niagara experienced first-hand the erosion of heavy industry in North America. The recession of 2008 shuttered plants for major employers such as General Motors, John Deere, and Cadbury Schweppes, the region's last major fruit processor. Their closures undermined Niagara's blue-collar image. Residents who sought to preserve a positive collective identity sought to revise their region's narrative. Some looked to the past, while others sought a new identity in prized cultural outputs and in environmental sustainability. The wine industry was a ready resource for them all. Agriculture had its roots in colonial Niagara, and some folks still recalled the golden age of Canada's fruit belt. Indeed, agriculture's potency as a heritage narrative rivalled that of manufacturing. The wineries recognized its appeal and mobilized it in their branding efforts. When selling themselves, they articulated a coherent and desirable vision that was reinforced through each element of their branding.

The heart of the wine country brand is its close association with a local history of tender-fruit farming and rural life. The sustainability of any heritage narrative requires that it remain relevant, seemingly authentic, and resonant in the imaginations of those who embrace it.[138] Relevance and a sense of authenticity can, however, slip away if the authors' actions betray their words. By claiming a local heritage narrative as their own, particularly by invoking the values of small-scale family production, the wineries had to nurture those values in ways that rang true with residents. Had they not done so, they would have chanced alienating them. The wineries' pursuit of affluent consumers heightened that risk by introducing social class as an element of local wine culture, potentially creating a perceived barrier to participation in grape and wine events.

The wine country brand was well established by 2020. Winery advertising, pamphlets, and websites had been updated over time and revised to match changing tastes in the written word, photography, and design. Notably, the image of the ideal consumer began to skew younger after 2015. Coincidentally, the tight rebranding of St. Catharines's fall festival seemed to loosen as "Grape and Wine" came back into official usage, and local craft beers appeared on festival menus. Still, the local grape and wine industry continued to reference rustic authenticity to woo Niagara residents as well as the affluent urbanites of Toronto and New York State.[139]

Both adaptations point to forms of mnemonic practice. They are instances in which institutions, corporations, groups, and individuals signalled that vineyards and wineries constituted mnemonic resources and products in Niagara. The industry's successful linkage of local agricultural tradition and wine had indeed resonated – or in Ahmed's words, had become "sticky" – among residents. Subsequently, industry leaders appeared willing to relent to some popular expressions of that tradition.[140] As noted at the start of this chapter, it is important to assess how an intangible asset benefits an entire community when it is deployed in place branding. This chapter described the development of the brand. In the next chapter, we explore Niagarans' responses to it.

Chapter Six

Residents Engage the Niagara Wine Industry

Our wine heartland ... The home of Ontario's first wine pioneers, Niagara-on-the-Lake combines rich Loyalist history with innovative agricultural techniques ... The main attraction, however, is still the award-winning wineries.

– Wine Country Ontario, *Travel Guide 2016*[1]

Q: Is wine part of Niagara's agricultural heritage?
A: Yes. It's a big part of the community. It represents livelihood and family pasts. And it's a local celebration of local achievement. Mostly because of Grape and Wine.

– Survey participant, 2012[2]

The COVID-19 pandemic that swept the world in 2020 hit the Niagara region hard. The virus took a physical and social toll on residents. The region's economy – and especially the tourism sector – was also ravaged.[3] The federal government closed Canada's borders to tourist traffic with the United States while the provincial government placed restrictions on tourist sites, restaurants, and non-essential shopping. During these closures, the Grape and Wine Festival's traditional community events were also cancelled. Residents expressed their dismay through conversations and social media. Sara Nixon shared their views. A public historian at the St. Catharines Museum and Welland Canals Centre, she noted the festival was "perhaps the most visible symbol of citizenship in the city." In a blog on the museum's website, she then put residents' dismay in perspective:

Local celebrations like the Niagara Grape and Wine Festival are knit together with the spirit of the community. Such public displays are only as much as the meaning and feeling that people [give?] to them – feelings of community pride, identity, and belonging can only be cultivated by the people who share in the festivities ... I hope that reading this brief history of the festival can bring back a few memories and recollections, and even more, I hope you can stir up a certain feeling or sense connected to Grape and Wine – that is the essence of the community spirit fostered at such public celebrations.[4]

The marketing agencies working for the region's wine industry could not have conceived a better appeal. The industry had long tied its brand to a "Niagara" identity rooted in local histories of tender-fruit farming, genealogy, and the continuing prevalence of family farms. As described in the previous chapter, the industry's marketing was disseminated through winery shops and multiple media platforms. If grapes and wine offered residents meaningful mnemonic resources, then residents may have recognized them as such and used them in ways that confirm the narrative. In this chapter, we explore their responses to the industry's efforts.

When rural areas pursue economic development by capitalizing upon local heritage, they often follow a four-stage process. According to Christopher Ray, the first stage requires that a suitable local attribute be identified for development. This attribute might be the local culture, the landscape, a specific narrative (which can be historical or entirely fictional), or any combination of such things. Second, that attribute and its qualities must be sold to non-residents through a marketing strategy. Third, attribute developers must promote the product to residents. And finally, residents must internalize – or take for granted – the attribute and its narrative as an "authentic" element of regional culture.[5] Regional officials and entrepreneurs in Niagara followed the first three stages of this model: they identified grapes, wineries, and wines as potential assets; they sold the region as a desirable destination; and they tied grapes and wine to Niagara's major heritage narratives. Had they, however, convinced Niagara residents to assume these associations? Our surveys suggested they had done so. The place brand possessed the same characteristics as successful mnemonic products: the weight of institutional backing, intertextual messaging, the ability to move people, and narrative resonance.

The role of brand marketing in regional economic development has been widely debated. One school of thought, commonly associated with supply-side thinking, argues that a place brand should demonstrate "self-congruity," or, in other words, it should reflect some existing characteristic or self-representation of a place.[6] Fostering such a brand is a difficult task. Marketers may face not only consumer resistance to brand messages, but local resistance as well. Maja Konecnik and Frank Go note that a brand application, "without sensitive inclusion and consideration of the significance of public space may result in a commercial orientation which runs the risk of spoiling the identity characteristics such as social relationships, history, and geography, and by extension may destroy an area's sense of place."[7] This concern is echoed among scholars who decry the threat of "overtourism" wherein place making leverages specific local heritage narratives to please visitors but in so doing displaces residents. Hugues Séraphin and colleagues, for example, contend that the deleterious effects of overtourism in particular can be linked to place branding.[8] The deployment of cultural assets, lack of diverse representation in the planning process, and "touristification" become key points of conflict.[9]

A second school of thought, associated with demand-side approaches, argues that the coherence of a place may be lost through the branding process.[10] At issue in this debate is how marketers balance market demand with a place's attributes.[11] Heather Skinner suggests that a geographical place cannot be created and managed like a new product because it already exists; it is already located in a particular set of beliefs, attitudes, and practices that comprise local culture.[12] Recent place-branding efforts seem to suggest otherwise. Cities across North America, Europe, and Oceania have done just this as they embraced the creative economy theory championed by urban studies scholar Richard Florida. His writings provide practical guidelines for reorganizing and managing both the tourist and resident experience of cities in order to attract "creative" or knowledge workers and the firms that hire them.[13] Case studies suggested that reimagining cities in this manner demonstrably altered the ways in which residents lived in their own neighbourhoods.[14]

Place-branding strategies are, then, potentially harmful even as many scholars argue they are benign if not beneficial for residents of a region.[15] Jeannette Hanna, a marketing communication scholar, argues the following:

> The key to disarming local skepticism of place-based brands is authenticity. By pinpointing the most meaningful cultural and heritage differentiators of a place, cultural mapping breathes credibility and genuine character into the community's brand story. Immersing the brand proposition in the "real deal" reconnects people to their roots, boosts local pride, and engages supporters in re-envisioning their future.[16]

Such literature argues that successful place branding requires wide public consultation. That may be true, but it also requires the distillation of a diverse community into a single brand. In many locations, this is impossible. Moreover, the exercise is not designed to connect individuals to a place, create community, or foster an inclusive identity. It is designed to sell local assets in a global market. A place brand must resonate with external audiences; as such, it remains subject to the whims of consumer tastes. The development of any one specific place brand will probably appeal most to local economic and political stakeholders. Hence, a branding exercise may not express a local identity as much as it may generate one.[17]

As a marketing strategy or as a cultural trope, the "rural idyll" – like that described in the last chapter – is a complex social construction.[18] Its long history reveals a rich set of shared characteristics: a simple life, well lived; rustic locations that feature communities securely attached to place; the presence of farm animals; and beautiful landscapes containing alluring vistas.[19] The term can be used in various ways. Mark Shucksmith, for example, outlines the ways the term has been deployed: as a nostalgic yearning for a pre-industrial past, as a charming pastoral setting in the present, or as a utopian vision for a sustainable future.[20] All three may overlap.[21]

The definition of what constitutes "rural" is up for debate.[22] Statistics Canada categorizes municipalities by population size and proximity to urban amenities. The agency defines "rural" as those populations "living outside centres of 1,000 people" and "rural and small town" as those "living outside the commuting zone of large urban centres."[23] According to its 2016 population estimates, the Niagara region's twelve municipalities included one large population centre (> 100,000), three medium urban centres (30,000–99,999), and eight small population centres (1,000–29,999).[24] By definition, then, there were no officially designated "rural areas" in the region.[25] This technical curiosity belied the fact that there was considerable agricultural land in Niagara. Below the escarpment the fruit belt remained home to orchards and vineyards despite suburban growth. Above the escarpment, most of the land was designated as good, general-purpose agricultural land. A 2017 report quantified this distribution: Niagara was comprised of 64 per cent agricultural land, 21 per cent wilderness or rural forest, and only 15 per cent urban land.[26] In response to the continued problem of suburban encroachment on rural lands, the provincial government retained two pieces of legislation – the *Greenbelt Act* and the *Places to Grow Act* – designed to limit sprawl and to concentrate urban growth.[27]

In economic terms, agriculture remained significant to regional prosperity. In 2016, for example, the regional gross domestic product (GDP) for the Niagara census area was $16.4 billion. Local agriculture had a GDP impact of $1.41 billion and employed almost twenty thousand people.[28] However, the wine country brand elides the fact that most of the region is not suitable for wine-quality viticulture. Nor are vineyards the most significant agricultural crop. The regional government reported gross farm receipt data aggregated by segment in 2016. Greenhouse cultivation was the strongest segment, accounting for 49.1 per cent of the region's farm receipts with 204 hectares of space under glass. The strength of this segment was later bolstered when investors sought year-round capacity for cannabis production after that crop was legalized for recreational use in 2018.[29] After greenhouses, the second-largest segment was poultry and egg farming (18.4 per cent). These farmers took inspiration from the wineries by hosting an annual PoultryFest in Smithville from 2000 to 2017.[30] The third-largest segment was fruit and nut tree farming (16.7 per cent), whose farm receipts did not include the value-added income produced separately by the wineries.[31] Within this segment, grape growers operated the greatest number of farms (386 farms, or 61.9 per cent) and had the greatest number of hectares under cultivation (6,366 hectares, or 63.2 per cent).[32]

Farms across the region had remained viable by adapting to changing economic conditions.[33] Farmers employed three strategies. First, some sought to diversify their production. Between 2011 and 2016, for example, the largest increases in farm receipts were in hog and pig farming and sheep and goat farming rather than in some of the more traditional food and fibre products. Second, larger farms pursued greater production through intensification. Such investments changed the shape of Niagara's farms. After 1990, the total number of farms and the aggregate farmland in production both decreased while average farm size, farm employment, gross farm receipts, and agricultural export volumes all increased.[34] By contrast, some smaller farms favoured multifunctional approaches that featured "extensification." This approach is defined by greater care in food production, sustainable farming practices, and entrepreneurial moves into processing and retailing.

Agritourism is among the most important extensification adaptations. Proponents of agritourism contend that farmers and their local communities reap economic benefits – as well as improvements in overall quality of life – by developing amenities for tourists.[35] In Niagara, wineries may be factories in a rural setting, but they also serve as tourist attractions. For many small operations, limited production volumes and prohibitive distribution costs inhibit sales through third

parties.[36] Hence, many wineries rely on urban visitors to achieve prof-
itability. Such dependence is not unusual. Studies suggest that the
success of some wine regions in Italy, California, Australia, and New
Zealand hinges upon nearby urban populations.[37] Scholars such as
Peter Howland and John Overton characterize these regions as "metro-
rural idylls" or "metropolitan wine regions."[38]

Visitors to Niagara's wineries fit this pattern. A series of market
research studies from 2003 to 2020 revealed a consistently proximate,
urban clientele for local wine. A 1999 survey of winery visitors found
that 87.7 per cent had arrived by personal vehicles during day trips or
weekend getaways. Regardless of their mode of transportation, the vast
majority were Ontario residents (76.2 per cent); the next largest group
came from New York State (5.4 per cent).[39] A similar study conducted in
2003–4 identified the Greater Toronto Area as the single largest source of
all visitors (roughly 30 per cent).[40] Later surveys echoed these results. A
2015 study conducted at three Niagara wineries reported that 82.6 per
cent of visitors were Ontario residents, while another 3.9 per cent came
from New York State. Again, the Greater Toronto Area was the single
largest source, representing 44.3 per cent of all visitors.[41] In none of these
studies do Niagara residents make up more than 17 per cent of visitors.

As consumers, winery visitors tend to be well educated and wealthy.[42]
They also tend to enjoy a variety of regional amenities. An Ontario
Ministry of Tourism report published in 2008 classified Niagara tourists into three main groups based on the experiences they sought while
in the region: "wine," "natural wonders," and "casinos."[43] Of the three
groups, wine tourists were most likely to seek all three experiences. Barbara Carmichael argues that "for most visitors wine tourism is a touring
experience"; this group is keen to explore the region for unique sights,
shops, and fine dining.[44] Donald Getz and Graham Brown suggest these
experiences are linked by a desire for authentic, personal, and perhaps intimate moments. Wine tourists, they argue, seek a place where
"romantic dreams and cultural yearnings can be fulfilled. The model is
also a clarion call for destination organizations, the wine industry, the
community and cultural-heritage institutions to collaborate in creating
product and communicating appropriate benefits."[45] Parallel studies
reinforce that a charming rural setting that allows an escape from the
city is an important visitor motivation.[46]

Niagara's wine region, then, is supported by its physical attributes,
but also by its propinquity to major metropolitan areas.[47] From the falls,
to battlefield parks, beaches, and cottages, to the fruit belt and blossoms,
and now to wine, the Niagara place brand has served as an "urban"
extension of the countryside rather than as intrinsically rural landscape.[48]

Indeed, the Niagara winescape emerged out of an ongoing and recursive relationship between brand representation and visitor practice. The combination has produced a hedonic landscape featuring affluence and taste that encourages the conspicuous consumption of a rural setting and its products.[49] Writing of other wine regions, Overton and Murray contend,

> A scene of ordered vineyards, dotted with wineries that offer tastings and restaurants and hotels that offer a pleasant relief from city life, and festivals that celebrate wine consumption perhaps more than they do the traditions of grape harvests, all contribute to a romantic construction of wine regions as sites of middle class consumption.[50]

As the wine industry became established and Toronto housing prices became unaffordable to many families by 2010, Niagara became a destination for amenity-based, peri-rural migration. Amid later COVID-induced flights to the country, one advertorial in *Toronto Life* magazine proclaimed,

> City dwellers looking to press re-start are now driving beyond the burbs and setting down roots in communities where they can reconnect with nature and afford houses, while not decamping too far from T.O. – it's gotta be close enough to pop into town for a Jay's game. Within an easy drive of Toronto, there's one area that checks all our boxes for the perfect (literal) greener pasture: Niagara.[51]

This suburbanization of rural Niagara continued apace.[52]

The wine country brand, then, invested the regional landscape with a narrative and set of values that linked the wineries' visibility to tangible and intangible aspects of the region's past. The wine industry and its allies in local government mobilized the brand to attract tourists, new residents, and corporate investors. Existing residents were, as we have noted, always exposed to these efforts. The brand circulated among them much like the region's many other mnemonic narratives. Residents could have accepted the brand narrative as readily as tourists; they could have sought similar experiences, consumption opportunities, and any accompanying cultural capital to be gained from such participation. Much like those other mnemonic narratives, residents were free to embrace, repeat, and embody the wine country narrative, or not.[53]

Our early surveys suggested that residents did embrace the branding effort. Participants in 2005 and 2008 reported that wine was a significant marker of regional identity. These surveys had sought to measure the impact of local mnemonic narratives marked by material commemorations. The strength of residents' identification with wine prompted us to consider how the industry was perceived: as a legacy of pioneer farms and the fruit belt, or as a purely contemporary narrative. Consequently, we decided to probe how residents had adopted the grape and wine narrative.

Niagara Residents Talk Wine

The 2009 survey of St. Catharines residents began with one simple question: If you had to choose one thing that identifies Niagara, what would it be? Of 219 participants, 29.7 per cent referred to the wine industry and vineyards, while another 9.6 per cent referred to farming, tender fruit, and other agricultural produce. By contrast, only 19.6 per cent referred to Niagara Falls, and only 1.4 per cent referred to national historical narratives such as the War of 1812 or the Freedom Trail. One person surprised herself, stating, "Wine [*pause*]. That's not what I thought I would have said. I'm not even a wine lover." Another asserted, "Our agriculture is the best thing to share with outsiders. We identify with the fruit and wines. It is the image and substance of local existence, it's all around." Moreover, an overwhelming majority of participants associated wine with local heritage narratives (over 82 per cent); even more associated wine with specifically *agricultural* heritage narratives (88.8 per cent). To be blunt, the industry's reimagining of Niagara as "wine country" reanimated narratives of the region's agricultural past for participants.

St. Catharines is Niagara's largest city, but its residents could not be expected to speak for the region as a whole. As public officials and local journalists could attest, there were marked differences in perspective among its various communities. These varied perspectives emerge from very real differences in the regional geography and economy (see map 0.1). The most pertinent divide for the tourism industry was the Niagara Escarpment, which runs east–west through the entire region. The City of Niagara Falls, of course, sits at its edge. It is worth repeating that below the escarpment, to the north, is the microclimate favourable to soft fruit and grapes. Above the escarpment, to the south, are farms producing a variety of grains, vegetables, and livestock. A mix of manufacturing industries are found both north and south of the escarpment. The Niagara place brand had been mostly linked with tourism centred on the falls. Since the mid-1970s, Niagara-on-the-Lake and the wine industry have joined the falls as major visitor destinations, and both are north of the escarpment. Hence, to understand how this new brand was received across the region, both north and south of the escarpment, it was important to conduct a survey beyond St. Catharines.

In 2012 we selected four centres that serve as hubs for their own surrounding villages and countryside. Two of these centres, Grimsby and
Virgil, sit north of the Niagara Escarpment. Both are surrounded by
farmland, but they also function as relatively affluent bedroom communities. Many Grimsby residents, for example, commute to work in
Hamilton and cities further west. Virgil, meanwhile, was developing
into a community hub for residents of Niagara-on-the-Lake after the
Old Town became dominated by the tourist trade for much of the year.[54]
The two other centres, Port Colborne and Welland, are located south of
the escarpment. Both were small cities outside the viticulture microclimate and removed from the region's most common tourist haunts.
Their local economies were built around the Welland Canal and manufacturing.[55] At the time of our research, both cities were struggling with
long-term declines in heavy industry and the after-effects of the global
economic recession between 2007 and 2009.

As described in chapter 1, this survey followed the same procedures
used in our St. Catharines surveys. We made one important change:
all participants were screened to ensure that they were residents of the
centre in which the interviews took place. In chapter 1 we also briefly
outlined the results. We noted, for example, that most residents in all
four centres drew upon the present for regional identifiers (table 1.2).

There was nonetheless one important difference: participants living
north of the escarpment were more inclined to name grapes and wine
as a regional identifier; participants living south of the escarpment were
more inclined to name Niagara Falls. After that, residents of Welland
and Port Colborne identified wine, agriculture, and the landscape and
climate in almost equal measure. These responses seem to reflect conditions particular to each centre. Virgil, despite recent residential growth,
retained its self-image as a rural hamlet. In 2020, a brief drive from
the centre of town in any direction left one surrounded by farms. By
contrast, Grimsby was a growing town and the other two centres were
small cities. They may have been surrounded by farms but agriculture
did not define them. Similarly, Grimsby and St. Catharines sit on Lake
Ontario while Port Colborne is on Lake Erie: not one participant mentioned their access to North America's Great Lakes. Instead, residents
of all three cities identified the positive social aspects of contemporary
Niagara: family life, friendly neighbours, good schools and athletic
facilities, and diverse shopping opportunities. In the midst of declining
industrial fortunes, they also noted some of the negative social aspects
of the region, such as unemployment, outmigration, alcohol or drug
dependency, and civic corruption. St. Catharines was the only centre
to note arts and culture as an important identifier. Perhaps its residents

Table 6.1. If you had to choose one thing that identifies Niagara, what would it be?

Responses	Percentage of responses				
	St. Catharines (n = 219)	Grimsby (n = 37)	Virgil (n = 52)	Welland (n = 39)	Port Colborne (n = 43)
Niagara Falls	19.6	24.3	3.8	28.2	48.7
Wine industry and vineyards	29.7	43.2	26.9	15.4	7.0
Agriculture (not grapes)	9.6	13.6	53.8	12.8	7.0
Landscape and climate	13.2	10.8	11.7	10.3	7.0
Positive social aspects	5.5	5.4	–	7.7	4.6
Tourism	5.0	–	3.8	2.6	4.7
Historical aspects	1.4	2.7	–	5.1	2.3
Negative social aspects	5.0	–	–	10.3	4.7
Welland Canal	1.4	–	–	5.1	14.0
Arts and culture	9.6	–	–	–	–
Industry	–	–	–	2.6	–

Notes: St. Catharines, 2009; Grimsby, Virgil, Welland, and Port Colborne, 2012.

were thinking of the city's status as the largest centre in the region and home to arts groups, an art gallery, a performing arts centre, a college, and a university, among other things.

Overall, residents north of the escarpment identified markers that were very closely aligned with the region's contemporary place brand. Participants did not name the falls or wine exclusively, but the most common alternatives were allied elements that evoke the falls and wine in marketing materials: the landscape, its climate, and agriculture. Residents south of the escarpment identified the same markers and elements but less frequently. Beyond the predictable public or commercial markers of the region, residents in the south also identified aspects of their own experiences of life in their hometowns.

To put the 2012 results in context, we sought to understand the participants' relationship with the wine industry. There were some eighty-five local wineries that year. We suspected that proximity to them would shape people's thinking, in that residents who live or work among them would be more inclined to highlight the industry than residents who live elsewhere. Indeed, the top-of-mind responses suggest that the farther one is from vineyards and wineries, the more wine recedes as a marker of regional identity (table 6.1 and map 6.1). As such, we asked participants if wine was in any way part of their daily routines. The data in table 6.2 indicate that it was for a clear majority north of the escarpment but not for a majority south of the escarpment. Participants

Table 6.2. Are wine or wine activities part of your daily routine?

	Percentage of responses			
Responses	Grimsby (n = 37)	Virgil (n = 52)	Welland (n = 39)	Port Colborne (n = 43)
Yes	56.8	75.0	25.6	41.9
No	43.2	25.0	71.8	55.8
No answer	–	–	2.6	2.3

in Welland were least likely to report that wine was a part of their daily lives. These results echo the results of the first question regarding top-of-mind responses.

Participants who responded in the positive were asked to identify how wine fit into their daily routines (table 6.3). This was posed as an open-ended question. We offered no prompts to guide the participants' thoughts and recorded every answer given by each participant. The five most common responses were remarkably consistent. Many residents north and south of the escarpment shared one thing in common: they were habitual consumers of wine. Many residents in the north also acknowledged that they lived near wineries and that wineries and vineyards were visible elements of the landscape. Interestingly, the physical visibility of wineries was also reported south of the escarpment, particularly in Port Colborne. That city is roughly thirty kilometres from the closest winery. Perhaps these participants had in mind the places where they worked and shopped more than where they resided. Nonetheless, it was clear that when many participants thought about wine's place in their daily lives, it was due to their enjoyment of wine itself or their proximity to wineries and vineyards.

After probing the participants' quotidian experience of the wine industry, we sought to understand how they actively engaged with it. This, too, was an open-ended question; we offered no prompts and did not limit responses. Once again, participants north and south of the escarpment commonly reported that they enjoyed wine (table 6.4). The frequency of this response was higher for this question than for the previous one. (The difference suggests that most participants drink wine, but not on a daily basis.) After consumption, the next most popular response was visiting wineries, and the third most common was attending wine-related events. Consumption, visiting wineries, and attending events was more common among those who lived closest to the wineries, but many residents south of the escarpment also told us that they, too, enjoy winery tours, shops, restaurants, and events.

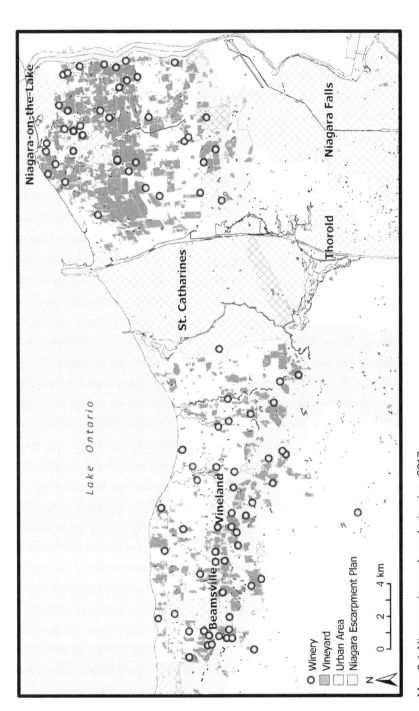

Map 6.1. Niagara vineyards and wineries, 2017.

Sources: Cartography by Sharon Janzen. Data from Niagara Escarpment Commission; Agriculture and Agri-food Canada, Natural Resources Canada, VGA Wines, Niagara Peninsula Conservation Authority, Esri Canada. Map prepared by Brock University Map, Data & GIS Library (2022).

Table 6.3. How do wine or wine activities fit into your daily routine?

	Percentage of responses			
Responses	Grimsby (n = 21)	Virgil (n = 39)	Welland (n = 10)	Port Colborne (n = 18)
Consume wine	42.9	16.9	60.0	61.9
Grow grapes	9.5	3.4	–	–
Make wine	–	–	–	–
Work at winery	9.5	5.1	30.0	9.5
Live near winery	9.5	33.9	–	–
Visibility of wine and wineries	19.1	37.3	10.0	28.6
Other	9.5	3.4	–	–

Table 6.4. In what ways do you participate in wine and wine-related activities?

	Percentage of responses			
Responses	Grimsby (n = 40)	Virgil (n = 93)	Welland (n = 33)	Port Colborne (n = 39)
Consume wine	42.5	32.3	42.4	33.3
Grow grapes	2.5	1.1	3.0	5.1
Make own wine	2.5	1.1	6.1	7.7
Visit winery	42.5	26.9	24.2	35.9
Attend wine events	5.0	33.3	18.2	7.7
Other	5.0	5.3	6.1	10.3

Almost half of the participants in Virgil and Grimsby reported that they visited wineries and/or attended their events. Many of the wineries in those centres had shops and/or restaurants that were open year-round. As such, residents may have viewed the wineries as local amenities that blended into the towns' retail landscapes. Quite simply, the wineries offered convenient places to shop and dine that were close to home. Similarly, a majority of Virgil residents reported that they attended wine-related events. Their village was surrounded by wineries that hosted events throughout the year such as cooking classes, concerts, and the annual Icewine Festival. Hence, the wineries' proximity allowed residents north of the escarpment to treat them as routine outings rather than special excursions.

It is one thing to understand how wine was incorporated into residents' lives and daily routines, but quite another to understand how it was incorporated into their conception of Niagara's past. To explore this issue, we asked four questions regarding participants' engagement

Table 6.5. Resident engagement with local agriculture and opinions regarding local agricultural heritage

Statements	Percentage of participants in agreement with each statement			
	Grimsby (n = 37)	Virgil (n = 52)	Welland (n = 39)	Port Colborne (n = 43)
I participate in agricultural activities.	89.2	96.2	89.7	88.4
Agriculture is significant to the region's heritage.	97.3	90.4	74.4	93.0
Wine is significant to the region's heritage.	100.0	75.0	82.1	83.7
Wine is part of the region's agricultural heritage.	91.9	75.0	84.6	83.7

with agricultural activities and their thoughts on agriculture's signifi-cance to the region's heritage. On a five-point Likert scale, participants north and south of the escarpment were unequivocal in their assess-ment that agriculture was a significant aspect of the region's heritage (table 6.5). This confirmed our broad assumption that Niagarans were aware of the region's agricultural past. Though Welland and Port Colborne featured industrial economies, the entire region remained typified by small centres and family-owned farms at the time of our surveys.[56] To underscore this point, participants north and south of the escarpment stated that they participated in some form of agricultural or agriculture-related activity. Most prevalent among these activities were positive interactions at farmers' markets, roadside produce stands, and fall fairs, but participants also shared many stories of summer jobs picking fruit or working on a family member's farm.

The third and fourth questions asked participants to consider the significance of wine in "local heritage" and then in "local agricultural heritage." The results were similar to the first and second questions (table 6.5). The vast majority of participants perceived wine to be a significant part of the region's heritage. These data confirmed our assumptions regarding residents north of the escarpment in "wine country." We were surprised, however, by the responses collected south of the escarpment. Few participants in Welland or Port Colborne named wine as a top-of-mind identifier for Niagara, and few lived or worked among vineyards and wineries. Nonetheless, they were just as inclined to accept wine as part of the region's heritage as resi-dents to the north. They were also just as likely to believe wine was significant to the region's agricultural heritage. Numerous residents north of the escarpment recalled the tender-fruit orchards that had been replaced by grape vines. As one such individual noted, "I have liked to drive to see the blossoms since I was a little girl. I'd love to see it stay that way." Many of these participants accepted that wine had some heritage value, but a minority in Grimsby were reluctant to accept wine as an element of the region's historic fruit belt. We will return to this point below.

These findings raise an important question about "visibility." It would be reasonable to assume that visibility can be ascribed to the material reality of vineyards and wineries. North of the escarpment, they were spacious, inviting, and growing in number between 2005 and 2016. As one Virgil resident noted, "We are surrounded by vineyards, we pass wineries daily. I always see the Wine Route signs, but they just seem like part of the landscape. I hardly notice anymore." This participant's easy transition from the landscape to the brand is telling. We believe that residents experienced a second kind of visibility associated with media representations. The wine industry and local government disseminated the wine country brand, laden with pastoral narratives and vineyard imagery, through multiple media platforms and channels. The industry achieved not just material visibility in the physical landscape, but also a high profile in its ideational representations.

The last component of the survey asked participants how they learned about wine and wine-related activities. This, too, was an open-ended question; we offered no prompts and did not limit responses. Participants were generous with their answers, which were aggregated into three categories: marketing and retail campaigns, passive sources, and the mass media (table 6.6). Three patterns emerged.

First, wineries were the most commonly cited source of information about wine. Four participants, for example, noted that they actively seek out knowledge from nearby wine professionals. One stated bluntly, "I still prefer to go to the source." This participant seemed to speak for many participants north and south of the escarpment. More than one-third of participants to the south of the escarpment were willing to travel to a winery even though they did not live near one. In the winery setting, residents would be immersed in the full brand experience, replete with each company's narrative and style. If the experience is positive, then the impact of this brand experience would be reinforced by the winery's other marketing efforts, such as its brochures, websites, and festival appearances.

Second, participants cited family and acquaintances as important sources of information. We refer to this as "word of mouth." Given

Table 6.6. Residents' sources of knowledge about wine and wine-related activities

	Percentage of responses			
Sources of knowledge	Grimsby (*n* = 67)	Virgil (*n* = 113)	Welland (*n* = 54)	Port Colborne (*n* = 60)
Wine marketing total, %	29.9	32.8	37.7	30.1
Brochures	7.5	4.4	–	3.3
LCBO*	4.5	6.2	5.6	3.3
Wineries	14.9	19.5	26.5	23.5
Festivals	3.0	2.7	5.6	–
Passive sources total, %	28.4	31.9	30.7	24.9
Word of mouth	19.4	24.8	14.8	15.0
Employment	1.5	1.8	13.0	3.3
Visibility	7.5	5.3	2.9	6.6
Mass media total, %	41.7	35.3	31.6	45.0
Newspapers	19.4	14.1	13.0	13.3
Magazines	4.5	–	3.7	–
Radio	–	7.1	–	6.7
Television	10.3	1.7	5.6	6.7
Internet	7.5	12.4	9.3	18.3

* The LCBO is the Liquor Control Board of Ontario. It is a provincial Crown corporation that holds a quasi-monopoly over the sale of alcoholic beverages in the province through its chain of retail stores. Ontario-based producers are permitted to sell their own products at their own places of manufacture and at a limited number of third-party locations.

the number of participants who enjoyed wine and visited wineries, it appears that a wine culture had developed among some residents of the region. Conversations with friends, neighbours, and co-workers could have provided trusted recommendations when looking for a special bottle, restaurant, or outing. As one person explained, "Just being in the area means that you are exposed to it all the time. I really don't consider how I learn it, it seems that I just do." Marketing professionals understand the power of word of mouth. If a visitor has a positive experience at a winery, that visitor can become a brand ambassador simply by sharing that experience with acquaintances.[57] After 2006, social media platforms emerged on the Internet that expanded the visitor's sphere of influence to a much larger audience through posted comments, tweets, "likes," and star ratings.[58] The influence of true word of mouth appears to be reinforced locally by the proximity of the wineries and the employment of residents in related industries. Several participants noted that they had gained knowledge working in local tourism and hospitality services, especially restaurants.

The third most common source of information was newspapers. Participants who identified newspapers as sources named five specific local titles: the daily St. Catharines *Standard*, the *Welland Tribune*, the *Fort Erie Times*, and the *Hamilton Spectator*, and the weekly *Niagara This Week*.[59] These papers covered the local wine industry in multiple ways. On a weekly basis, readers could expect reviews of recent vintages and winery restaurants, news of industry events, or profiles of up-and-coming growers, wine makers, and chefs. This coverage framed the wine industry in a positive light; it was feel-good, human-interest material about an industry that was boosting the region's economy and profile, a stance echoed in the same papers' coverage of the War of 1812 anniversary. Less often, stories also covered the industry's internal conflicts and its vulnerability to national and international policy regimes. Ultimately, both sets of stories maintained familiar local narratives and positively positioned Niagara as an international wine region. The messages were also ubiquitous insofar as the same topics were covered by multiple outlets. Participants did not name specific outlets beyond the newspapers already mentioned, but other mass media were identified in a general fashion, particularly the Internet. One exception was a familiar Ontario magazine, *Food and Drink*. As one participant claimed, "All I know comes from that magazine."

The example of *Food and Drink* raises a final issue: the impact of advertising on perceptions of the region. The magazine is a custom-content advertising vehicle for the Liquor Control Board of Ontario (LCBO), a retail chain owned by the Government of Ontario. The magazine is available for free in all LCBO outlets. It is modelled after high-quality food magazines; its pages are glossy and thick with fashion-conscious imagery. All its content is crafted to market alcoholic beverages, be it editorial matter or branded advertising. Every issue features at least one Niagara wine with an article synchronized with the vintner's own marketing. During our study *Food and Drink* was complemented by the *Wine Country Ontario Travel Guide*. This booklet was published annually by the Wine Council of Ontario and distributed throughout the province. Although it included other Ontario wine regions such as Prince Edward County and the Lake Erie North Shore, the booklet devoted most of its pages to the Niagara region. Its look was strikingly similar to *Food and Drink*. The written content promised unique and memorable experiences among family farms and heritage settings while its photography portrayed rural idylls, vineyards, and affluent wine drinkers. As noted at the opening of this chapter, the tone of this industry material – widely available and free of charge – was readily echoed by participants in our surveys. Such content found intertextual echoes in other wine, food, and travel magazines that highlighted Niagara destinations.

Table 6.7. To what extent do you learn about wine and wine-related activities from advertising?

Responses	Percentage of responses			
	Grimsby ($n = 37$)	Virgil ($n = 52$)	Welland ($n = 39$)	Port Colborne ($n = 43$)
Significant contribution	54.1	32.7	20.5	44.2
Minor contribution	35.1	51.9	61.5	39.5
No contribution	10.8	15.4	10.3	2.3
Negative contribution	–	–	2.6	–
No answer	–	–	5.1	14.0

Our last question asked participants to gauge the significance of advertising to their personal knowledge of wine and the industry. Their responses revealed that a majority of participants living north and south of the escarpment considered advertising to be a commonplace source of their information (table 6.7). A Welland resident stated, "You can't help but notice the signs, the LCBO advertisements, the advertisements in the *Globe and Mail*." However, there was no consensus among centres regarding the importance of this source. It was rated to be a significant source among a majority of participants in Grimsby, and a minor one among those in Virgil and Welland, while participants in Port Colborne were almost equally divided. Perhaps the most significant conclusion to be drawn from this data is that more than 80 per cent of respondents acknowledged that advertising contributed in some measure to their knowledge of wine.

Overall, participants' responses in 2012 indicated wide popular knowledge of the region's agricultural history. An overwhelming proportion of participants, north and south of the escarpment, viewed agriculture in general to be a significant aspect of local heritage. Participants also indicated that wine was a significant part of that heritage. For most of them, agriculture and wine were linked. The transition from fruit farming to viticulture to wine production was, for some, easily accommodated. As one participant told us of grapes, wine, and agriculture, "They go hand in hand." Another stated that "Wine is part of the agricultural heritage. It was inspired by grapes grown here." Such comments echoed the claims found in wine marketing.

The mass media play an ambiguous role in the production of heritage narratives. George Lipsitz contends the narratives disseminated by national and international media companies can loosen the bond of local experience and dislocate local memory.[60] At the same time, such media can foster connections with experiences and memories well beyond the local. Various authors have demonstrated how print, broadcast, film, and digital media have contributed to the constitution of remembered events, heritage sites, and national heritage narratives.[61] The work of historian Alison Landsberg is germane here. Following Lipsitz, Landsberg points to the ability of mass communication to open avenues to the past for those who did not live it. She writes,

> Prosthetic memories are neither purely individual nor entirely collective but emerge at the interface of individual and collective experience. They are privately felt public memories that develop after an encounter with mass cultural representations of the past, when new images and ideas come into contact with a person's own archive of experience.[62]

As Landsberg notes, the commodification of particular narratives enables them to circulate at previously unthinkable scales and "enables transmission of memories to those who have no 'natural' or biological claim to them."[63] Thus the work of wine industry entrepreneurs, place-branding marketers, and local governments had successfully replicated the work of memory entrepreneurs who champion traditional nation-building narratives.[64]

This process is not so different from Ray's four-step process to create a successful place brand for rural areas.[65] Place brand managers operate, in effect, as memory entrepreneurs. Their work is inspired by select narratives that tell of a place's meaningful characteristics. This may include recourse to narratives highlighting the formative events in a place's past. Like other memory entrepreneurs, they claim special significance for these narratives by enumerating the beliefs and values that animated the key historical actors. These narratives, beliefs, and values are among the place's intangible assets. Place brand managers may hope these beliefs and values still animate the community today, but primarily they have goods and services to sell. Place brand managers turn to the media – any medium – to keep the narrative accessible and alive. Material commemorations may take the form of heritage buildings or preserved landscapes. By contrast, the mass media deliver a diverse package of information to a wider audience. This is, of course, crucial for tourism. Residents, too, will likely engage this information. If it is positive and fits with residents' notions of self-identity, it may become added to their stores of identity-building resources and may, eventually, assume the status of a mnemonic product in and of itself. Even then, memory entrepreneurs may ask individuals to make manifest their personal endorsement of the narrative by participation in a set form of observance. If this observance becomes ritualized, through festivals or anniversaries, then the place brand managers' work may become self-perpetuating within the community itself.

Hegemony or Engagement

The development of feelings in common does not imply residents of a branded community are dupes of brand managers. Participants in the 2012 survey shared the positive beliefs and values linked with the wine country brand, and the Grape and Wine Festival featured in their comments regarding the industry and local heritage. However, it is also telling that the vast majority never referred to the festival by its new name (the Niagara Wine Festival). As one noted, "Grape and Wine … provides an outlet for cultural events. It identifies the region and evolves with the community." And another: it "has been going on forever. Most of us grew up with it. It's just part of our lives." Annual rituals like the festival facilitate a link between autobiographical memory and historical memory narratives, much as Remembrance Day does for military commemoration. The regional government understands this point:

> Niagara, being one of the oldest settled parts of the province, is particularly rich in history. Our agricultural heritage in urban families is sustained today through parents or grandparents who grew up on farms, through celebrations of agriculturally based festivals such as Thanksgiving, and in the ability of urban families to escape to their roots, in the farmlands and rural communities that are still accessible from our cities.[66]

Extending this line of thought, vineyards and wineries may function in the same way that monuments do. They can appear timeless and unchanging and therefore impinge only on the periphery of residents' consciousness as they tend to their daily routines.

Participants did not share simple perceptions of the regional landscape despite the confluence of their individual experiences and the branding effort. Their answers revealed a broad knowledge of the wine industry and its place in Niagara. Indeed, some participants insisted the industry should be distinguished from agriculture because it represented a break with the past. One participant explained, "I have mixed feelings. This is the place for it, but when they ripped out the peach and cherry trees and other soft fruit trees we lost food diversity." Another stated, "The wine industry keeps people working more than fruit trees. But I still prefer fruit trees." Several participants echoed those sentiments. More ominously, one stated, "It's taking over. You no longer hear 'agriculture' and think peaches or apples or cherries. You think grapes and wine. I haven't decided if that is a dangerous thing yet." Consciousness of this newness often triggered memories of lost orchards. Several participants echoed the sentiments of one who stated, "I just feel bad that we have to get rid of a lot of the fruit and processing plants. It's unfair to farmers. It seems like wine is the only solution around here, for good or bad." Or this: "Agriculture is significant only because of the wine. Tender fruits have gone down the drain and agriculture wouldn't exist if wine didn't exist." Both comments reinforce the point that grapes and wine had much greater visibility than greenhouse cultivation and poultry farming.

A more common response viewed the transition from fruit to wine as evolutionary rather than revolutionary: "It's evolved into this. Niagara was first tender fruits. Then wine emerged. The climate is perfect for wine. It's an industry that makes a lot of money and because of this has become the main part of Niagara heritage." In other words, some participants saw an almost natural evolution in market adaptation and crop selection as reason enough to include wine in local agricultural heritage despite the industry's relatively recent ascendance. In sum, as much as these participants supported the industry's goals and achievements, they were not passive observers of the local landscape or the industry's claims.

While many participants identified the wine industry as a threat to orchards, only three identified urban development as one. Tender fruit production in Niagara peaked in 1951. Since then, the amount of land converted from orchards to vineyards has been matched by the urbanization of Niagara's landscape. Indeed, among the areas subject to the greatest urban sprawl up to the period of our research were tender-fruit lands in St. Catharines, Grimsby, and Virgil. Most residents who live in centres north of the escarpment own homes that were built on former orchards.[67]

When participants shared their thoughts on the region's "heritage," we understood their notions of that concept influenced their answers to our questions. Many participants identified "heritage" with something intrinsic to a society or place due to natural or environmental factors. Hence, one person told us that "Wineries are here. It's in the soil." Another opined, "Wine – there are wineries everywhere. The environment is perfect for wine." Others understood "heritage" as a synonym for "history," and they often referenced local toponomy to support wine's inclusion in local narratives. Referring to a village near Grimsby, one participant asserted, "Grapes have been here for many years … Look at Vineland!" Still others linked "heritage" with personal rather than regional narratives: "I grew up with it. Wine is in my soul, it is part of my heritage." The majority, however, saw wine as an important, contemporary cultural activity: "It seems that wine is part of local heritage. It's local. My family drinks wine every day, as do our friends. It's part of the culture." Some participants clearly aligned "heritage" with widespread social recognition or established reputation. For example, one person told us that "Wine is one of the things we're known for, something to be proud of, and we deserve to be proud of it, because we make great wine." Another said, "It brings tourists, it's our image. But we do make some of the best wine. We've found our niche, especially with the ice wine." Yet another linked reputation and "heritage" with economic significance: "Yes, because it draws people here. It's one of the two things that bring people here. The falls is first."

Those who aligned heritage explicitly with history sometimes questioned wine's role in regional narratives. These participants tended to reflect on the length of time required for an artefact or practice to become "traditional." One participant noted quite succinctly, "Wine is a part of our local heritage, but not in the historical meaning of the term ... It's a recent phenomenon." Another told us that the industry "is relatively new, so it's not part of our heritage yet. It's only been around for thirty years. It's too soon to be part of our heritage." One contrasted the wine industry with agricultural traditions and drew a sharp distinction between them: "Grape growing is a part of our agricultural heritage, but wineries now are different. Farmers have been here longer. The wineries have only been here a short time. They are outsiders." Another bluntly declared that wine makers were not "real farmers."

Some participants offered more ambiguous responses to our questions. Participants, for example, were aware there was wine before the industry's sudden evolution in the 1970s. One observed that "it has become prominent in the last twenty-five years, but it started well before that." On this point, one participant distinguished between the past and the present in her conception of the industry's local influence: "It's grand what they've done with Ontario wines. Because of the improvements, we've been able to make it a part of our identity. I think 'heritage' is a tricky word because it brings up history, but 'identity' or 'an identifier' is probably more accurate for me."

Participants were well aware of the regional government's recent branding campaign and the wineries' own advertising. When answering questions regarding local heritage, 29 participants commented on these efforts. They had mixed feelings. Nineteen were generally positive towards the brand, and these respondents were split fairly evenly between those living above ($n = 10$) and below ($n = 9$) the escarpment.

The remaining 10 were indifferent to, or critical of, the rebranding. One participant articulated this tension by noting that wine is "part of our identity, although it borders on gimmicky sometimes because the idea is so marketable."

Participants who approved of the brand saw two main benefits. First, participants acknowledged the wine industry provided investment and employment benefits in a weak economy. For example, one participant told us, "it puts people to work in an economically depressed region. The wine festival brings all sorts of people to the region that wouldn't be here otherwise." And similarly, "The wine is important at a global scale. There's beginning to be a stronger connection between wine and the region ... [It] is good for marketing [the region] and good with the tourists." Participants seemed impressed that wine could provide a positive economic niche that residents could support. One participant declared, "I'm very proud of it. It makes you feel good that we recognize it and promote it. VQA and local wine is promoted across Canada, and it is a good product."

The last quotation points to the second benefit suggested by participants: that the wine industry's growth had bolstered a sagging local identity affected by deindustrialization and unemployment. One person told us, "Wine making certainly raised the appeal of Niagara. It makes it more sophisticated." Another echoed this sentiment but added, "we still have a long way to go. We need to shake this underdog or lower-class image." Another simply seemed surprised by the wineries' developing reputations: "As I was growing up, I felt the need to go away, then as I talk to you about heritage, I realize there is a lot of good stuff here. You become humbled. Good ol' Grimsby isn't that bad."

Several participants echoed these sentiments and opined that regional promoters could do more. One person said, "Labels can be improved upon. Grimsby does not have a lot of wineries or advertising. Highway 8 [a two-lane regional road running from St. Catharines to Hamilton] needs to be more user-friendly and designed better. It could be a new Niagara Parkway." Another stated that wine is "not at the forefront. Unless it's right in front of you. We need to educate and spark interest in it to participate. People all across the peninsula should embrace it." Some participants living south of the escarpment were particularly keen on tying their local economies to the wine industry. One said, "We need to incorporate wine into Canal Days [an annual festival in Port Colborne]. There are lots of wineries around the peninsula. They need to make a bigger presence around the peninsula. For example, they could come to the Welland Food Fair. Don't make people come to Niagara-on- the-Lake."

Only six participants were overtly critical of the brand. Perhaps it is telling that five of them resided south of the escarpment. Two participants noted the affected nature of the grape and wine brand and questioned its ability to represent the region in an authentic way. As one commented, the brand was "Bullshit ... We want to be a place [that we're not] ... everything suffers because of it ... I know it's B.S. so I don't listen to it. It becomes propaganda." This comment, and others like it, contrasted the upscale image of Niagara in most wine marketing with the speaker's perception of the region. Two comments were blunt in this regard: "People don't drink wine. This is a region of beer drinkers. It's not part of people's lives like it is in other places. In Europe even men drink wine." And, "Wine caters to a rich demographic." These critics perceived Niagara as a blue-collar community upholding a "traditional" or "unpretentious" masculinity that may be stereotypically white, Anglo-Saxon, and Protestant. The difference between their own self-conceptions and their readings of wine advertisements were simply too great. They did not see themselves or their communities reflected in the brand.[68]

We believe these responses, taken altogether, indicate the wine country brand succeeded in complex ways among Niagara residents. The wine industry was the most common, top-of-mind identifier for residents north of the escarpment, and it joined a select list of identifiers south of it. It did so because it offered a useful resource to individuals when constructing a personal or local identity. The ubiquitous brand narrative – promoted with intertextual repetition through multiple media channels, events, and word of mouth – moved residents towards positive associations linking the landscape of everyday life with the region's agricultural heritage. This was true both north and south of the escarpment.

A Final Taste

We offer two final thoughts. Place brands can never reflect the full breadth of place-based heritage narratives. Participants demonstrated that the lived reality of any region is too complex to reduce to a single slogan or image.[69] The Niagara region, like every other, exists as a political entity that is riven with its own diverse constituencies, with their own needs and priorities. In this case, brand managers and regional officials latched on to one industry in one part of the region to revive the local economy. To do so, they promoted that industry as the face of the entire region and invited residents to review their understanding of local heritage narratives. As with any mnemonic construction, there were detractors. Some participants did not equate wine making with pioneer agriculture or soft-fruit farming. Others did not see themselves included in the portrayal of wine culture.

Second, place brand managers should be mindful of all audiences with access to their messages. Residents are not beholden to local branding efforts. And, as marketing and communication scholar Natàlia Lozano-Monterrubio reminds us, social media can change the way destinations are researched by potential tourists.[70] Information available online includes the contributions of locals as well as professional marketers. If a destination's brand marketing is not consistent with these other sources of information, potential tourists will soon find out. More pointedly, if a brand runs counter to residents' own beliefs and values, or if brands serve to diminish a local experience, residents will not serve as brand ambassadors. They may, in fact, work to disrupt these branding efforts.[71]

Chapter Seven

Conclusion

KNOW IT: Niagara's iconic waterfall and world-class wine are just two of the region's stories ... Take a closer look at the region you thought you knew.
– Niagara Economic Development, website home page, July 2018[1]

The Regional Municipality of Niagara established a new economic development agency in 2018. With it came a new mandate, a new slogan, and a new website. Landing on its home page, visitors were greeted with the bold declaration quoted above. The page also featured a photograph that reinforced the statement: it showed a confident engineer examining a precision industrial part in a spotless workspace. That was all. There was no waterfall, rustic landscape, grape vine, or exposed brick wall bearing signs crafted from reclaimed planks. Instead, the message and image evoked advanced manufacturing and international trade. Geography and stories were still important. The region's selling points had long been its rich physical resources and its colonial history; now they included its proximity to American markets and its promising technological future. Positioning Niagara in these terms made one point clear: the wine industry was now an established identifier for the region. It required no further explanation. The agency quite simply equated the profile of the wine industry with that of the falls.

Memory entrepreneurs have staged hundreds of commemorations throughout the Niagara region. We understand their purpose. Mnemonic products contribute to identity formation at local, provincial, and national levels. The narratives and values they promote are evident in an inventory of the region's material commemorations. Using Eviatar Zerubavel's concept of the commemogram, we plotted the time periods marked by these plaques.[2] Narratives surrounding the War of 1812 are, by far, the most celebrated. They are not only more numerous, but also the largest, most visible, and most publicized. They are also among the most recent. As we noted, the bicentennial of the war featured prominently in the federal government's well-funded effort to rewrite Canadian foundation narratives. With this new investment, memory entrepreneurs at all levels eagerly added to the stock of material markers. This included the construction of new monuments, the renewal of existing sites, and an increased use of heroic names in local toponymy.

The motivations of memory entrepreneurs may be well known, but the responses of local residents to their efforts are not. Are local residents proud of celebrated Niagarans and their roles in Canadian nation building, major engineering works, or the Underground Railroad? Do they identify with them? Do they internalize values and assumptions captured in stone? We were not the first to consider these types of questions, but our research differs from similar projects. We did not ask participants how they engage with the past; we asked participants how they think about their communities, then compared their responses to narratives marked by local mnemonic products. We also pursued a constructivist approach to our surveys, adjusting our questions as we progressed to test our assumptions, our tools, and our results.[3] Given this process, participants' sustained identification of grapes and wine as a significant heritage identifier encouraged us to make an unpredictable transition from monuments to wineries. That said, future research into Niagara's mnemonic products and practices requires further methodological and conceptual innovations. Our surveys, for example, provided answers to our initial questions; they represent popular, and mostly white, engagement with local mnemonic traditions. However, they may not account for the differences in identity markers that potentially emerge from close attention to the intersections of ethnicity, class, gender, and/or sexuality.[4]

Many of the region's mnemonic products celebrate people, places, and events of a distant and receding past. Nearly two decades of research and five surveys brought one clear conclusion: traditional narratives and the mnemonic products and practices that support them did not loom near the top of participants' consciousness. Without prompts to consider the past, our research participants demonstrated that Niagara's major historical narratives have had a limited impact. The bicentennial of the War of 1812 provided a rare opportunity to test the efficacy of memory entrepreneurs' efforts while they remained fresh in residents' minds. However, our survey following the bicentennial produced very familiar results: participants did not draw upon the War of 1812 when we asked about regional identity. The petition to remove the Watson Monument provided a similar opportunity. Debate raged through the summer of 2020. Perhaps a survey would have revealed an engaged public conversant with Alexander Watson's story. On social media, however, many supporters and opponents of removal admitted they knew little of the North-West Resistance or Watson's role in it. Just as tellingly, public commentary on the issue ended within days of council's decisive meeting. Let us be blunt: few contemporary experiences ignited residents' interest in the most commonly marked narratives.

Despite these findings, we contend that memorials retain some mnemonic potency. The Watson Monument, for example, was erected by a community eager to claim its place in the nation. On one level, it commemorates a specific soldier, Alexander Watson. In this context, the monument failed at its task. St. Catharines residents have largely forgotten Watson and the reason why he was memorialized at city hall. On a second level, the monument commemorates military service and the ideal of the citizen-soldier. In this context, the Watson Monument and other memorials marking subsequent wars still convey a message to which individuals attend. Set in a wide intertextual net of cognitive and emotional prompts and citizenship codes, service memorials continue to evoke a shared sense of nationalism, sacrifice, and honour. The power structures that maintain this net employ language and rituals linking European customs with North American innovations. Watson's death was located not simply in the contemporary moment, but in an evolving history of military service to the Crown throughout the empire. These rituals had honoured Isaac Brock and the dead of Ridgeway; they would later honour the Canadians who died in South Africa, Europe, and elsewhere. The performance of these rituals every 11 November is a mnemonic practice that also supports the narrative. So long as there are residents willing to engage these narratives and perpetuate these rituals, such monuments have the potential to remain relevant. Thus, even though 94 per cent of our survey participants in 2005 knew nothing about Watson or the North-West Resistance, many clearly understood the symbolic purpose of the monument.

The memory entrepreneurs who shaped the Watson Monument's placement and removal were St. Catharines city officials, the local militia, and social justice advocates. The memory entrepreneurs who rooted the wine country brand in local agriculture were Niagara regional officials and the wine industry. Their narrative was not communicated through plaques, classrooms, textbooks, and Heritage Minutes, but through branding expressed in labels, advertising, architecture, and tours. The narrative has two strands. One celebrates the innovation and chutzpah necessary to grow wine-quality grapes in Niagara. The second links this new industry to the pioneering agriculture of the first settlers. Innovation is highly prized, but the local heritage of soft-fruit farming lends an aura of authenticity to modern winery operations. It harnesses consumer desires for locally sourced goods and sustainable consumption to the romance of small-scale production and personal contact with charismatic vintners.

The wineries' mnemonic products consist of rural Niagara's natural and built landscapes. They provide visual place holders for narratives of soft-fruit farming and viticulture. In some cases, wineries have restored heritage buildings to house their operations.[5] Even without such buildings, however, they still benefit from the visual appeal of fertile farmlands. Their vineyards occupy fields that have been cultivated by settler farmers for two centuries. Moreover, the simple visibility of the wineries serves as a constant reminder of the narratives absorbed via advertising, visits, or word of mouth. A viewer may, then, perceive the vineyards and wineries as symbols of a timeless rural Niagara.

The wine industry also has its annual rituals. In mnemonic terms, a wine festival can function in ways like Remembrance Day. Clearly, November is a time when Canadians are asked to contemplate the tragic human cost of military engagement and the bravery of those who serve. Wine festivals, by contrast, are marketing events that foster revelry. For most tourists, a festival may be a one-time experience that is never revisited. For residents, however, a wine festival can be an annual experience that renews the individual's connection to the local community. Like fall fairs throughout Ontario, the annual Grape and Wine Festival invites Niagarans to celebrate the local harvest and participate in agricultural traditions that seem genuinely local. When done well, the festival is an anticipated event on the local calendar that is marked with family and friends. As a mnemonic practice, then, the festival can achieve something that is culturally meaningful for the local community while it builds consumer knowledge of winery brands. Participants in our surveys, from 2009 to 2016, indicated that this had been true in the past. Grape and Wine, more than any one winery, was the heart of their Niagara.

Just as the Watson Monument's form and narrative provoked a variety of responses, so, too, did the Wine Country brand. Not all survey participants equated vineyards with orchards or bought into the winery mystique. As we noted, attempts to brand places inevitably must focus on a few select regional attributes. In this case, various industry and government agencies sought increased integration of the grape and wine brand into the region's own wider marketing to promote economic growth. The brand was imprinted on the region's landscapes. Wine routes criss-cross the region. Where they pass through cities and small towns, municipal crews have refurbished streetscapes to attract urban, middle-class consumers in search of small-town charm. While this brand extension was not fully realized at the time of writing, vineyards and wineries had become increasingly symbolic of Niagara.[6]

The lived experience of Niagara has, then, become commodified.[7] These changes may affect residents' relationships with their own spaces. Some appreciate the changes prompted by the wine industry: the echoes of the region's agricultural heritage, the opportunity for local outings, and the potential for economic growth. Others, however, view them as inauthentic or intimidating.[8] Such residents may feel less engaged with local sites of identity and, by extension, may feel their sense of regional citizenship diminish. We described the tension surrounding the gentrification of the Grape and Wine Festival. Many residents believed this process sanitized their fall fair and masked the realities of its host city.[9] We also note that new residents attracted by Niagara's brand promise, a rural idyll, found themselves at odds with the dust, noise, and smells associated with agricultural operations.[10] Finally, the character of local farms was also changing. The conversion of agricultural lands to urban uses, as well as investments in agricultural intensification, had some farmers concerned.[11] While the average farm size was increasing through the 2010s, the number of farms was shrinking. Further, farm business succession – the movement of farms between generations – was decreasing.[12] These trends suggested family farms were being replaced by "agri-businesses focused on economies of scale and corporate policies."[13] The corporate inflection of the Niagara identity revealed that feelings of loss or alienation were genuine possibilities as a community's civic culture was deployed for private gain.

Be that as it may, the wine industry's branding resonated with a majority of survey participants. Traditional narratives of refugee settlement, military service, and feats of engineering simply did not. Both the branding and the traditional narratives are marked, ritualized, and encountered in similar ways, but that does not mean residents absorb them in similar ways. When, for example, Niagara was touched by the War in Afghanistan, many Niagarans honoured the dead by attending the funeral of Corporal Albert Storm, keeping Remembrance Day observances, and engaging in related mnemonic practices. Perhaps knowledge of these practices prompted the regret expressed by survey participants when they could not identify Watson's story. Regardless, while participants took lessons from the past, they lived in the present. This point was made abundantly clear by our commemograms. When participants prioritized wine over traditional narratives, then, we do not believe they were ignorant or indifferent to those narratives. Rather, we believe they found an appealing, usable past in wine that animated an appealing, usable present.

Scholars of culture remind us that individuals try to render their lives meaningful as best they can. To do so, they draw upon whatever symbolic resources are at their disposal to gain information, insight, or inspiration. Such resources form the basis of belonging. Individuals will select elements to construct coherent narratives about themselves that are consistent with their place and time.[14] These narratives will necessarily involve a consideration of pasts available to them – be they individual, familial, or communal.

For any one narrative to be widely useful, three factors come into play: accessibility, which can be fostered through articulate and authoritative sponsors; intertextual ubiquity through wide and varied dissemination; and resonance with contemporary experience. It is worth remembering that a narrative does not have to be officially sanctioned, academically tested, or even remotely true. To wit, the efficacy of interpersonal communication and passive learning was a recurring theme in our findings. Word of mouth, social media, marketing campaigns, and the landscape itself – both the natural and the built environment – conveyed significant mnemonic cues that could reinforce or undermine narratives distributed through channels conventionally considered more authoritative. This point comes into play whenever narratives are designed for specific constituencies and yet become available to all.

The wineries' narrative is useful to residents because it meets the test of authoritative sponsorship, ubiquity, and resonance. Linking the wineries to the history of tender fruit and agricultural heritage has naturalized wineries' location in the rural landscape. As our surveys indicated, this linkage confers an aura of legitimacy upon the residents' gaze, much as it may the tourists' gaze. It helps that the narrative is appealing. The wineries can portray themselves as the inheritors of Niagara's soft-fruit traditions, rescuing lost orchards through their conversion to vineyards. Their protected status in the provincially designated Greenbelt adds official sanction to their existing historic value. The vineyards and wineries therefore provide tangible reassurance that the region's agriculture is sustainable, and that its pastoral vistas will be maintained as the world beyond the region continues to change.[15]

More prosaically, residents do not have to accept the connection to the soft-fruit narrative to appreciate the wineries. The industry represents itself in ways that evoke a particular aesthetic and sensuality in contemporary terms: it can be elegant, tasteful, and edifying, or stylish, sexy, and fun. Either way, residents may see the industry as a success story in which they can participate. North of the escarpment, ready accessibility to wineries makes them convenient amenities for residents seeking not just wine but gifts, restaurants, and unique places to impress visiting family and friends.

We believe that authoritative sponsorship, ubiquity, and resonance also explain the weak relationship between Niagara's "official" mnemonic arcs and residents' "vernacular" understanding of the region. John Bodnar offered these terms to distinguish between the historical narratives promoted by official bodies and the narratives that the public retain and embrace.[16] While our survey participants knew that major events had occurred in the region, they did not commit the details to memory. Rather, as Anthony Giddens has noted, such narratives are consigned to a realm that still has be watched over, but with minimal care.[17] Material commemorations, in particular, had no impact on our participants' consciousness unless their narratives were recalled in contemporary experiences and were supported through intertextual referencing. Laura Secord, for example, was known as much for the chocolatier that bears her name as for her own patriotic efforts during the War of 1812. Without such relevance or intertextual support, many narratives were simply too inconsequential to daily life to be actively retained.

We end with a suggestion regarding mnemonic products occupying public space. In a provocative book, Bonnie Honig suggests that public things are at the foundation of democracies.[18] In her words,

> Without public things, we have nothing or not much to deliberate about, constellate around, or agonistically contest ...
>
> Public things are part of the "holding environment" of democratic citizenship; they furnish the world of democratic life. They do not take care of our needs only. They also constitute us, complement us, limit us, thwart us, and interpellate us into democratic citizenship.[19]

This argument encompasses public mnemonic products and the spaces they inhabit. They emerge from relations of power. Their creation and maintenance are often tautological; memory entrepreneurs have a specific purpose in erecting them. The removal of such products from public space evokes a similar power.[20] At stake in both cases are claims on the public gaze and, ultimately, the resources residents may employ to construct identity and belonging.

Editing a plaque or removing a monument does little to eliminate structural or systemic marginalization. Alex Barker rightly notes that removing monuments does not erase history; the past is not so easily rectified or reconciled.[21] According to Honig's logic, an event – like removing a monument – must become a sustainable movement.[22]

Sustainable social change can only emerge through repeated public engagement with the troubling narratives themselves. Thus, removal of the Watson Monument may have brought the racist roots of the North-West Resistance home to local residents, but were residents also prompted to act on the findings of the Truth and Reconciliation Commission and to engage other social inequities? Despite the recent reframing of mnemonic landscapes across Canada, many Indigenous communities throughout the country still lacked basic infrastructure and continued to face systemic and structural discrimination.[23] In Niagara, a 2020 report contended that Indigenous persons composed 24.3 per cent of the region's homeless residents, despite making up only 3 per cent of its entire population.[24]

Some scholars argue that public mnemonic products can facilitate reconciliation and healing.[25] James E. Young once described the concept of "counter-monuments": "brazen, painfully self-conscious memorial spaces conceived to challenge the very premises of their being."[26] Their forms subvert the triumphal form of most monuments. They often connote weakness and complexity rather than strength and single-mindedness.[27] The Louis Riel statue at St. Boniface College, Winnipeg, and the American Vietnam Veterans Memorial in Washington, DC, provide examples. Erika Doss as well as William Logan and Keir Reeves have since argued that memorials evoking shame, anger, and trauma are an increasingly important facet of democratic nationhood as they undermine traditional, monolithic, and whiggish national narratives.[28] A public reorientation of mnemonic products may bear witness to the past and prod viewers via their cognitive and emotional responses to take responsibility for a troubling past.[29] In the case of Germany, Young writes,

> In fact, the best German memorial to the Fascist era and its victims may not be a single memorial at all, but simply the never to be resolved debate over which kind of memory to preserve, how to do it, in whose name, and to what end. Instead of a fixed figure for memory, the debate itself – perpetually unresolved amid ever-changing conditions – might be enshrined.[30]

The goal is to foster healing through public dialogue.[31] Can a mnemonic product accomplish this? Perhaps. Sustained dialogue, however, requires that its narrative be sponsored, ubiquitous, and resonant. It must be in public view through all the means of public communication available to contemporary society. Within this intertextual net, however, the mnemonic product will remain only one source of information among many, and available only to those to whom it remains visible. Any new or recontextualized mnemonic product would require a culturally and institutionally sustained effort to overcome the kinds of public indifference or economic, cultural, and political cleavages that the Watson Monument revealed.

In sum, memory making sits at the interstices of mnemonic products and the minds of the people who engage them. Memory entrepreneurs can champion specific narratives, but residents clearly do not internalize every narrative marked in the landscape. In Niagara, many residents paid no more than a passing glance to mnemonic products. Thus, a mnemonic landscape may serve only as a symbolic reduction of a complex narrative that has been learned through an intertextual set of resources. An earlier statement bears repeating: a monument may function as a prompt to remember, but it cannot prompt what was never known.

Notes

Introduction

1 "Welcome to the Salem Chapel Sanctuary of History," Salem Chapel BME Church Harriet Tubman Underground Railroad National Historic Sites, accessed 11 June 2018, http://salemchapelbmechurch.ca/index.html. Local ambivalence to the plight of African Americans was clearly articulated in this editorial from a St. Catharines–based paper: "The Emancipation Proclamation," *The Constitutional*, 8 January 1863, 2. See also Daniel G. Hill, "Early Black Settlement in the Niagara Peninsula," in *Immigration and Settlement in the Niagara Peninsula*, ed. John Burtniak and Patricia G. Dirks (St. Catharines: Brock University, 1981), 65–80; Dann J. Broyld, "'Justice Was Refused Me, I Resolved to Free Myself': John W. Lindsay. Finding Elements of American Freedoms in British Canada, 1805–1876," *Ontario History* 109, no. 1 (2017): 27–59, https://doi.org/10.7202/1039198ar.

2 St. Catharines descendants of Underground Railroad travellers prefer the term "Freedom Seekers" to "escaped slaves."

3 See Jeffery Olick, *The Sins of the Fathers: Germany, Memory, Method* (Chicago: University of Chicago Press, 2016); Astrid Erll, *Memory and Culture: A Semiotic Model*, trans. Sara B. Young (Houndsmills, UK: Palgrave McMillan, 2011); Anna Green, "Can Memory Be Collective?," in *The Oxford Handbook of Oral History*, ed. Donald A. Ritchie (New York: Oxford University Press, 2011), 96–111.

4 Maurice Halbwachs, *On Collective Memory*, ed. and trans. Lewis A. Coser (Chicago: University of Chicago Press, 1992).

5 Olick, *Sins of the Fathers*, 43. Emphasis in the original.

6 Laurajane Smith, *Uses of Heritage* (London: Routledge: 2006).

7 Smith, 46. Emphasis in the original.

8 Olick, *Sins of the Fathers*.

9   Pierre Nora, "Between Memory and History: *Les Lieux de Mémoire*," *Representations* 26 (April 1989): 7–24, https://doi.org/10.2307/2928520.

10 Brian S. Osborne, "Constructing Landscapes of Power: The George Etienne Cartier Monument, Montreal," *Journal of Historical Geography* 24, no. 4 (October 1998): 431–58, https://doi.org/10.1006 /jhge.1998.0090; Brian S. Osborne, "Landscape, Memory, Monuments, and Commemoration: Putting Identity in Its Place," *Canadian Ethnic Studies* 33, no. 3 (Fall 2001): 39–76.

11 "Census Profile, 2016 Census: St. Catharines-Niagara [Census Metropolitan Area] Ontario," Statistics Canada, accessed 28 February 2020, https:// www12.statcan.gc.ca/census-recensement/2016/dp-pd/prof/details/page.cf m?Lang=E&Geo1=CMACA&Code1=539&Geo2=PR&Code2=35&SearchText =st.%20catharines%20niagara&SearchType=Begins&SearchPR=01&B1=All& TABID=1&type=0.

12 Christin Köber and Tilmann Habermas, "How Stable Is the Personal Past? Stability of Most Important Autobiographical Memories and Life Narratives across Eight Years in a Life Span Sample," *Journal of Personality and Social Psychology* 113, no. 4 (October 2017): 608–26, https://psycnet .apa.org/doi/10.1037/pspp0000145. See also Jefferson Singer, Pavel Blagov, Meredith Berry, and Kathryn Oost, "Self-Defining Memories, Scripts, and the Life Story: Narrative Identity in Personality and Psychotherapy," *Journal of Personality* 81, no. 6 (December 2013): 569–81, https://doi.org/10.1111/jopy.12005.

13 Rauf Garagozov, "Painful Collective Memory: Measuring Collective Memory Affect in the Kabakh Conflict," *Peace and Conflict: Journal of Peace Psychology* 22, no. 1 (February 2016): 28, https://psycnet.apa.org /doi/10.1037/pac0000149.

14 Halbwachs, *On Collective Memory*. See also, Olick, *Sins of the Fathers*.

15 Alexandru Cuc, Yasuhiro Ozura, David Manier, and William Hirst, "On the Formation of Collective Memory: The Role of a Dominant Narrator," *Memory and Cognition* 34, no. 4 (June 2006): 752–63, https://doi .org/10.3758/bf03193423.

16 Michel Foucault, *Power/Knowledge: Select Interviews and Other Writings, 1972–1977*, ed. Colin Gordon (New York: Pantheon Books: 1980).

17 Edward Said, "Invention, Memory, and Place," *Critical Inquiry* 26, no. 2 (Winter 2000): 179.

18 Elizabeth Jelin, *State Repression and the Labors of Memory* (Minneapolis: University of Minnesota Press, 2003).

19 Jennifer R. Nájera, "Remembering Migrant Life: Family Collective Memory and Critical Consciousness in the Midcentury Migrant Stream," *Oral History Review*, 45, no. 2 (Summer/Fall 2018): 211–31, https://doi .org/10.1093/ohr/ohy037.

20 Luigi Cajani, "Historians between Memory Wars and Criminal Laws: The Case of the European Union," *Yearbook of the International Society of History Didactics / Jahrbuch Der Internationalen Gesellschaft Für Geschichtsdidaktik* 28/29 (2009): 39–55.

21 E.B. Perlman, "The Role of an Archivist in Shaping Collective Memory on Kibbutz: Through Her Work on the Photographic Archive," *Journal of Visual Literacy* 30, no. 1 (2011): 1–18, https://doi.org/10.1080/23796529. 2011.11674682; Zinaida Manzuch, "Archives, Libraries and Museums as Communicators of Memory in the European Union Project," *Information Research: An International Electronic Journal* 14, no. 2 (June 2009): 400–24; Richard Harvey Brown and Beth Davis-Brown, "The Making of Memory, the Politics of Archives, Libraries and Museums in the Construction of National Consciousness," *History of the Human Sciences* 11, no. 4 (November 1998): 17–32, https://doi.org/10.1177/095269519801100402.

22 William Ocasio, Michael Mauskapf, and Christopher W.J. Steele, "History, Society, and Institutions: The Role of Collective Memory in the Emergence and Evolution of Societal Logics," *Academy of Management Review* 41, no. 4 (October 2016): 676–99, https://doi.org/10.5465/amr.2014.0183; Ali Usman Qasmi, "Identity Formation through National Calendar: Holidays and Commemorations in Pakistan," *Nations & Nationalism* 23, no. 3 (July 2017): 620–41, https://doi.org/10.1111/nana.12310; Scott Boehm, "Privatizing Public Memory: The Price of Patriotic Philanthropy and the Post-9/11 Politics of Display," review of *The Price of Freedom: Americans at War*, by David Allison, Howard Morrison, Dik Daso, Barton Hacker, Jennifer Jones, *American Quarterly* 58 no. 4 (December 2006): 1147–66.

23 Margaret Conrad, Kadriye Ercikan, Gerald Friesen, Jocelyn Letourneau, Delphine Muise, David Northrup, and Peter Seixas, *Canadians and Their Pasts* (Toronto: University of Toronto Press, 2013).

24 Kenneth E. Foote and Maoz Azaryahu, "Toward a Geography of Memory: Geographical Dimensions of Public Memory and Commemoration," *Journal of Political & Military Sociology* 35, no. 1 (Summer 2007): 125–44.

25 Eviatar Zerubavel, *Time Maps: Collective Memory and the Social Shape of the Past* (Chicago: University of Chicago Press, 2003), 3.

26 L. Churchill, J.K. Yamashiro, and Henry L. Roediger III, "Moralized Memory: Binding Values Predict Inflated Estimates of the Group's Historical Influence," *Memory* 27, no. 8 (September 2019): 1099–109, https://doi.org/10.1080/09658211.2019.1623261; Liljana Siljanovska, "Mass Media and Cultural Memory: Idealization of Values," *Kultura (Skopje)* 4, no. 7 (2014): 113–21.

27 David Maples, *Heroes and Villains: Creating National History in Contemporary Ukraine* (Budapest: Central European University Press, 2007); Sam Edwards, *Allies in Memory: World War II and the Politics of Transatlantic*

Commemoration, c. 1941–2001 (Cambridge: Cambridge University Press, 2015); Ana Claudia Marques, "Founders, Ancestors, and Enemies: Memory, Family, Time, and Space in the Pernambuco Sertão," *Journal of the Royal Anthropological Institute* 19, no. 4 (December 2013): 716–33, https://doi.org/10.1111/1467-9655.12061.

28 Charles W.J. Withers, "Place, Memory, Monument: Memorializing the Past in Contemporary Highland Scotland," *Cultural Geographies* 3, no. 3 (July 1996): 325–44, https://doi.org/10.1177/147447409600300304; Elizabeth Crooke, "Confronting a Troubled History: Which Past in Northern Ireland's Museums?," *International Journal of Heritage Studies* 7, no. 2 (June 2001): 119–36, https://doi.org/10.1080/713772347.

29 Sibylle Puntscher, Christoph Hauser, Karin Pichler, and Gottfried Tappeiner, "Social Capital and Collective Memory: A Complex Relationship," *Kyklos* 67, no. 1 (February 2014): 117, https://doi.org/10.1111/kykl.12046.

30 Eleftherios Klerides and Michalinos Zembylas, "Identity as Immunology: History Teaching in Two Ethnonational Borders of Europe," *Compare: A Journal of Comparative and International Education* 47, no. 3 (May 2017): 416–33, https://doi.org/10.1080/03057925.2017.1292847.

31 See Teresa Brennan, *The Transmission of Affect* (Ithaca, NY: Cornell University Press: 2004).

32 Kaitlin Murphy, *Mapping Memory: Visuality, Affect, and Embodied Politics in the Americas* (New York: Fordham University Press, 2018).

33 Kathleen Stewart, *Ordinary Affects* (Durham, NC: Duke University Press, 2007), 2. Emphasis in the original.

34 Margaret Wetherell, *Affect and Emotion: A New Social Science Understanding* (Los Angeles: Sage, 2012), 3.

35 Laurajane Smith and Gary Campbell, "The Elephant in the Room: Heritage, Affect and Emotion," in *A Companion to Heritage Studies*, ed. William Loban, Máiréad Nic Craith, and Ullrich Kockel (Chichester, UK: Wiley-Blackwell, 2015), 443–60, https://doi.org/10.1002/9781118486634.ch30. See also Owen J. Dwyer and Derik H. Alderman, *Civil Rights Memorials and the Geography of Memory* (Athens: University of Georgia Press, 2008); Conrad et al., *Canadians and Their Pasts*.

36 Tim Cresswell, *Maxwell Street: Writing and Thinking Place* (Chicago: University of Chicago Press, 2019), 94.

37 Joan Nogué and Jordi de San Eugenio Vela, "Geographies of Affect: In Search of the Emotional Dimension of Place Branding," *Communication and Society* 31, no. 4 (December 2018): 28, https://doi.org/10.15581/003.31.4.27-42.

38 Sara Ahmed, "Affective Economies," *Social Text* 22, no. 2 (Summer 2004): 117–39, https://doi.org/10.1215/01642472-22-2_79-117.

39 Rumi Sakamoto, "Mobilizing Affect for Collective War Memory: Kamikaze Images in Yūshūkan," *Cultural Studies* 29, no. 2 (March 2015): 158–84, https://doi.org/10.1080/09502386.2014.890235.

40 Tuuli Lähdesmäki, "Politics of Affect in the EU Heritage Policy Discourse: An Analysis of Promotional Videos of Sites Awarded with the European Heritage Label," *International Journal of Heritage Studies* 22, no. 8 (2017): 711, https://doi.org/10.1080/13527258.2017.1317649.

41 Eviatar Zerubavel, *Taken for Granted: The Remarkable Power of the Unremarkable* (Princeton, NJ: Princeton University Press, 2018), 22, quoting Karsten Hundeide, "The Tacit Background of Children's Judgments," in *Culture, Communication, and Cognition: Vygotskian Perspectives*, ed. James V. Wersch (Cambridge: Cambridge University Press, 1985), 311. Emphasis in the original.

42 C. Thi Nguyen, "Monuments as Commitments: How Art Speaks to Groups and How Groups Think in Art," *Pacific Philosophical Quarterly* 100, no. 4 (December 2019): 971–94, https://doi.org/10.1111/papq.12279.

43 Ahmed, "Affective Economies."

44 See Ahmed, "Affective Economies"; Sakamoto, "Mobilizing Affect for Collective War Memory"; Carol X. Zhang, Honggen Xiao, Nigel Morgan, and Tuan Phong Ly, "Politics of Memories: Identity Construction in Museums," *Annals of Tourism Research* 73 (November 2018): 116–30, https://doi.org/10.1016/j.annals.2018.09.011.

45 Zerubavel, *Taken for Granted*, 14.

46 Manuel Castells, *The Information Age: Economy, Society and Culture*, vol. 2, *The Power of Identity* (London: Blackwell, 1997), 7.

47 Anthony Giddens, *Modernity and Self-Identity: Self and Society in the Late Modern Age* (Stanford, CA: Stanford University Press, 1991), 188.

48 See Foucault, *Power/Knowledge*; Anthony Giddens, *The Constitution of Society: Outline of the Theory of Structuration* (Berkeley: University of California Press, 1984).

49 Conrad et al., *Canadians and Their Pasts*.

50 Smith and Campbell, "The Elephant in the Room."

51 Michael Billig, *Banal Nationalism* (London: Sage, 1995).

52 Billig, 41.

53 See the Toppled Monuments Archive, available at https://www.toppledmonumentsarchive.org.

54 John Bodnar, *Remaking America: Public Memory, Commemoration, and Patriotism in the Twentieth Century* (Princeton, NJ: Princeton University Press, 1992).

55 Roy Rosenzweig and David Thelen, *The Presence of the Past: Popular Uses of History in American Life* (New York: Columbia University Press, 1998); Paul Ashton and Paula Hamilton, *History at the Crossroads: Australians and*

the Past (Sydney: Halstead Press, 2010); Conrad et al., *Canadians and Their Pasts*

56 Ashton and Hamilton, *History at the Crossroads*; Conrad et al., *Canadians and Their Pasts.*

57 Apologies to Anthony Giddens, *Modernity and Self-Identity.*

58 Zerubavel, *Time Maps.*

1. Assessing Public Engagement with Historical Narratives in Niagara

1 Edward W. Miller, excerpt from "On the Erection of a Monument on the Battlefield of Lundy's Lane," *The Tribune*, 2 August 1895, posted by Andrew Porteus to Niagara Falls Poetry Project, 14 December 2017, https://niagarapoetry.ca/2017/12/14/miller/.

2 Robert Malcomson, *A Very Brilliant Affair: The Battle of Queenston Heights, 1812* (Toronto: Robin Bass Studio, 2003).

3 Stephen Strauss, "Greatest Canadian List Pared Down," *Globe and Mail*, 18 October 2004, https://www.theglobeandmail.com/arts/greatest -canadian-list-pared-down/article1142200.

4 Keith Basso, *Wisdom Sits in Places: Landscape and Language among the Western Apache* (Albuquerque: University of New Mexico Press, 1996).

5 Zinaida Manzuch, "Archives, Libraries and Museums as Communicators of Memory in the European Union Project," *Information Research: An International Electronic Journal* 14, no. 2 (June 2009): 400–24; Richard Harvey Brown and Beth Davis-Brown, "The Making of Memory, the Politics of Archives, Libraries and Museums in the Construction of National Consciousness," *History of the Human Sciences* 11, no. 4 (November 1998): 17–32, https://doi.org/10.1177/095269519801100402; Kirk Savage, *Monument Wars: Washington D.C., the National Mall and the Transformation of the Memorial Landscape* (Berkeley: University of California Press, 2009); Sanford Levinson, *Written in Stone: Public Monuments in Changing Societies* (Durham, NC: Duke University Press, 1998); Sergiusz Michalski, *Public Monuments: Art in Political Bondage, 1870–1997* (London: Reaktion Books, 1998); Robert Hayashi, "Transfigured Patterns: Contesting Memories at the Manzanar National Historic Site," *Public Historian* 25, no. 4 (Fall 2003): 51–71, https://doi.org/10.1525/tph.2003.25.4.51; Adam H. Domby, "Captives of Memory: The Contested Legacy of Race at Andersonville National Historic Site," *Civil War History* 63, no. 3 (September 2017): 253–94, https://doi.org/10.1353/cwh.2017.0037; Kate Hawkey, "Whose History Is This Anyway: Social Justice and a History Curriculum," *Education, Citizenship and Social Justice* 10, no. 3 (November 2015): 187–98, https:// doi.org/10.1177/1746197915583938; Keffrelyn Brown, "Race, Racial

Cultural Memory and Multicultural Curriculum in an Obama 'Post-Racial U.S.,'" *Race, Gender and Class* 18, nos. 3–4 (2011): 123–34; Jakob Bronec, "Transmission of Collective Memory and Jewish Identity in Post-War Jewish Generations through War Souvenirs," *Heritage* 2, no. 3 (September 2019): 1785–98, https://doi.org/10.3390/heritage2030109.

6 Michalski, *Public Monuments.*

7 Michalski, 27.

8 Erika Doss, *Memorial Mania: Public Feeling in America* (Chicago: University of Chicago Press, 2010).

9 M. Christine Boyer, *The City of Collective Memory: Its Historical Imagery and Architectural Entertainments* (Cambridge, MA: MIT Press, 1996); David Birdsell and Leo Groarke, "Toward a Theory of Visual Argument," *Argumentation and Advocacy* 33, no. 1 (Summer 1996): 1–10; Anthony Blair, "The Possibility and Actuality of Visual Arguments," *Argumentation and Advocacy* 33, no. 1 (Summer 1996): 23–39; Cara Finnegin, "Recognizing Lincoln: Image Vernaculars in Nineteenth Century Visual Culture," *Rhetoric and Public Affairs* 8, no. 1 (Spring 2005): 31–58, https://doi.org/10.1353/RAP.2005.0037; Cara Finnegin, "The Naturalistic Enthymeme and Visual Argument: Photographic Representation in the 'Skull Controversy,'" *Argumentation and Advocacy* 37, no. 3 (Winter 2001): 133–9, https://doi.org/10.1080/00028533.2001.11951665.

10 James Young, "The Counter-Monument: Memory against Itself in Germany Today," *Critical Inquiry* 18, no. 2 (Winter 1992): 270, https://doi.org/10.1086/448632.

11 Savage, *Monument Wars*; Hamzah Muzaini and Brenda Yeoh, "Memory-Making 'from Below': Rescaling Remembrance at the Kranji War Memorial and Cemetary, Singapore," *Environment and Planning A: Economy and Space* 39, no. 6 (June 2007): 1288–1305, https://doi.org/10.1068/a3862.

12 Rebecca Clare Dolgoy and Jerzy Elżanowski, "Working through the Limits of Multidirectional Memory: Ottawa's Memorial to the Victims of Communism and National Holocaust Monument," *Citizenship Studies* 22, no. 4 (May 2018): 433–51, https://doi.org/10.1080/13621025.2018.1462507; Joanne Chianello, "Feds to Shell Out $4M More on Victims of Communism Memorial," *CBC News*, 19 April 2021, https://www.cbc.ca/news/canada/ottawa/victims-communism-memorial-budget-1.5993395.

13 Elizabeth Jelin, *State Repression and the Labors of Memory* (Minneapolis: University of Minnesota Press, 2003); Erin Kaipainen, "Graffiti, Memory and Contested Space: Mnemonic Initiatives following Trauma and/ or Repression in Buenos Aires, Argentina" (master's thesis, Brock University, 2007); Rodrigo Navarrete Sánchez and Ana Maria López, "Scratching Behind the Walls: Graffiti and Symbolic Political Imagination at Cuartel San Carlos (Caracas, Venezuala)," in *Memories from the Darkness:*

Archaeology of Repression and Resistance in Latin America, Contributions to a Global Historical Archaeology, ed. Pedro Funari, Andres Zarankin, and Melissa Salerno (New York: Springer, 2009), https://doi.org/10.1007/978-1-4419-0679-3_8.

14 Ntsikelelo B. Breakfast, Gavin Bradshaw, and Richard Haines, "Attacks on South African Monuments: Mediating Heritage in Post-Conflict Society," *Africa's Public Service Delivery and Performance Review* 6, no. 1 (December 2018): e1–e12, https://doi.org/10.4102/apsdpr.v6i1.184.

15 Jurij FikFak, "Cultural and Social Representations on the Border: From Disagreement to Coexistence," *Human Affairs* 19, no. 4 (December 2009): 350–62, https://doi.org/10.2478/v10023-009-0049-1.

16 Carola Lentz, "Ghanaian 'Monument Wars': The Contested History of the Nkrumah Statues," *Cahiers d'Études africaines* 57, no. 3 [227] (2017): 551–82, https://doi.org/10.4000/etudesafricaines.20822.

17 Brian S. Osborne, "Corporeal Politics and the Body Politic: The Representation of Louis Riel in Canadian Identity," *International Journal of Heritage Studies* 8, no. 4 (December 2002): 308, https://doi.org/10.1080/1352725022000037209. See also H.V. Nelles, *The Art of Nation-Building* (Toronto: University of Toronto Press, 1999).

18 Clifton Hood, "An Unusable Past: Urban Elites, New York City's Evacuation Day, and the Transformations of Memory Culture," *Journal of Social History* 37, no. 4 (Summer 2004): 884–913, https://doi.org/10.1353/jsh.2004.0050. See also Eviatar Zerubavel, *Time Maps: Collective Memory and the Social Shape of the Past* (Chicago: University of Chicago Press, 2003); N.N. Korzh, "Representation of Historical Knowledge in Collective Memory," *Journal of Russian and Eastern European Psychology* 39, no. 3 (May 2001): 69–83, https://doi.org/10.2753/RPO1061-0405390369; Sharon Macdonald, "Undesirable Heritage: Fascist Material Culture and Historical Consciousness in Nuremberg," *International Journal of Heritage Studies* 12, no. 1 (January 2006): 9–28, https://doi.org/10.1080/13527250500384464; Sarah J. Purcell, "Commemoration, Public Art, and the Changing Meaning of the Bunker Hill Monument," *Public Historian* 25, no. 2 (Spring 2003): 55–71, https://doi.org/10.1525/tph.2003.25.2.55; Yvonne Whelan, "The Construction and Deconstruction of a Colonial Landscape: Monuments to British Monarchs in Dublin before and after Independence," *Journal of Historical Geography* 28, no. 4 (October 2002): 508–33, https://doi.org/10.1006/jhge.2002.0441.

19 Osborne, "Corporeal Politics and the Body Politic," 308.

20 Robert Musil, "Monuments," in *Posthumous Papers of a Living Author*, trans. Peter Wortsman (Brooklyn: Archipelago Books, 2006), 64.

21 Sarah Rankin and David Crary, "Countries Seek 'New History' as Figures Are Reexamined after George Floyd's Death," *Global News*, 12 June 2020,

last updated 23 June 2020, https://globalnews.ca/news/7058287/us
-statues-monuments-racism/; Jon Quelly, "New Interactive Map Details
67 Confederate Monuments (and Counting) Removed since George Floyd
Murder," *Common Dreams*, 10 July 2020, https://www.commondreams.
org/news/2020/07/10/new-interactive-map-details-67-confederate
-monuments-and-counting-removed-george.

22 See plates 1–15 in R. Cole Harris, ed., *The Historical Atlas of Canada: From the
Beginning to 1800*, vol. 1 (Toronto: University of Toronto Press, 1987). See also
Helen Hornbeck Tanner, ed., *Atlas of Great Lakes Indian History* (Norman:
University of Oklahoma Press, 1987); Olive Patricia Dickason, *Canada's First
Nations: A History of Founding Peoples from Earliest Times*, 2nd ed. (Toronto:
Oxford University Press, 1997); Bruce Trigger, *Children of Aataentsic: A
History of the Huron People to 1660* (Montreal: McGill-Queen's University
Press, 1988); Conrad E. Heidenreich, *Huronia: A History and Geography of the
Huron Indians, 1600–1650* (Toronto: McClelland and Stewart, 1971).

23 D. Brian Deller, Christopher Ellis, and Merle Franklin, "The Rogers Site:
An Early PaleoIndian Site in the Niagara Peninsula Region of Ontario,"
Archaeology of Eastern North America 46 (2018): 103–34.

24 J.V. Wright, "Iroquoian Agricultural Settlement," in Harris, *Historical
Atlas of Canada*, plate 12; Ronald F. Williamson and Ronald I. MacDonald,
Legacy of Stone: Ancient Life on the Niagara Frontier (Toronto: East End
Books, 1998), 1–25, 132–40.

25 See, for example, Mary Jackes, "The Mid Seventeenth Century Collapse
of Iroquoian Ontario: Examining the Last Burial Place of the Neutral
Nation," in *Vers une anthropologie des catastrophes : 9ᵉ Journées anthropologie
de Valbonne*, ed. Luc Buchet, Catharin Rigede, Isabelle Séguy, and Michel
Signoli (Antibes: Éditions APDCA/INED, 2008), 347–73; Marian White,
"On Delineating the Neutral Iroquois of the Eastern Niagara Peninsula of
Ontario," *Ontario Archaeology*, no. 17 (1972): 62–74.

26 Betty E. Eley and Peter H. von Bitter, *Cherts of Southern Ontario* (Toronto:
Royal Ontario Museum, 1989).

27 Williamson and MacDonald, *Legacy of Stone*, 133.

28 Helen Hornbeck Tanner, "The Iroquois Wars, 1641–170," in *Atlas of Great
Lakes Indian History*, ed. Helen Hornbeck Tanner, cartography by Miklos
Pinther (Norman: University of Oklahoma Press, 1987), 29–35; Peter
Schmalz, *The Ojibwa of Southern Ontario* (Toronto: University of Toronto
Press, 1991); Leroy V. Eid, "The Ojibwa-Iroquois War: The War the Five
Nations Did Not Win," *Ethnohistory* 26, no. 4 (Autumn 1979): 297–324,
https://doi.org/10.2307/481363; Victor Konrad, "An Iroquois Frontier:
The North Shore of Lake Ontario during the Late Seventeenth Century,"
Journal of Historical Geography 7, no. 2 (April 1981): 129–44, https://doi
.org/10.1016/0305-7488(81)90116-X.

29 Trigger, *Children of Aataentsic*, 398–402; Helen Hornbeck Tanner, "Distribution of Late Prehistoric Cultures c. 1400–1600," in Tanner, *Atlas of Great Lakes Indian History*, 24–8.

30 Frank H. Severance, *An Old Frontier of France: The Niagara Region and Adjacent Lakes under French Control*, vol. 1 (New York: Dodd, Mead, 1917), 13–22; J.V. Wright, "Iroquoian Agricultural Settlement."

31 Jackes, "Mid Seventeenth Century Collapse of Iroquoian Ontario," 349, following P.F.X. de Charlevoix, *History and General Description of New France*, trans. from the Paris ed. of 1744, vol. 1, trans. John Gilmary Shea (Chicago: Loyola University Press, 1870), 265; Reuben Gold Thwaites, ed., *The Jesuit Relations and Allied Documents*, vol. 7 (Cleveland, OH: Burrows Brothers, 1897), 223.

32 Alun Hughes, "On the Meaning of Niagara," in *History Made in Niagara*, ed. Michael Ripmeester, David Butz, and Loris Gasparotto (St. Catharines: Elbow Island Publishing, 2019), 263–74.

33 George R. Stewart, *Names on the Land: A Historical Account of Place-Naming in the United States* (Boston: Houghton Mifflin Company, 1967), 83; Trigger, *Children of Aataentsic*; Robert S. Allen, *His Majesty's Indian Allies: British Indian Policy in the Defence of Canada, 1774–1815* (Toronto: Dundurn, 1992), 12–38; Jackes, "Mid Seventeenth Century Collapse of Iroquoian Ontario"; Tanner, "Iroquois Wars, 1641–170," 29–35.

34 Robert West Howard, *Thundergate: The Forts of Niagara* (Englewood Cliffs, NJ: Prentice Hall, 1968); Michael Ripmeester, "The Development of a British Landscape at Niagara, 1759–1765," in *Historic Landscape Preservation*, ed. Nancy Pollock-Ellwand (Waterloo: University of Waterloo Heritage Resources Centre, 2002).

35 Severance, *An Old Frontier of France*, 1:23–35, 2:275–350; John N. Jackson, *St. Catharines Ontario: Its Early Years* (Belleville: Mika, 1976), 61–82; Canada, *Treaty Texts–Upper Canada Land Surrenders, Niagara Treaty of 1781, No. 381* (head of Lake Ontario, August 1764), and *Between the Lakes Purchase and Collins Purchase, No. 3* (Lincoln County, 7 December 1792), retrieved from the Government of Canada, last modified 7 March 2016, accessed 23 March 2022, https://www.rcaanc-cirnac.gc.ca/eng/13703721 52585/1581293792285.

36 See Donald B. Smith, *Sacred Feathers: The Reverend Peter Jones (Kahkewaquonaby) and the Mississauga Indians* (Lincoln: University of Nebraska Press, 1987); Donald B. Smith, "The Dispossession of the Mississauga Indians: A Missing Chapter in the Early History of Upper Canada," in *Historical Essays on Upper Canada: New Perspectives*, ed. Bruce Wilson and J.K. Johnson (Ottawa: Carleton University Press, 1989), 23–51.

37 Jeffrey Shantz, "Struggle for the Land: Racism, Class, and the Six Nations Land Reclamation," *New Politics* 12, no. 1 (Summer 2008): 83–8; Theresa

McCarthy, *In Divided Unity: Haudenosaunee Reclamation at Grand River* (Tucson: University of Arizona Press, 2016).

38 John Leslie and Ron Maguire, eds., *The Historical Development of the Indian Act*, 2nd ed. (Ottawa: Ministry of Indian and Northern Affairs, 1979), 1–12; Elizabeth Elbourne, "Broken Alliance: Debating Six Nations' Land Claims in 1822," *Cultural and Social History 9*, no. 4 (December 2012): 497–525, https://doi.org/10.2752/147800412X13434063754445.

39 William Denevan, "After 1492: Nature Rebounds," *Geographical Review 106*, no. 3 (July 2016): 381–98, https://doi.org/10.1111/j.1931-0846.2016.12175.x.

40 R. Louis Gentilcore, "The Beginnings of Settlement in the Niagara Peninsula (1782–1792)," *Canadian Geographer 7*, no. 2 (June 1963): 72–82, https://doi.org/10.1111/j.1541-0064.1963.tb00315.x; Alun Hughes, "Lines on the Land," in Ripmeester, Butz, and Gasparotto, *History Made in Niagara*, 17–20; Daniel G. Hill, "Early Black Settlement in the Niagara Peninsula," in *Immigration and Settlement in the Niagara Peninsula: Proceedings, Third Annual Niagara Peninsula History Conference, Brock University, 25–26 April 1981*, ed. John Burtniak and Patricia G. Dirks (St. Catharines: Brock University, 1981): 65–80.

41 John Jackson, *Names across Niagara* (St. Catharines: Vanwell Publishing, 1989); Andrew F. Burghardt, "The Origin and Development of the Road Network of the Niagara Peninsula, Ontario, 1770–1851," *Annals of the Association of American Geographers 59*, no. 3 (September 1969): 417–40, https://doi.org/10.1111/j.1467-8306.1969.tb00683.x.

42 Colin Read, "The Short Hills Raid of June, 1838, and Its Aftermath," *Ontario History 68*, no. 2 (1968): 93–115.

43 Robert Malcolmson, *Burying General Brock: A History of Brock's Monuments* (Niagara-on-the-Lake: Friends of Fort George, 1996).

44 Peter Vronsky, *Ridgeway: The American Fenian Invasion and the 1866 Battle that Made Canada* (Toronto: Allen Lane/Penguin, 2011).

45 Alun Hughes, "Terrorist Attacks on the Welland Canal," in Ripmeester, Butz, and Gasparotto, *History Made in Niagara*, 145–60.

46 See Peter Gideon Prins, "Group Preferences for Rural Amenities and Farmland Preservation in the Niagara Fruit Belt" (master's thesis, University of Waterloo, 2005); John N. Jackson, *Management of Rural Resources the Niagara Peninsula* (St. Catharines: Brock University, 1985); Hugh Gayler, *Niagara's Changing Landscapes* (Ottawa: Carleton University Press, 1994).

47 See John N. Jackson, *The Welland Canals and Their Communities: Engineering, Industrial, and Urban Transformation* (Toronto: University of Toronto Press, 1997).

48 Wendy Haaf, "Canada's Top Retirement Destinations: St. Catharines, ON," *Good Times: Canada's Magazine for Successful Retirement*, 6 January 2020, https://goodtimes.ca/canadas-top-retirement-destinations-st -catharines-on/.

49 Adam Bisby, "Go Train Fuels Niagara Housing Boom," *National Post*, 4 July 2019, A1; Allen Benner, "Unemployment Rate Increases in Niagara," *Niagara Falls Review*, 10 July 2017, https://www. niagarafallsreview.ca/news/niagara-region/2017/07/10/ unemployment-rate-increases-in-niagara.html; Allan Benner, "Niagara's Living Wage Reaches $18.12 Hour," *The Standard* (St. Catharines), 5 November 2019, A3.

50 M. Christine Boyer, *The City of Collective Memory*; Paul Butterfield, "Monuments and Memories: What History Can Teach the Architects at Ground Zero," *New Republic*, 3 February 2003, 27–32; Miguel de Oliver, "Historical Preservation and Identity: The Alamo and the Production of a Consumer Landscape," *Antipode* 28, no. 1 (January 1996): 1–23, https://doi.org/10.1111/j.1467-8330.1996.tb00669.x; Erika Doss, "Death, Art, and Memory in the Public Sphere: The Visual and Material Culture of Grief in Contemporary America," *Mortality* 7, no. 1 (March 2002): 63–82, https://doi.org/10.1080/13576270120102553; Sarah Bennett Farmer, "Oradour-sur-Glane: Memory in a Preserved Landscape," *French Historical Studies* 19, no. 1 (Spring 1995): 27–47, https://doi. org/10.2307/286898; Nuala Johnson, "Cast in Stone: Monuments, Geography, and Nationalism," *Environment and Planning D: Society and Space* 13, no. 1 (February 1995): 51–65, https://doi.org/10.1068/ d130051; Sharon Macdonald, "Undesirable Heritage: Fascist Material Culture and Historical Consciousness in Nuremberg," *International Journal of Heritage Studies* 12, no. 1 (January 2006): 9–28, https://doi. org/10.1080/13527250500384464; Brian S. Osborne, "From Native Pines to Diasporic Geese: Placing Culture, Setting Our Sites, Locating Identity in a Transnational Canada," *Canadian Journal of Communication* 31, no. 1 (March 2006): 147–75, https://doi.org/10.22230/cjc.2006v31n1a1781; Brian S. Osborne, "Landscape, Memory, Monuments, and Commemoration: Putting Identity in Its Place," *Canadian Ethnic Studies* 33, no. 3 (Fall 2001): 39–76; Purcell, "Commemoration, "; Robert Shipley, *To Mark Our Place: A History of Canadian War Memorials* (Toronto: NC Press, 1987); Whelan, "The Construction and Deconstruction of a Colonial Landscape," 508–33.

51 Zerubavel, *Time Maps*.

52 Zerubavel, 27.

53 Wayne Cook has since passed away and his website has been taken down. Elements of the website are preserved on the Wayback Machine

at https://archive.org/. A complementary site acknowledges Cook's efforts and provides its own list of historical plaques in Niagara. See Alan Brown's Ontario's Historical Plaques, accessed 26 June 2018, https://www.ontarioplaques.com/.

54 Mark Nuttall, *Arctic Homeland: Kinship, Community and Development in Northwest Greenland* (Toronto: University of Toronto Press, 1992).

55 A plaque that celebrates a tool that labourers used during the construction of the second Welland Canal.

56 The twenty-sixth mayor of St. Catharines and founder of the St. Catharines *Standard*. His name adorns a park, an arena, and a bridge.

57 See Jonathan Vance, *Death So Noble: Memory, Meaning and the First World War* (Vancouver: UBC Press, 1997); Shipley, *To Mark Our Place*.

58 See Roy Rosenzweig and David Thelen, *The Presence of the Past: Popular Uses of History in American Life* (New York: Columbia University Press, 1998); Margaret Conrad, Kadriye Ercikan, Gerald Friesen, Jocelyn Letourneau, Delphine Muise, David Northrup, and Peter Seixas, *Canadians and Their Pasts* (Toronto: University of Toronto Press, 2013); Paul Ashton and Paula Hamilton, *History at the Crossroads: Australians and the Past* (Sydney: Halstead Press, 2010).

59 Paul Litt, "The Apotheosis of the Apothecary: Retailing and Consuming the Meaning of a History Site," *Journal of the Canadian Historical Association/Revue de la Société historique du Canada*, 10, no. 1 (1999): 297–321, https://doi.org/10.7202/030517ar; Alan Gordon, "Heritage and Authenticity: The Case of Ontario's Sainte-Marie-among-the-Hurons," *Canadian Historical Review* 85, no. 3 (September 2004): 507–31, https://doi.org/10.3138/CHR.85.3.507.

60 Ian McKay, *The Quest of the Folk: Antimodernism and Cultural Selection in Twentieth-Century Nova Scotia* (Montreal: McGill-Queen's University Press, 1994).

61 See Rosenzweig and Thelen, *The Presence of the Past*; Conrad et al., *Canadians and Their Pasts*.

62 Laurajane Smith and Gary Campbell, "The Elephant in the Room: Heritage, Affect and Emotion," in *A Companion to Heritage Studies*, ed. William Loban, Máiréad Nic Craith, and Ullrich Kockel (Chichester, UK: Wiley-Blackwell, 2015), 443–60, https://doi.org/10.1002/9781118486634.ch30.

63 We are indebted to Stewart Hoover, Lynn Schofield Clark, and Diane Alters, *Media, Home, and Family* (New York: Routledge, 2004).

64 Charmayne Highfield and Jayne E. Bisman, "The Road Less Travelled: An Overview and Example of Constructivist Research in Accounting," *Australasian Accounting, Business and Finance Journal* 6, no. 5 (2012): 3–22.

65 D.L. Stewart, "Researcher as Instrument: Understanding 'Shifting' Findings in Constructivist Research," *Journal of Student Affairs Research*

and Practice 47, no. 3 (September 2010): 291–306, https://doi
.org/10.2202/1949-6605.6130.

66 William Cannon Hunter, Namho Chung, Ulrike Gretzel, and Chulmo
Koo, "Constructivist Research in Smart Tourism," *Asia Pacific Journal
of Information Systems* 25, no. 1 (March 2015): 105–20, https://doi
.org/10.14329/apjis.2015.25.1.105.

67 See Hoover, Clark, and Alters, *Media, Home, and Family*.

68 All percentages have been rounded to the nearest integer.

69 Clare J.A. Mitchell, R. Greg Atkinson, and Andrew Clark, "The Creative
Destruction of Niagara-on-the-Lake," *Canadian Geographer* 45, no. 2
(June 2001): 285–99, https://doi.org/10.1111/j.1541-0064.2001.tb01489.x.

70 The respondent was referencing statements made in André Picard,
"Fattest or Fittest?," *Globe and Mail*, 21 July 2001, A1.

71 Karim M. Tiro, "Now You See It, Now You Don't: The War of 1812 in
Canada and the United States," *Public Historian* 35, no. 1 (February 2013):
87–97, https://doi.org/10.1525/tph.2013.35.1.87.

72 TNS Canadian Facts, *Canadians' Knowledge & Perception of the War of
1812: Final Report Submitted to Department of Canadian Heritage* (Toronto:
TNS Canadian Facts, 2011); Steven Chase, "Harper's 1812 Overture,"
Globe and Mail, 5 October 2011, A4; Canadian Press, "Few Canadians
Aware of War of 1812," *CBC News*, 29 August 2012, https://www.
cbc.ca/news/politics/few-canadians-aware-of-war-of-1812-research-
suggests-1.1176690.

73 Canadian Press, "Few Canadians Aware of War of 1812."

74 John Moffat, "Policing History: Burke's Constabulary and Forensic
Functions of Rhetoric and Controversy over the War of 1812
Commemoration," *Rhetor: Journal of the Canadian Society for the Study of
Rhetoric* 6 (2016): 18–35; Claire T. Sjolander, "Through the Looking Glass:
Canadian Identity and the War of 1812," *International Journal* 69, no. 2
(June 2014): 152–67, https://doi.org/10.1177/0020702014527892.

75 Sjolander, "Through the Looking Glass," 164.

76 Scott Staring, "Harper's History," *Policy Options* 34, no. 2 (2013): 42–8.

77 Heritage Canada, *Evaluation of the Commemoration of the Bicentennial of
the War of 1812: July 2011 to March 2014*, document no. CH7-29/2015E
([Ottawa]: Heritage Canada, [2015]), 2–3.

78 Moffat, "Policing History"; Sjolander, "Through the Looking Glass."

79 The main daily papers were the St. Catharines *Standard*, the *Welland
Tribune*, and the *Niagara Falls Review*. On Postmedia's conservatism,
see Jeffrey Simpson, "Conservatives Pedalling Furiously but Going
Nowhere," *Globe and Mail*, 25 March 2011, A9; Jessica Chin, "Postmedia,
Sun Front Pages Replaces with Full-Page Political Ads," *Huffington Post
Canada*, 19 October 2015, https://www.huffpost.com/archive/ca

/entry/postmedia-sun-front-pages-replaced-with-full-page-political
-ads_n_8326634.

80 Heritage Canada, *Evaluation of the Commemoration of the Bicentennial of the War of 1812*, 5.

81 Paul Forsyth, "All Eyes on Thorold for 1812 Bash," *Niagara This Week*, 19 July 2013, 1.

82 "Lundy's Lane Sign Honours Historic 1814 Battle," Niagara Falls Tourism, July 17, 2014, https://www.niagarafallstourism.com/blog/lundys-lane -sign-honours-historic-1814-battle/.

83 Scott Rosts, "Landscape of Nations Ready for Unveiling: Native Memorial to Open Sunday at Queenston Heights," *Town Crier* (Niagara-on-the-Lake), 28 September 2016, 1; Paul Forsyth, "'Stunning' First Nations Peace Monument to Be Unveiled Oct. 7," *Niagara This Week*, 1 October 2017, 1.

84 "Brock Parkway a Reflection of Niagara: A Drive Along the Former Hwy. 405 Is Now a Teachable Moment, Transport Minister Says," *Niagara This Week*, 25 October 2006, SC00.

85 Paul Forsyth, "Region Supports Sir Isaac Brock Way Name Change," *Niagara This Week*, 4 September 2014, 1.

86 Melinda Cheevers, "Niagara on Fire: Commemorating 200 Years since the Burning of the Town," *Town Crier* (Niagara-on-the-Lake), 5 December 2013, 1.

87 Melinda Cheevers, "Marking the End of the War of 1812: Bicentennial Wraps Up with Treaty of Ghent Ratification Commemoration," *Town Crier* (Niagara-on-the-Lake), 19 February 2015, 4.

88 The Niagara region had no local television station during the period under study, although Cogeco (a cable television company) maintained a local access channel for its subscribers. The closest Canadian station was in Hamilton, Ontario, but many residents lived closer to American stations in Buffalo, New York.

89 Canadian Media Directors' Council, *Media Digest 2014/2015* (Toronto: CDMC and Marketing Magazine, 2014), 69 (supplement to *Marketing Magazine* 119, no. 10 [2014]); Canadian Media Directors' Council, *Media Digest 2015/2016* (Toronto: CMDC and Marketing Magazine, 2015), 121. During the period under study, Canadian newspaper research shifted from measures of circulation (i.e., hard copies distributed) to readership. "Readership" incorporates two measures: (1) the rate at which single copies of the print edition are read by multiple individuals, and (2) the number of unique readers who only read the digital edition.

90 W. Gamson and A. Modigliani, "Media Discourse and Public Opinion on Nuclear Power: A Constructionist Approach," *American Journal of Sociology* 95, no. 1 (1989): 1–37.

91 Enric Xicoy, Cristina Perales-García, and Rafael Xambó, "Shaping Public Opinion for Confrontation: Catalan Independence Claims as Represented

in Spanish, Catalan, Valencian, and Basque Editorials," *American Behavioral Scientist* 61, no. 9 (2017): 1042.

92 See Gamson and Modigliani, "Media Discourse and Public Opinion on Nuclear Power."

93 Eric Klinenberg, *Heat Wave: A Social Autopsy of a Disaster in Chicago* (Chicago: University of Chicago Press, 2003).

94 Jessica Murphy, "Tories Plan National Bash for War of 1812," *The Standard* (St. Catharines), 11 October 2011, A1 and A5; Brett Clarkson, "Canada Loads Up to Celebrate War with a Bang," *The Standard*, 12 October 2011, A1 and A4.

95 Forsyth, "'Stunning' First Nations Peace Monument to Be Unveiled Oct. 7."

96 John Law, "1812 Legacy Council Comes to a Close," *The Standard* (St. Catharines), 26 February 2015, A4.

97 Grant Lafleche, "War of 1812 *fought right here*," *The Standard* (St. Catharines), 18 June 2011, A1 and A4.

98 Shawn Jeffords, "Onley Kicks Off 1812 Celebrations," *The Standard* (St. Catharines), 3 January 2012, A2.

99 Tiro, "Now You See it, Now You Don't," 92.

100 Michael Billig, *Banal Nationalism* (London: Sage, 1995).

101 See Boyer, *The City of Collective Memory*; Birdsell and Groarke, "Toward a Theory of Visual Argument"; Anthony Blair, "Visual Arguments"; Finnegin, "Recognizing Lincoln,"; Finnegin, "The Naturalistic Enthymeme."

2. A War Memorial as a Mnemonic Device

1 [Lewis Redman Ord], *Reminiscences of a Bungle, by One of the Bunglers, and Two Other Northwest Rebellion Diaries* (Toronto: Grip, 1887), 28–9. See also R.C. Macleod, ed., *Reminiscences of a Bungle, by One of the Bunglers, and Two Other Northwest Rebellion Diaries* (Edmonton: University of Alberta Press, 1983), xliii–xliv.

2 "Comment and Criticism," *Canadian Militia Gazette*, 1, no. 7 (23 June 1885): 49.

3 Grant Lafleche, "Final Farewell," *The Standard* (St. Catharines), 9 December 2006, A1, A2.

4 Grant Lafleche, "Final Farewell: Hundreds Expected to Attend Funeral," *The Standard* (St. Catharines), 7 December 2006, A4; Grant Lafleche, "Students to View Soldier's Funeral Procession," *The Standard*, 8 December 2006, A4.

5 Snezhana Demitrova, "'Taming the Death': The Culture of Death (1915–18) and Its Remembering and Commemorating through First World War Soldier Monuments in Bulgaria (1917–44)," *Social History* 30, no. 2 (2005): 175–94. See also Thomas Laqueur, "Names, Bodies and the Anxieties of Erasure," in *Social and Political Body*, ed. T.R. Schatzki and W. Natter (New York: Guildford, 1996), 123–41.

6 Robert J. Harding, "Glorious Tragedy: Newfoundland's Cultural Memory of the Attack at Beaumont Hamel, 1916–1925," *Newfoundland & Labrador Studies* 21, no. 1 (2006): 3–40.

7 D.N. Sprague, *Canada and the Métis, 1869–1885* (Waterloo: Wilfrid Laurier University Press, 1988); Frits Pannekoek, *A Snug Little Flock: The Social Origins of the Red River Resistance of 1869–70* (Winnipeg: Watson and Dyer, 1991); Blair Stonechild and Bill Waiser, *Loyal til Death: Indians and the North-West Rebellion* (Markham: Fifth House, 2010); Ron Bourgeault, "The Struggle for Class and Nation: The Origin of the Métis in Canada and the National Question," in *1492–1992: Five Centuries of Imperialism and Resistance*, ed. Ron Bourgeault, Dave Broad, Lorne Brown, and Lori Foster (Halifax: Fernwood, 1992), 153–87; Allan Greer, "Settler Colonialism and Beyond," *Journal of the Canadian Historical Association/Revue de la Société historique du Canada* 30, no. 1 (2019): 61–86, https://doi.org/10.7202/1070631ar.

8 On the Red River resistance, see Sprague, *Canada and the Métis*; Pannekoek, *A Snug Little Flock*; J.M. Bumsted, *Trials and Tribulations: The Red River Settlement and the Emergence of Manitoba, 1811–1870* (Winnipeg: Great Plains Publications, 2003). On the Orange Order, see Cecil Houston and William J. Smyth, "The Orange Order and the Expansion of the Frontier in Ontario, 1830–1900," *Journal of Historical Geography* 4, no. 3 (July 1978): 251–64, https://doi.org/10.1016/0305-7488(78)90264-5; Hereward Senior, *Orangeism: The Canadian Phase* (Toronto: McGraw-Hill Ryerson, 1972); J.M. Bumsted, "Why Shoot Thomas Scott? A Study in Historical Evidence," in *Thomas Scott's Body and Other Essays on Early Manitoba History* (Winnipeg: University of Manitoba Press, 2000), 197–209; Ian Radforth, *Royal Spectacle: The 1860 Visit of the Prince of Wales to Canada and the United States* (Toronto: University of Toronto Press, 2004), 164–205.

9 Myrna Kostash, ed., *The Frog Lake Reader* (Edmonton: NeWest, 2009); Garry Radison, *Kā-pēpāmahchakwēw – Wandering Spirit: Plains Cree War Chief* (Yorkton, SK: Stone Ridge, 2009); S.A. Martin, "Johnson, Theresa Mary (Gowanlock)," in *Dictionary of Canadian Biography*, vol. 7, *1891–1900* (Toronto: University of Toronto/Université Laval, 1990), 478.

10 Desmond Morton, *The Last War Drum: The North-West Campaign of 1885* (Toronto: Hakkert, 1972); Bob Beal and Rod Macleod, *Prairie Fire: The North-West Rebellion of 1885* (Edmonton: Hurtig, 1984); Sprague, *Canada and the Métis, 1869–1885*; Desmond Morton, *A Military History of Canada*, 4th ed. (Toronto: McClelland and Stewart, 1999); John L. Tobias, "Canada's Subjugation of the Plains Cree, 1879–1885," *Canadian Historical Review* 64, no. 4 (December 1983): 519–48, https://doi.org/10.3138/CHR-064-04-04; Stonechild and Waiser, *Loyal til Death*.

11 "The Killed," *Winnipeg Free Press*, 14 May 1885, 1; "Editorial," *Thorold Post*, 15 May 1885, 2; "The Official List," *Welland Telegraph*, 15 May 1885,

8; "Private Watson," *Hamilton Spectator*, 19 May 1885, 1; C.P. Mulvaney, *Canada's Northwest Rebellion* (Toronto: Hovey, 1885), 242. On Winnipeg's railway boom, see Alan F.J. Artibise, ed., *Gateway City: Documents on the City of Winnipeg, 1873–1913* (Winnipeg: Manitoba Record Society/University of Manitoba Press, 1979), 55–60; R.G. MacBeth, *The Making of the Canadian West* (Toronto: Briggs, 1898).

12 A.N. Mowatt, "Decoration Day," *The Standard* (St. Catharines), 11 May 1946, 12; Morton, *Last War Drum*, chap. 4.

13 George T. Denison, *Soldiering in Canada: Recollections and Experiences* (Toronto: Morang, 1900), 261–6; Thomas Flanagan, *Riel and the Rebellion: 1885 Reconsidered* (Saskatoon: Western Producer Prairie, 1983); Phillipe R. Mailhot and Douglas M. Sprague, "Persistent Settlers: The Dispersal and Resettlement of the Red River Métis, 1870–1885," *Canadian Journal of Ethnic Studies* 17, no. 2 (1985): 1–30; Don McLean, *1885: Métis Rebellion or Government Conspiracy?* (Winnipeg: Pemmican, 1985); Stonechild and Waiser, *Loyal til Death*, 46–64.

14 "Riel's Collapse," *Welland Tribune*, 22 May 1885, 5.

15 "What Will Be Done with Riel?," *Thorold Post*, 22 May 1885, 2.

16 "Editorial," *Sentinel and Orange & Protestant Advocate*, 25 March 1886, 4.

17 "What Will Be Done with Riel?," 2.

18 "Riel's Collapse," 5.

19 Brock University Archives & Special Collections, *Niagara Lodges of the Loyal Orange Association Collection*, RG57, Box 5, LOL 844 Merritton, minute book 1883–1885, 2 September 1885. See also Box 3, LOL 1334 Fonthill, minute book, 21 October 1885.

20 Douglas Owram, "The Myth of Louis Riel," *Canadian Historical Review* 63, no. 3 (September 1982): 315–36, https://doi.org/10.3138/CHR-063-03-01; Susan Mann Trofimenkoff, *The Dream of Nation: A Social and Intellectual History of Quebec* (Toronto: Macmillan, 1982); Carl Berger, *The Sense of Power: Studies in the Ideas of Canadian Imperialism* (Toronto: University of Toronto Press, 1970), 233–58; A.I. Silver, "Ontario's Alleged Fanaticism in the Riel Affair," *Canadian Historical Review* 69, no. 1 (March 1988): 21–50, https://doi.org/10.3138/CHR-069-01-02; Paul Maroney, "Lest We Forget: War and Meaning in English Canada, 1885–1914," *Journal of Canadian Studies*, 32, no. 4 (Winter 1997): 108–24, https://doi.org/10.3138/jcs.32.4.108.

21 Lyrics by Lawrence Buchan quoted in George Broughall, *The 90th on Active Service* (Winnipeg: George Bishop, 1885); Bruce Tascona and Eric Wells, *Little Black Devils: A History of the Royal Winnipeg Rifles* (Winnipeg: Royal Winnipeg Rifles/Frye Publishing, 1983).

22 "The Rebellion," *Winnipeg Free Press*, 18 May 1885, 4.

23 "City Notes," *Winnipeg Free Press*, 19 May 1885, 1; "Casualties at Batoche," *Winnipeg Sun*, 3 July 1885, 19; "The Casualties," *The Globe* (Toronto),

15 May 1885, 2; "North West Rebellion," *Montreal Star*, 18 May 1885, 1; "Our Fallen Heroes," *Toronto Mail*, 15 September 1886, 8.

24 Paul Litt, Ronald F. Williamson, and Joseph W.A. Whitehorne, *Death at Snake Hill: Secrets from a War of 1812 Cemetery* (Toronto: Dundurn Press, 1993); Adam J. Barker, "Deathscapes of Settler Colonialism: The Necro-Settlement of Stoney Creek, Ontario, Canada," *Annals of the American Association of Geographers*, 108, no. 4 (2018): 1134–49; Stephanie Spar, "Understanding Conflict through Burial: Neural Network Analysis of Death and Burial in the War of 1812," *Ontario Archaeology*, nos. 89–90 (2010): 58–68. Natalie Southworth, "Harbour Project Runs into Dead of War of 1812," *Globe and Mail*, 15 January 2000, A23; Philip Longworth, *The Unending Vigil: A History of the Commonwealth War Graves Commission, 1917–1984*, 2nd ed. (London: Leo Cooper / Martin Secker & Warburg, 1985), preface; William M. Taylor, "War Remains: Contributions of the Imperial War Graves Commission and the Australian War Records Section to Material and National Cultures of Conflict and Commemoration," *National Identities* 17, no. 2 (2015): 217–40; Graham Oliver, "Naming the Dead, Writing the Individual," in *Cultures of Commemoration: War Memorials, Ancient and Modern*, ed. Polly Low, Graham Oliver, and P.J. Rhodes (Oxford: Oxford University Press, 2012), 113–34; Lee Jackson, *Dirty Old London: The Victorian Fight against Filth* (New Haven, CT: Yale University Press), 105–33.

25 James Stevens Curl, *The Victorian Celebration of Death* (Newton Abbot, UK: David & Charles, 1972), 27–53; Thomas R. Forbes, "By What Disease or Casualty: The Changing Face of Death in London," in *Health, Medicine and Mortality in the Sixteenth Century*, ed. Charles Webster (Cambridge: Cambridge University Press, 1979), 118–39; Julian Litten, *The English Way of Death: The Common Funeral since 1450* (London: Robert Hale, 1991), 143–72; Nigel Llewellyn *The Art of Death: Visual Culture in the English Death Ritual, c. 1500–1800* (London: Reaktion, 1991), 109–21.

26 Curl, *The Victorian Celebration of Death*, 1–26; Sarah Tarlow, "Romancing the Stones: The Graveyard Boom of the Later 18th Century," in *Grave Concerns: Death and Burial in England 1700–1850*, ed. Margaret Cox (York: Council for British Archaeology/Pennine, 1998), 33–43; Ralph Houlbrooke, *Death, Religion and the Family in England, 1480–1750* (Oxford: Clarendon, 1998), 331–71. See also Chris Brooks, *Mortal Remains: The History and Present State of the Victorian and Edwardian Cemetery* (Exeter, UK: Wheaton, 1989), 1–76; Pat Jalland, *Death in the Victorian Family* (Oxford: Oxford University Press, 1996), 284–99; Llewellyn, *The Art of Death*, 109–21.

27 Jalland, *Death in the Victorian Family*, 17–38. See also Litten, *The English Way of Death*, 143–72.

28 Llewellyn, *The Art of Death*, 28.

29 G.P. de T. Glazebrook, *Life in Ontario: A Social History* (Toronto: University of Toronto Press, 1968), 28; Susan Smart, *A Better Place: Death and Burial in Nineteenth-Century Ontario* (Toronto: Dundurn, 2011), 83–98; Ross D. Fair, "Model Farmers, Dubious Citizens: Reconsidering the Pennsylvania Germans of Upper Canada, 1786–1834," in *Beyond the Nation? Immigrants' Local Lives in Transnational Cultures*, ed. Alexander Freund (Toronto: University of Toronto Press, 2012), 79–106; Catherine Paterson, "The Heritage of Life and Death in Historical Family Cemeteries of Niagara, Ontario" (PhD diss., McMaster University, 2013). On the American experience, see David Charles Sloane, *The Last Great Necessity: Cemeteries in American History* (Baltimore: Johns Hopkins University Press, 1991).

30 Paterson, "The Heritage of Life and Death," 32–3; Brian Young, *Respectable Burial: Montreal's Mount Royal Cemetery* (Montreal: McGill-Queen's University Press, 2003), 6; Gladys Humphrey and Bob Humphrey, *Victoria Memorial Park Cemetery c. 1793–1862* ([Toronto]: [Ontario Genealogical Society, Toronto Branch], [1998]); Andrew King, "Ottawa's Park of the Dead," *Ottawa Rewind* (blog), 16 January 2017, ottawarewind. com/2017/01/16/ottawas-park-of-the-dead; Christopher Rouse, "Workers Unearth Grave," *Kingston Whig-Standard*, 26 September 1992, 6A; Janina Enrile, "Skeleton Park," *The Journal* (Kingston), 28 October 2011, 3.

31 Gary Laderman, *The Sacred Remains: American Attitudes toward Death, 1793–1883* (New Haven, CT: Yale University Press, 1996), 27–46; Susan-Mary Grant, "Patriot Graves: American Identity and the Civil War Dead," *American Nineteenth Century History* 5, no. 3 (2004): 74–100; Russ Castronovo, *Necro Citizenship: Death, Eroticism, and the Public Sphere in the Nineteenth-Century United States* (Durham, NC: Duke University Press, 2001).

32 Drew Gilpin Faust, *This Republic of Suffering: Death and the American Civil War* (New York: A.A. Knopf, 2008), 135. See also Richard Huntington and Peter Metcalfe, *Celebrations of Death: The Anthropology of Mortuary Ritual* (Cambridge: Cambridge University Press, 1979), 184–211.

33 Faust, *This Republic of Suffering*, 61–101.

34 Abraham Lincoln, draft of the Gettysburg Address (Nicolay copy), Abraham Lincoln Papers at the Library of Congress, Manuscript Division, accessed 8 June 2023, https://www.loc.gov/exhibits/gettysburg-address/exhibition-items.html.

35 Edwin Black, "Gettysburg and Silence," *Quarterly Journal of Speech* 80, no. 1 (1994): 21–36; Martha Watson, "Ordeal by Fire: The Transformative Rhetoric of Abraham Lincoln," *Rhetoric & Public Affairs* 3, no. 1 (2000): 33–48; Michael Leff and Jean Goodwin, "Dialogic Figures and Dialectical Argument in Lincoln's Rhetoric," *Rhetoric & Public Affairs* 3, no. 1 (2000): 59–69; Ian Finseth, "The Civil War Dead: Realism and the Problem of Anonymity," *American Literary History* 25, no. 3 (2013): 535–62. The

Commonwealth War Graves Commission adopted a similar stance when it formed during the First World War. See Philip Longworth, *The Unending Vigil: A History of the Commonwealth War Graves Commission, 1917–1984* (London: Leo Cooper/Secker and Warburg, 1985), 1–28.

36 Huntington and Metcalfe, *Celebrations of Death*, 153–83; Helen Rappaport, *A Magnificent Obsession: Victoria, Albert, and the Death that Changed the British Monarchy* (New York: St Martin's, 2011), 86–104.

37 Litten, *English Way of Death*, 173–94; Llewellyn, *Art of Death*, 101–8; Houlbrooke, *Death, Religion and the Family*, 255–80.

38 "The Funeral of His Late Royal Highness the Prince Consort," *The Globe* (Toronto), 10 January 1862, 1; Smart, *A Better Place*, 63–82; Rappaport, *A Magnificent Obsession*.

39 The four men were William Gladstone, Horatio Nelson (Viscount Nelson), Arthur Wellesley (Duke of Wellington), and Robert Napier (Baron Napier of Magdala).

40 Robert Malcolmson, *Burying General Brock: A History of Brock's Monuments* (Niagara-on-the-Lake: Friends of Fort George, 1996).

41 Adam Shortt, *Lord Sydenham* (London: Oxford University Press, 1926), 343–4; "Latest from Quebec," *The Globe* (Toronto), 2 August 1865, 2; "Latest from Quebec," *The Globe*, 3 August 1865, 2; *Quebec Mercury*, 2 August 1865, quoted in Maurice Brodeur, ed., *Le Premier Père de la Confédération des Provinces Canadiennes* (1927; Montmagny, QC: Lieu historique du Canada Maison Sir Étienne-Paschal-Taché, 2004), 17–18; Yves Hébert, *Étienne-Paschal Taché, 1795–1865 : Le militaire, le médecin, et l'homme politique* (Quebec City: Éditions GID, 2006), 271–82; Fennings Taylor, *The Hon. Thos. D'Arcy McGee: A Sketch of His Life and Death* (Montreal: Lovell, 1868), 43–60; David A. Wilson, *Thomas D'Arcy McGee*, vol. 2, *The Extreme Moderate, 1857–1868* (Montreal: McGill-Queen's University Press, 2011).

42 David Mills, *The Idea of Loyalty in Upper Canada, 1784–1850* (Montreal: McGill-Queen's University Press, 1988), 12–33; George Sheppard, *Plunder, Profit, and Paroles: A Social History of the War of 1812 in Upper Canada* (Montreal: McGill-Queen's University Press, 1994).

43 James A. Wood, *Militia Myths: Ideas of the Canadian Citizen Soldier, 1896–1921* (Vancouver: UBC Press, 2010), 1–29; J. Mackay Hitsman, *Safeguarding Canada, 1763–1871* (Toronto: University of Toronto Press, 1968); Jane Errington, *The Lion, the Eagle, and Upper Canada: A Developing Colonial Ideology*, 2nd ed. (Montreal: McGill-Queen's University Press, 2012), 166–84; John Garner, *The Franchise and Politics in British North America, 1755–1867* (Toronto: University of Toronto Press, 1969).

44 Nicholas Penny, "'Amor Publicus Posuit': Monuments for the People and of the People," *Burlington Magazine* 129, no. 1017 (1987): 793–800; Mary

P. Ryan, "Democracy Rising: The Monuments of Baltimore, 1809–1842," *Journal of Urban History* 36, no. 2 (2010): 127–50.

45 J.H.L., "Note 189. The Welsford-Parker Monument in Halifax, Nova Scotia," *Journal of the Society for Army Historical Research* 8, no. 32 (1929): 129–31.

46 William D'arcy, *The Fenian Movement in the United States, 1858–1886* (Washington, DC: Catholic University of America Press, 1947); Peter Vronsky, *Ridgeway: The American Fenian Invasion and the 1866 Battle that Made Canada* (Toronto: Allen Lane, 2011).

47 Brian R. Gabrial, "The Second American Revolution: Expressions of Canadian Identity in News Coverage at the Outbreak of the United States Civil War," *Canadian Journal of Communication* 33, no. 1 (February 2008): 21–37, https://doi.org/10.22230/cjc.2008v33n1a1872.

48 "Funeral," *St. Catharines Journal*, 13 June 1866, 2; Robert A. Macbeth, "Mewburn, Frank Hamilton," in *Dictionary of Canadian Biography*, vol. 15 (Toronto: University of Toronto/Université Laval, 2003).

49 Curl, *Victorian Celebration of Death*, 1–26; Houlbrooke, *Death, Religion and the Family*, 255–94; Jalland, *Death in the Victorian Family*, 210–24; Smart, *A Better Place*, 83–98.

50 "Funeral," *St. Catharines Journal*, 13 June 1866, 2. See also Alexander Somerville, *Narrative of the Fenian Invasion of Canada* (Hamilton: Lyght, 1866), 114–15. Mewburn's grave remains in the churchyard at St John's Church in Niagara Falls. The Queen's Own Rifles erected a new gravestone beside the original stone in 2011.

51 Peter Vronsky, *Ridgeway*, 211–14.

52 Denison, *Soldiering in Canada*, 293–301; Charles A. Boulton, *Reminiscences of the North-West Rebellions* (Toronto: Grip, 1886), 197–8; Desmond Morton, *Ministers and Generals: Politics and the Canadian Militia, 1868–1904* (Toronto: University of Toronto Press, 1970).

53 Frederick Middleton, *Suppression of the Rebellion in the North West Territories of Canada, 1885* (1886; Toronto: University of Toronto Press, 1948), 53.

54 Denison, *Soldiering in Canada*, 261–6.

55 Canada, *House of Commons Debates*, 5th Parliament, 3rd Session, vol. 3, 28 May 1885, 2169; George Beauregard, *Le 9me Bataillon au Nord-Ouest (journal d'un militaire)* (Quebec City: Joseph G. Gingras, 1886), 39–40; W.H. Jackson, *Report on the Matters in Connection with the Suppression of the Rebellion in the North-West Territories in 1885* (Ottawa: Department of Militia and Defence/MacLean, Roger and Co., 1887), 27.

56 Macbeth, *The Making of the Canadian West*, 174–5; "Caron, Jean Sr. (1833–1905)," in *Veterans and Families of the 1885 Northwest Resistance*, ed. Lawrence J. Barkwell (Saskatoon: Gabriel Dumont Institute, 2011), 71–3.

57 Beauregard, *Le 9me Bataillon*, 32–3, 38.

58 *The Riel Rebellion 1885* (Montreal: Witness Printing House, [1885]), 23.

59 Parks Canada, *Battle of Tourond's Coulee/Fish Creek National Historic Site of Canada: Management Plan* (Ottawa: Crown Printer, 2007).

60 George Goulet and Terry Goulet, "The Saskatchewan Uprising and the St. Laurent Nuns," in Barkwell, *Veterans and Families*, 32–42; H.P. Dwight to A.P. Caron, telegram, 16 May 1885, reproduced in *Telegrams of the North West Campaign*, ed. Desmond Morton and Reginald H. Roy (Toronto: Champlain Society, 1972), 287.

61 Canada, *House of Commons Debates*, 5th Parliament, 3rd Session, vol. 3, 21 May 1885, 2029.

62 H.P. Rusden, "Notes on the Suppression of the North West Insurrection," in *Reminiscences of a Bungle ... and Two Other Northwest Rebellion Diaries*, ed. R.C. Macleod (Edmonton: University of Alberta Press, 1983), 293; H.P. Dwight to A.P. Caron, telegram, 19 May 1885, reproduced in Macleod, *Telegrams of the North West Campaign*, 301–2.

63 "Editorial," *Canadian Military Gazette*, 23 June 1885, 49.

64 "The Soldiers' Graves," *Toronto Mail*, 16 May 1885, 4; "To Be Buried at Home," *Hamilton Spectator*, 23 May 1885, 1; Tascona and Wells, *Little Black Devils*, 39.

65 "Bodies of the Dead" and "From Winnipeg," *Hamilton Times*, 22 May 1885, 1; Keith A. Foster, "Moose Jaw: The First Decade, 1882–1892" (master's thesis, University of Regina, 1978); Mary H. Bishop, *Moose Jaw: A History in Words and Pictures* (Lunenburg, NS: MacIntyre Purcell, 2017).

66 "Massacre of Ten Whites at Frog Lake," *The Globe* (Toronto), 10 April 1885, 2; "Big Bear's Captives," *The Globe*, 23 June 1885, 2; "Local News," *The Globe*, 14 July 1885, 6; Theresa Gowanlock and Theresa Delaney, *Two Months in the Camp of Big Bear* (Parkdale: The Times, 1885); Sarah Carter, five interview segments, in Kostash, *The Frog Lake Reader*, 129–33.

67 Jackson, *Report on the Matters*, 27.

68 Beauregard, *Le 9me Bataillon*, 89–90; Macbeth, *Making of the Canadian West*, 176.

69 "Arrival of the Dead," *Winnipeg Free Press*, 23 May 1885, 1.

70 "Letter from Rev. Mr. Gordon," *Winnipeg Free Press*, 23 May 1885, 1.

71 Jackson, *Report on the Matters*, 31.

72 Macbeth, *Making of the Canadian West*, 206–7; Beauregard, *Le 9me Bataillon*, 88.

73 Denison, *Soldiering in Canada*, 329.

74 "Reception at Toronto," *Daily British Whig* (Kingston, ON), 21 July 1885, reprinted in *The Riel Rebellion 1885*, ed. Nick Mika and Helma Mika (Belleville: Mika, 1972), 176–7; Maroney, "Lest We Forget," 108–24; David A. Sutherland, "Halifax's Encounter with the North-West Uprising of 1885," *Journal of the Royal Nova Scotia Historical Society* 13 (2010): 63–79; Ian Radforth, "Celebrating the Suppression of the North-West Resistance of 1885: The Toronto Press and the Militia Volunteers," *Histoire sociale/Social History* 67, no. 95 (November 2014): 601–39, https://doi.org/10.1353/his.2014.0042.

75 Susan G. Davis, *Parades and Power: Street Theatre in Nineteenth-Century Philadelphia* (Berkeley: University of California Press, 1986); Mary Ryan, "The American Parade: Representations of the Nineteenth-Century Social Order," in *The New Cultural History*, ed. Lynn A. Hunt (Berkeley: University of California Press: 1989), 131–53; Peter G. Goheen, "Parading: A Lively Tradition in Early Victorian Toronto," in *Ideology and Landscape in Historical Perspective*, ed. Alan R.H. Baker and Gideon Biger (Cambridge: Cambridge University Press, 1992), 330–51; Peter G. Goheen, "Negotiating Access to Public Space in Mid-Nineteenth Century Toronto," *Journal of Historical Geography* 20, no. 4 (October 1994): 430–49, https://doi .org/10.1006/jhge.1994.1033; Robert Cupido, "Public Commemoration and Ethnocultural Assertion: Winnipeg Celebrates the Diamond Jubilee of Confederation," *Urban History Review* 38, no. 2 (Spring 2010): 64–74, https://doi.org/10.7202/039675ar.

76 "From St. Catharines," *Hamilton Spectator*, 29 May 1885, 1.

77 "Military Funeral," *Welland Telegraph*, 29 May 1885, 1.

78 "Military Funeral," 1.

79 "From St. Catharines," *Hamilton Spectator*, 30 May 1885, 1; "St. Catharines," *Thorold Post*, 5 June 1885, 3; "Private Watson," *Welland Tribune*, 5 June 1885, 1.

80 William Westfall, *Two Worlds: The Protestant Culture of Nineteenth-Century Ontario* (Montreal: McGill-Queen's University Press, 1989); Glazebrook, *Life in Ontario*, 78; Alan Lauffer Hayes, *Anglicans in Canada: Controversies and Identity in Historical Perspective* (Champaign: University of Illinois Press, 2004); Peter B. Moore, *St. Thomas Saints: The First Hundred Years* (St. Catharines: St. Thomas Church/Rannie, 1976); Grand Orange Lodge of British America, *Canadian Orange Lodge Officers 1888* (Toronto, 1888).

81 W.R. Harris, *The Catholic Church in the Niagara Peninsula, 1626–1895* (Toronto: Briggs, 1885); Lynne Marks, *Revivals and Roller Rinks: Religion, Leisure, and Identity in Late-Nineteenth-Century Small-Town Ontario* (Toronto: University of Toronto Press, 1996), 119. See also Ruth Bleasdale, "Class Conflict on the Canals of Upper Canada in the 1840s," *Labour/Le Travail* 7 (Spring 1981): 9–39, https://doi.org/10.2307/25140020; W. Thomas Matthews, "The Myth of the Peaceable Kingdom: Upper Canadian Society during the Early Victorian Period," *Queen's Quarterly* 94, no. 2 (1987): 383–401; Roberta M. Styran and Robert R. Taylor, *This Great National Object: Building the Nineteenth-Century Welland Canals* (Montreal: McGill-Queen's University Press, 2012), 264–88.

82 "Private Watson," *Welland Tribune*, 5 June 1885, 1. See also "From St. Catharines," *Hamilton Spectator*, 30 May 1885, 1; "A Tribute to the Dead," *Winnipeg Free Press*, 30 May 1885, 1.

83 "Regimental Notes," *Canadian Militia Gazette*, 2 July 1885, 81; "Lieu. Kippen's Funeral," *Toronto Mail*, 29 May 1885, 1; *Weekly British Whig*

(Kingston, ON), 23 July 1885, reprinted in Mika and Mika, *The Riel Rebellion 1885*, 176–7; *The Gazette* (Montreal), 21 July 1885, reprinted in Mika and Mika, *The Riel Rebellion 1885*, 178–80; "Gleanings," *Canadian Militia Gazette*, 7 July 1885, 70; "From Toronto" and "The Funeral of Lieut. Fitch," *Hamilton Spectator*, 28 May 1885, 1; "The Burial of Private Moore [*sic*]," *Hamilton Spectator*, 30 May 1885, 1; "At Rest," *The Globe* (Toronto), 2 June 1885, 6. In 1990, a monument to honour Les Voltigeurs de Québec was unveiled in Quebec City. The figure of a soldier is dressed in the unit's 1885 uniform. "Monuments in Place Georges-V across from the Armoury," Voltigeurs de Québec, 5 November 2018, https://manegemilitaire.ca/en/monuments-in-place-george-v-across-from-the-armoury.

84 Morton, *Military History of Canada*, 99–106.

85 City of St. Catharines, *City Council Minutes*, 8 June 1885, mflm 171–315; "Neighbourhood News: St. Catharines," *Hamilton Spectator*, 2 June 1885, 2; "Neighbourhood News: St. Catharines," *Hamilton Spectator*, 24 June 1885, 2; "Neighbourhood News: St. Catharines," *Hamilton Spectator*, 11 July 1885, 2; "St. Catharines," *The Globe* (Toronto), 4 August 1885, 2; "St. Catharines," *Thorold Post*, 17 September 1886, 2; "Comment and Criticism," *Canadian Militia Gazette*, 30 June 1885, 57; "Gleanings," *Canadian Militia Gazette*, 9 February 1886, 319.

86 "Comment and Criticism," *Canadian Militia Gazette*, 17 November 1885, 218; "A.W. Kippen Monument," *Perth (ON) Courier*, 10 June 1887, 2.

87 City of St. Catharines, *Victoria Lawn Cemetery, Registry of Interments 1856–1902* (St. Catharines: City of St. Catharines, n.d.): 115; City of St. Catharines, *Victoria Lawn Cemetery, Registry of Lots Sold 1856–1947* (St. Catharines: City of St. Catharines, n.d.), section W.

88 City of St. Catharines, *City Council Minutes*, 6 August 1886, mflm 171–315; "St. Catharines," *Thorold Post*, 20 November 1885, 2; "St. Catharines," *Thorold Post*, 13 August 1886, 2.

89 Huntington and Metcalfe, *Celebrations of Death*, 181–3. Russ Castronovo argues a similar point for the dead of the American Civil War. See Castronovo, *Necro Citizenship*.

90 "St. Catharines," *Winnipeg Free Press*, 14 September 1886, 1.

91 Contemporary accounts report that the arms of the Dominion appeared on each side. Perhaps they were in the original proposal, but no image of the monument shows them. See "A Northwest Hero," *True Witness and Catholic Chronicle* (Montreal), 22 September 1886, 3; "Regimental Notes," *Canadian Militia Gazette*, 30 September 1886, 502.

92 City of St. Catharines, *City Council Minutes*, 23 May 1887, mflm 171–316; "St. Catharines," *Hamilton Spectator*, 15 September 1886, 1.

93 "St. Catharines," *Thorold Post*, 3 September 1886, 2; "St. Catharines," *Thorold Post*, 10 September 1886, 3.

94 "St. Catharines," *Thorold Post*, 17 September 1886, 2.

95 "St. Catharines," *Welland Tribune*, 17 September 1886, 8.

96 "St. Catharines," *Thorold Post*, 17 September 1886, 2.

97 "St. Catharines," *Winnipeg Free Press*, 15 September 1886, 1; "Our Fallen Heroes," *Toronto Mail*, 15 September 1886, 8.

98 Mathew McRae, "Remembering Rebellion, Remembering Resistance: Collective Memory, Identity, and the Veterans of 1869–70 and 1885" (PhD diss., University of Western Ontario, 2018), 131–5. Our thanks to Dr. McRae for sharing an early draft of his research with us.

99 Walter Allward quoted in "The Noble Dead," *The Globe* (Toronto), 29 June 1896, 6. Allward also completed monuments for the War of 1812 (located in Toronto), Second Boer War (Windsor and Toronto), and First World War (Brantford, Peterborough, and Stratford).

100 Stonechild and Waiser, *Loyal til Death*, 226–7; "Battleford Hangings," *Saskatchewan Indian* 3, no. 7 (July 1972): 5.

101 Diane P. Payment, *The Free People – Li Gens Libre: A History of the Métis Community of Batoche, Saskatchewan* (Calgary: University of Calgary, 2009), 253–82; Carolyn Podruchny and Jesse Thistle, "A Geography of Blood: Uncovering the Hidden Histories of Métis People in Canada," in *Spaces of Difference: Conflicts and Cohabitation*, ed. Ursula Lehmkuhl, Hans-Jürgen Lüsebrink, and Laurence McFalls (Münster: Waxmann, 2016), 61–79; "Ouellette, Moïse Napoléon (1840–1911)," in Barkwell, *Veterans and Families*, 201.

102 Bill Waiser, *Saskatchewan: A New History* (Calgary: Fitzhenry & Whiteside, 2005), 21–40, 171–3; Mary Ellen Turpel-Lafond, "Maskêko-sâkahikanihk: One Hundred Years for a Saskatchewan First Nation," in *Perspectives of Saskatchewan*, ed. Jene M. Porter (Winnipeg: University of Manitoba Press, 2009), 76–104; Payment, *The Free People – Li Gens Libre*, 272–99; Podruchny and Thistle, "A Geography of Blood," 61–79. The Winnipeg monuments are placed at the provincial legislature, St. Boniface College, St. Boniface Museum, and Riel Memorial Park.

3. The Watson Monument through Time

1 Elizabeth Jelin, *State Repression and the Labors of Memory* (Minneapolis: University of Minnesota Press, 2003).

2 Iain Hay, Andrew Hughes, and Mark Tutton, "Monuments, Memory and Marginalization in Adelaide's Prince Henry Gardens," *Geographiska Annaler* 86b, no. 3 (2004): 201–16.

3 Rhiannon Johnson, "Métis Move Louis Riel Day Event," *CBC News*, 16 November 2017, https://www.cbc.ca/news/indigenous/ontario-metis -louis-riel-day-northwest-rebellion-monument-1.4404191.

4   Charlene Mires, *Independence Hall in American Memory* (Philadelphia: University of Pennsylvania Press, 2002).

5   Robert J. Harding, "Glorious Tragedy: Newfoundland's Cultural Memory of the Attack at Beaumont Hamel, 1916–1925," *Newfoundland & Labrador Studies* 21, no. 1 (January 2006): 3–40.

6   "David Watson," *The Standard* (St. Catharines), 7 August 1900, 1.

7   St. Catharines Board of Trade, *Annual Report of the St. Catharines Board of Trade for the Year 1900* (St. Catharines: St. Catharines Board of Trade, 1901); Edwin Poole, *St. Catharines Ontario Souvenir 1904* (St. Catharines: Print Shop, 1904); *The Standard, The Garden City of Canada, Anno Domini, 1907* (St. Catharines: Standard Printing, 1907).

8   "St. Catharines in Summer," *The Globe* (Toronto), 29 June 1889, 1; City of St. Catharines Museum, "Lord Stanley's Cadets 1891" (3165-N) and "Owls B.C. 1889" (N2357); James Munro advertisement in *St. Catharines City Directory 1910*, ed. J.M. Poole (Toronto: Poole, 1910), [ix].

9   William D'Arcy, *The Fenian Movement in the United States, 1858–1886* (Washington, DC: Catholic University of America Press, 1947); Peter Vronsky, *Ridgeway: The American Fenian Invasion and the 1866 Battle that Made Canada* (Toronto: Allen Lane, 2011).

10   "The Limeridge Action," *St. Catharines Constitutional*, 7 June 1866, 1; "The Military Hospital," *St. Catharines Constitutional*, 14 June 1866, 1; R.L. Rogers, *History of the Lincoln and Welland Regiment* (St. Catharines: Lincoln and Welland Regiment, 1954), 25–32.

11   Mewburn Road in Niagara Falls may have been named for John H. Mewburn, but this cannot be confirmed. The Mewburns were a prominent family in the area and several members, if not the entire family, could have been the nominal member.

12   Paul Maroney, "Lest We Forget: War and Meaning in English Canada, 1885–1914," *Journal of Canadian Studies* 32, no. 4 (Winter 1997): 108–24, https://doi.org/10.3138/jcs.32.4.108.

13   David W. Blight, "Decoration Days: The Origins of Memorial Day in North and South," in *The Memory of the Civil War in American Culture*, ed. Alice Fahs and Joan Waugh (Chapel Hill: University of North Carolina Press, 2004), 94–129; Martha E. Kinney, "If Vanquished I Am Still Victorious," *Virginia Magazine of History & Biography* 106, no. 3 (1998): 237–66; Drew Gilpin Faust, *This Republic of Suffering: Death and the American Civil War* (New York: A.A. Knopf, 2008), 211–49.

14   Maroney, *Lest We Forget*, 108–24.

15   "Regimental and Other News," *Canadian Militia Gazette*, 5 June 1890, 181; "Celebration of the Veterans of '66 To-day," *The Globe* (Toronto), 2 June 1894, 20; Maroney, *Lest We Forget*, 108–24.

16   See, for example, "Decoration Day," *The Globe* (Toronto), 3 June 1898, 3; "Tributes," *Toronto Star*, 2 June 1898, 2.

17 "Editorial," *Grip*, 10 June 1893, 356.

18 "The Old Flag," *The Standard* (St. Catharines), 21 July 1898, 4; "The Volunteer Movement," *St. Catharines Journal*, 23 July 1898, 1. See also "Niagara Volunteers," *The Globe* (Toronto), 28 April 1898, 2; "Veterans Reunion," *The Standard*, 21 July 1898, 1.

19 "Decoration Day," *The Standard* (St. Catharines), 28 July 1898, 5.

20 "Still Willing," *The Standard* (St. Catharines), 21 July 1898, 1.

21 "Volunteer Movement," *St. Catharines Journal*, 23 July 1898, 1.

22 "Veterans' Day," *Thorold Post*, 29 July 1898, 4.

23 "Veterans' Reunion," *The Standard* (St. Catharines), 21 July 1898, 4. See also "66 Veterans' Reunion," *Toronto Telegram*, 20 July 1898, 8.

24 "Decoration Day," *The Standard* (St. Catharines), 28 July 1898, 5; "Veterans' Reunion," *The Globe* (Toronto), 20 July 1898, 10; McRae, "Remembering Rebellion, Remembering Resistance," 123–9.

25 Carman Miller, *Painting the Map Red: Canada and the South African War, 1899–1902* (Montreal: McGill-Queen's University Press, 1993); Robert J.D. Page, "Canada and the Imperial Idea in the Boer War Years," *Journal of Canadian Studies* 5, no. 1 (February 1970): 33–49, https://doi.org/10.3138/jcs.5.1.33; Desmond Morton, *Canada at Paardeberg* (Ottawa: Balmuir/Canadian War Museum, 1986); Desmond Morton, *A Military History of Canada*, 4th ed. (Toronto: McClelland and Stewart, 1999).

26 Richard Price, *An Imperial War and the British Working Class* (London: Routledge and Kegan Paul, 1972); Robert Page, *The Boer War and Canadian Imperialism*, CHA Historical Booklet No. 44 (Ottawa: Canadian Historical Association, 1987).

27 Page, *The Boer War*; Millar, *Painting the Map Red*, 423–44.

28 "Given a Hearty Welcome," *The Standard* (St. Catharines), 30 July 1900, 1 and 4; "Were Given a Great Send-Off," *The Standard*, 1 October 1900, 1; "Canadians Lionized," *The Standard*, 6 December 1900, 1; "City and Vicinity," *The Standard*, 11 December 1900, 3; "The Veterans' Night," *The Standard*, 13 December 1900, 4; "The Heroes Received," *The Standard*, 26 December 1900, 1.

29 Miller, *Painting the Map Red*, 49–64, 109, 203–4, and 408.

30 Katie Pickles, *Female Imperialism and National Identity: Imperial Order Daughters of the Empire* (Manchester: Manchester University Press, 2002), 108–21. See also Millar, *Painting the Map Red*, 433–6; "South African War Memorials," Veterans Affairs Canada, last modified 14 February 2019, https://www.veterans.gc.ca/eng/remembrance/memorials/overseas/south-african-war; "Welcome to the Maple Leaf Legacy Project," Maple Leaf Legacy Project, accessed 9 June 2023, http://www.mapleleaflegacy.ca/wp.

31 *The Standard* (St. Catharines), 24 July 1900, 2; "A Memorial Tablet," *The Standard*, 24 December 1900, 2.

32 "Solemn Memorial Service," *The Standard* (St. Catharines), 9 July 1900, 1; Peter B. Moore, *St. Thomas Saints: The First Hundred Years* (St. Catharines: St. Thomas Church/Rannie, 1976), 29.

33 "Lieu. Birch," *The Standard* (St. Catharines), 19 July 1900, 1–2; "Dangerously Wounded," *The Standard*, 26 August 1901, 1; "Sgt. Black," *The Standard*, 6 September 1901, 1; "President McKinley," *The Standard*, 7 September 1901, 1; "Private A. Ratcliffe," *The Standard*, 26 September 1900, 1.

34 St. Catharines Museum and Archives, City of St. Catharines, *City Council Minutes*, 31 October 1900.

35 "City and Vicinity," *The Standard* (St. Catharines), 28 October 1901, 3.

36 "A Memorial Tablet," *The Standard* (St. Catharines), 24 December 1900, 2.

37 Canada, *Fifth Census of Canada 1911: Areas and Population by Provinces, Districts and Subdistricts*, vol. 1 (Ottawa: King's Printer, 1912), 80; "Prince of Wales and War Memorial Unveiled," *The Standard* (St. Catharines), 8 August 1927, 1; Grant Lafleche, "St. Catharines Suffered Losses of War in Silence," *The Standard*, 19 November 2016.

38 The commission also produced two stone monuments that provided a focal point within each cemetery. Generally, the Cross of Sacrifice was placed in smaller cemeteries and the Stone of Remembrance was placed in larger ones. The former was a stone cross bearing an inverted bronze sword, the whole resting on an octagonal stone base. After the war, several centres across Canada adopted this design for their civic war memorials. Reporters in St. Catharines seem to have borrowed the name of this monument for the wooden cross at city hall. Philip Longworth, *The Unending Vigil: A History of the Commonwealth War Graves Commission, 1917–1984* (London: Leo Cooper/Secker and Warburg, 1985); Julie Summers, *Remembered: The History of the Commonwealth War Graves Commission* (London: Merrell, 2007), 12–24; John Wolffe, *God and Greater Britain: Religion and National Life in Britain and Ireland 1843–1945* (London: Routledge, 1994), 213–53; Michèle Barrett, "Death and the Afterlife: Britain's Colonies and Dominions," *Race, Empire, and First World War Writing*, ed. Santanu Das (Cambridge: Cambridge University Press, 2010), 301–20.

39 "St. Catharines Joins Heartily," *The Standard* (St. Catharines), 11 November 1922, 1.

40 David Cannadine, "War and Death, Grief and Mourning in Modern Britain," in *Mirrors of Mortality: Studies in the Social History of Death*, ed. J. Whaley (London: Europa Publications, 1981), 187–242; Denise Thomson, "National Sorrow, National Pride: Commemoration of War in Canada, 1918–1945," *Journal of Canadian Studies*, 30, no. 4 (Winter 1995): 5–27, https://doi.org/10.3138/jcs.30.4.5.

41 "St. Kitts in Joyous Raptures," *The Standard* (St. Catharines), 11 November 1918, 1.

42 "Present Honor Flags," *The Standard* (St. Catharines), 16 November 1918, 3.

43 "Public Thanksgiving Service," *The Standard* (St. Catharines), 30 November 1918, 1.

44 Laura Wittman, *The Tomb of the Unknown Soldier, Modern Mourning, and the Reinvention of the Mystical Body* (Toronto: University of Toronto Press, 2011); Robert Shipley, *To Mark Our Place: A History of Canadian War Memorials* (Toronto: NC Press, 1987).

45 George K.B. Adams, quoted in "St. Catharines Joins Heartily," *The Standard* (St. Catharines), 11 November 1922, 1.

46 "St. Catharines IODE," *The Standard* (St. Catharines), 7 December 1936, 18 and 27.

47 Lewis W. Broughall, quoted in "Canadian Fallen Are Honored," *The Standard* (St. Catharines), 2 June 1919, 3.

48 See also Tonya Davidson, "Imperial Nostalgia, Social Ghosts, and Canada's National War Memorial," *Space and Culture* 19, no. 2 (May 2016): 177–91, https://doi.org/10.1177/1206331215623220.

49 Thomson, "National Sorrow, National Pride," 5–27; Melissa Davidson, "Acts of Remembrance: Canadian Great War Memory and the Public Funerals of Sir Arthur Currie and Canon F.G. Scott," *Études Canadiennes/Canadian Studies* 80 (2016): 109–27, https://doi.org/10.4000/eccs.688.

50 Ted Harrison, *Remembrance Today: Poppies, Grief, and Heroism* (London: Reaktion, 2012), 148–82; "Armistice Anniversary," *The Standard* (St. Catharines), 11 November 1921, 1; "Flower of Remembrance," *The Standard*, 4 November 1922, 4.

51 Shipley, *To Mark Our Place*; Thomson, "National Sorrow, National Pride," 5–27; Jonathan F. Vance, *Death So Noble: Memory, Meaning, and the First World War* (Vancouver: UBC Press, 1997), 198–225.

52 "Outlook of the Church," *The Globe* (Toronto), 14 September 1918, 11; "Women's Party Organized," *The Globe*, 25 September 1918, 6; Jeffrey Richards, *Imperialism and Music: Britain, 1876–1953* (Manchester: Manchester University Press: 2001), 152–3; Rose Collis, *Death and the City: The Nation's Experience, Told through Brighton's History* (Brighton, UK: Hanover Press, 2013), 170.

53 David Cannadine, *Aspects of Aristocracy: Grandeur and Decline in Modern Britain* (New Haven, CT: Yale University Press, 1994), 77–108; Vance, *Death So Noble*, 213–14.

54 Cannadine, *Aspects of Aristocracy*, 77–108; Harrison, *Remembrance Today*, 77–110.

55 "This City Pays Homage," *The Standard* (St. Catharines), 11 November 1919, 1.

56 "Second Anniversary," *The Standard* (St. Catharines), 11 November 1920, 1.

57 "War Memorials," *The Standard* (St. Catharines), 7 May 1919, 4.

58 "St. Catharines Joins Heartily," *The Standard* (St. Catharines), 11 November 1922, 1.

59 "Visit of the Vice Regal Party," *The Standard* (St. Catharines), 30 May 1918, 4; "City and Vicinity," *The Standard*, 31 May 1918, 2.

60 "Second Anniversary," *The Standard* (St. Catharines), 11 November 1920, 1; "Armistice Anniversary," *The Standard*, 11 November 1921, 1; "St. Catharines Joins Heartily," *The Standard*, 11 November 1922, 1; "Observance of Armistice," *The Standard*, 11 November 1923, 3; "Cease Fire," *The Standard*, 11 November 1924, 1–2 and 7; "Two Minutes," *The Standard*, 11 November 1925, 1; "Armistice Day," *The Standard*, 11 November 1926, 1–2; "Armistice Day," *The Standard*, 8 November 1927, 4; "Two Minutes of Silence," *The Standard*, 10 November 1927, 1.

61 "Decoration Day Plan," *The Standard* (St. Catharines), 28 May 1927, 24.

62 "Increased Attendance," *The Standard* (St. Catharines), 6 June 1927, 1 and 3.

63 Pickles, *Female Imperialism*, 108–21; "Memorial to Heroes," *The Standard* (St. Catharines), 9 December 1918, 1; "Memorial Hall Is Proposed," *The Standard*, 17 December 1918, 1; Mrs. Charles Porter and Mrs. David Craig, "Sixtieth Anniversary History, Municipal Chapter, IODE St. Catharines … 1912–1972" (unpublished manuscript, 1973), RG309, Series IV, f.3.9, IODE Earl Mountbatten Chapter fonds, Archives and Special Collections, Brock University.

64 Vance, *Death So Noble*, 202–11.

65 "Unveiling," *The Standard* (St. Catharines), 6 August 1927, 1 and 3.

66 "Impressive Parade," *The Standard* (St. Catharines), 6 June 1927, 4.

67 "When Governor General Visited," *The Standard* (St. Catharines), 14 May 1936, 1.

68 "War Shrine," *The Standard* (St. Catharines), 2 June 1938, 13; "Decoration Day," *The Standard*, 4 June 1938, 20; "Thousands Witness," *The Standard*, 6 June 1938, 9. See also Graham Phair, *Snapshots of the Homefront, 1939–1941* (St. Catharines: Vanwell, 2009).

69 Albert E. Coombs, *St. Catharines, Ontario: Historical Facts* (St. Catharines: City of St. Catharines, 1939); Coombs, *St. Catharines, Ontario: Historical Facts* (St. Catharines: City of St. Catharines, 1948).

70 Albert E. Coombs, *History of the Niagara Peninsula* (Montreal: Historical Foundation, 1950), 67.

71 Finding aid, RG575, IODE William Hamilton Merritt Chapter fonds, Archives and Special Collections, Brock University.

72 Melinda Cheevers, "Decoration Day Marked in St. Catharines," *Niagara This Week*, 5 June 2017, 24.

73 *The Standard, Beautiful and Busy St. Catharines* (St. Catharines: Standard Printing, 1921); City of St. Catharines, *The Garden City Bids You Welcome* (St.

Catharines: City of St. Catharines, [ca. 1925]); Niagara Peninsula Blossom Time
Committee, *Follow the Blossom Time Route* ([St. Catharines]: Niagara Peninsula
Blossom Time Committee, [1937]); Lincoln County, *Lincoln County Welcomes
You* ([St. Catherines]: Lincoln County, [ca. 1946]); City of St. Catharines, *The
Garden City of Canada, St. Catharines* (St. Catharines: City of St. Catharines,
[ca. 1947]); Patricia Jasen, *Wild Things: Nature, Culture, and Tourism in Ontario,
1790–1914* (Toronto: University of Toronto Press, 1995); Cecila Morgan,
*Creating Colonial Pasts: History, Memory, and Commemoration in Southern Ontario,
1860–1980* (Toronto: University of Toronto Press, 2015), 112–71.

74 "Which Way," *The Standard* (St. Catharines), 25 August 1936, 12; "City
    Hall," *The Standard*, 9 September 1936, 8.

75 "Monument at City Hall," *The Standard* (St. Catharines), 23 July 1954, 9.

76 See, for example, City of St. Catharines, *St. Catharines: Holiday in Canada's
    Garden Land* (St. Catharines: City of St. Catharines, [ca. 1946]); City of St.
    Catharines, *Welcome to St. Catharines Ontario* (St. Catharines: City of St.
    Catharines, [ca. 1960]).

77 R. Janet Powell and Barbara F. Coffman, *Lincoln County 1856–1956* (St.
    Catharines: Lincoln County, 1956), 92.

78 "Part of Funeral Procession," *The Standard* (St. Catharines), 18 January
    1936, 14; A.N. Mowatt, "Decoration Date," *The Standard*, 11 May 1946, 12;
    "Monument at City Hall," *The Standard*, 23 July 1954, 9; "Chilly Guard
    Duty," *The Standard*, 29 October 1965, 1; "Watson's Watch," *The Standard*, 6
    April 1971, 9; "He Still Stands on Guard!" *The Standard*, 8 October 1983, 8;
    Gail Robertson, "Century-Old Statue," *The Standard*, 30 July 1985, 9; Dennis
    Gannon, "Yesterday and Today," *The Standard*, 12 May 1997, B3; Monique
    Beech, "Decaying Statue Stirs Debate," *The Standard*, 16 November 2009, A1.

79 City of St. Catharines, *Council Minutes*, 31 March 1971; "Watson's Watch,"
    The Standard (St. Catharines), 6 April 1971, 9.

80 Lest We Forget [pseud.], "The Watson Statue," *The Standard* (St.
    Catharines), 16 April 1971, 5.

81 "A Short-lived Memorial," *The Standard* (St. Catharines), 1 June 1971, 4.

82 Norman Macdonald, telephone interview with Michael Ripmeester, 21
    October 2005; City of St. Catharines, *Council Minutes*, 26 May 1971, 2;
    "Third Restoration Firm," *The Standard* (St. Catharines), 29 April 1971, 9;
    Andrew Dreschel, "Has Time Tainted Hero's Renown?," *The Standard*, 22
    May 1982, 9.

83 Donald Swainson, "Rieliana and the Structure of Canadian History," *Journal
    of Popular Culture* 14, no. 2 (Fall 1980): 286–97, https://doi.org/10.1111
    /j.0022-3840.1980.1402_286.x; Carl Berger, *The Writing of Canadian History*,
    2nd ed. (Toronto: University of Toronto Press, 1986), 282–92; Desmond
    Morton, "Reflections on the Image of Louis Riel a Century After," in *Images
    of Louis Riel in Canadian Culture*, ed. Ramon Hathorn and Patrick Holland

(Lampeter, UK: Edwin Mellen, 1992), 47–62; Claude Rocan, "Images of Riel in Contemporary School Textbooks," in Hathorn and Holland, *Images of Louis Riel*, 93–126.

84 Monique Dumontet, "Essay 16: Controversy in the Commemoration of Louis Riel," in *Mnemographia Canadensis*, vol. 2, *Remember and See*, ed. D.M.R. Bentley (London, ON: Canadian Poetry Press, 1999), 89–111; Brian S. Osborne, "Corporeal Politics and the Body Politic: The Re-presentation of Louis Riel in Canadian Identity," *International Journal of Heritage Studies* 8, no. 4 (December 2002): 303–22, https://doi.org/10.1080/135272502 2000037209; Frances W. Kaye, *Hiding the Audience: Viewing Art and Arts Institutions on the Prairies* (Edmonton: University of Alberta Press, 2003), 185–226; Michael Dawson, *The Mountie from Dime Novel to Disney* (Toronto: Between the Lines, 1998), 109–43; Alan McCullough, "Parks Canada and the 1885 Rebellion/Uprising/Resistance," *Prairie Forum* 27, no. 2 (Fall 2002): 161–98; Diane P. Payment, *The Free People – Li Gens Libre: A History of the Métis Community of Batoche, Saskatchewan* (Calgary: University of Calgary, 2009), 282–99; Chris Andersen, "More than the Sum of Our Rebellions: Métis Histories beyond Batoche," *Ethnohistory* 61, no. 4 (Fall 2014): 609–33, https://doi.org/10.1215/00141801-2717795.

85 Chester Brown, *Louis Riel: A Comic Strip Biography* (Montreal: Drawn and Quarterly, 2003); John Coulter, *Riel: A Play in Two Parts* (Toronto: Ryerson, 1962); Harry Somers, Mavor Moore, and Jacques Languirand, *Louis Riel* [33⅓ rpm] (Centrediscs, 1985); Robert Williston, "Louis Riel," Museum of Canadian Music, accessed 31 January 2018, http://citizenfreak.com /playlists/190-louis-riel.

86 George Bloomfield, dir., *Riel* [video] (Canadian Broadcasting Corporation, 1979); Gil Cardinal, dir., *Big Bear* [video] (BFS Entertainment, 1998); Dominion Institute, prod., *The Retrial of Louis Riel* [video] (CBC Newsworld, 2003); Carolyn Strange, "Hybrid History and the Retrial of the Painful Past," *Crime, Media, Culture: An International Journal* 2, no. 2 (August 2006): 197–215, https://doi.org/10.1177/1741659006065419; Denise Boitea and David Stansfield, dirs., "Forming a Nation," episode 16 of *Origins: A History of Canada* [video] (TVOntario, 1986).

87 James Monro and Bill Maylone, dirs., *Making History: Louis Riel and the North West Rebellion of 1885* [video] (National Film Board of Canada, 1997); Mark Starowicz, prod., "Taking the West," episode 10 of *Canada: a People's History* [video] (Canadian Broadcasting Corporation, 2001); Chris Wattie, "87% Vote to Acquit Riel," *National Post*, 24 October 2002, A14.

88 John N. Jackson and Sheila M. Wilson, *St. Catharines: Canada's Canal City* (St. Catharines: St. Catharines Standard, 1992); Robert Shipley, Wesley B. Turner, and Richard M. Pearce, *St. Catharines: Garden on the Canal* (Burlington: Windsor, 1987); St. Catharines Chamber of Commerce, *Tour*

Operators Manual for St. Catharines, Ontario (St. Catharines: Chamber of Commerce, 1985); City of St. Catharines, *Economic Development and Tourism Services, St. Catharines: The Heart of Niagara* [video] (St. Catharines: Omnimedia, [ca. 2000]); "Advertising Works," *The St. Catharines Centennial Book* (St. Catharines: Lincoln Graphics, 1975).

89 James Wood, *Militia Myths: Ideas of the Canadian Citizen Soldier, 1896–1921* (Vancouver: UBC Press, 2010), 19–50.

90 R.L. Rogers, *History of the Lincoln and Welland Regiment* (Ottawa: Lincoln and Welland Regiment, 1954).

91 Dreschel, "Has Time Tainted Hero's Renown?," 9.

92 Robert R. Taylor, *Discovering St. Catharines' Heritage: The Old Town* (St. Catharines: St. Catharines Local Architectural Conservation Advisory Committee, 1981), 15; Robert R. Taylor, *Discovering St. Catharines' Heritage*, 2nd ed. (St. Catharines: Vanwell, 1992), 15. Taylor left the monument out of a historical photo book. See Taylor, *St. Catharines: Touring St. Catharines in a REO circa 1910–1920* (St. Catharines: St. Catharines Museum, 1992).

93 Gail Robertson, "Century-Old Statue," *The Standard* (St. Catharines), 30 July 1985, 9.

94 The Carousel Players, quoted in *Grand Old Lady: A Musical Revue* [libretto] (St. Catharines: Carousel Players, 1976).

95 David MacKenzie, correspondence with Russell Johnston (1 May 2018); Buffy Sainte-Marie, "The Universal Soldier," *It's My Way!* [LP] (Vanguard, 1964). The song observes that no general would succeed without the common foot soldier, noting, "He's the universal soldier / And he really is to blame."

96 "A Memorial Tablet," *The Standard* (St. Catharines), 24 December 1900, 2.

97 Knowles, *Inventing the Loyalists*; Vance, *Death So Noble*.

98 A.N. Mowatt, "Decoration Date," *The Standard* (St. Catharines), 11 May 1946, 12.

4. Residents Engage the Watson Monument

1 Details of the authors' 2008 survey are provided in chapter 1.

2 Caroline Grech, "Ceremony to Honour Heroics of Soldier Killed in Action: Statue in Front of City Hall Unknown to Many Local Residents," *The Standard* (St. Catharines), May 13, 2002, A5.

3 Truth and Reconciliation Commission of Canada, *Canada's Residential Schools: The Final Report* (Montreal: Truth and Reconciliation Commission of Canada/McGill-Queen's University Press, 2016); Phillip Morris, "As Monuments Fall, How Does the World Reckon with a Racist Past?," *National Geographic*, 29 June 2020, https://www.nationalgeographic.com/history/2020/06/confederate-monuments-fall-question-how-rewrite-history/.

4 Kenneth E. Foote and Maoz Azaryahu, "Toward a Geography of Memory: Geographical Dimensions of Public Memory and Commemoration," *Journal of Political & Military Sociology* 35, no. 1 (Summer 2007): 125–44; Eviatar Zerubavel, *Time Maps: Collective Memory and the Social Shape of the Past* (Chicago: University of Chicago Press, 2003), 3; Margaret Conrad, Kadriye Ercikan, Gerald Friesen, Jocelyn Letourneau, Delphine Muise, David Northrup, and Peter Seixas, *Canadians and Their Pasts* (Toronto: University of Toronto Press, 2013).

5 Brian S. Osborne, "Corporeal Politics and the Body Politic: The Representation of Louis Riel in Canadian Identity," *International Journal of Heritage Studies* 8, no. 4 (December 2002): 303–22, https://doi.org/10.1080/1352725022000037209; H.V. Nelles, *The Art of Nation Building* (Toronto: University of Toronto Press, 1999); Clifton Hood, "An Unusable Past: Urban Elites, New York City's Evacuation Day, and the Transformations of Memory Culture," *Journal of Social History* 37, no. 4 (Summer 2004): 884–913, https://doi.org/10.1353/jsh.2004.0050; Sarah J. Purcell, "Commemoration, Public Art, and the Changing Meaning of the Bunker Hill Monument," *Public Historian* 25, no. 2 (Spring 2003): 55–71, https://doi.org/10.1525/tph.2003.25.2.55.

6 Jan Assman and John Czaplicka, "Collective Memory and Cultural Identity," *New German Critique* 65 (Spring/Summer 1995): 125–33, https://doi.org/10.2307/488538.

7 Peter Stupples, "Visual Culture, Synthetic Memory and the Construction of National Identity," *Third Text* 17, no. 2 (June 2003): 127–39, https://doi.org/10.1080/09528820309661.

8 Eviatar Zerubavel, *Taken for Granted: The Remarkable Power of the Unremarkable* (Princeton, NJ: Princeton University Press), 22.

9 Elliot Turiel, *The Development of Social Knowledge: Morality and Convention* (Cambridge: Cambridge University Press, 1993); Zerubavel, *Taken for Granted*; Stupples, "Visual Culture."

10 John Berger, *Ways of Seeing* (London: British Broadcasting Corporation; Harmondsworth, UK: Penguin, 1972); Michael Baxandall, *Painting and Experience in Fifteenth Century Italy* (Oxford: Oxford University Press, 1972).

11 M. Christine Boyer, *The City of Collective Memory: Its Historical Imagery and Architectural Entertainments* (Cambridge, MA: MIT Press, 1996). For more on visual arguments see, Gillian Rose, *Visual Methodologies: An Introduction to Researching with Visual Materials*, 4th ed. (London: Sage Publications, 2016); David Birdsell and Leo Groarke, "Toward a Theory of Visual Argument," *Argumentation and Advocacy* 33, no. 1 (Summer 1996): 1–10; J. Anthony Blair, "The Possibility and Actuality of Visual Argument," *Argumentation and Advocacy* 33, no. 1 (Summer 1996): 23–39.

12 Zerubavel, *Taken for Granted*; Michael N. Billig, *Banal Nationalism* (London: Sage, 1995).

13 Sara Ahmed, "Affective economies," *Social Text* 22, no. 2 (Summer 2004): 117–39, https://doi.org/10.1215/01642472-22-2_79-117; Rumi Sakamoto, "Mobilizing Affect for Collective War Memory: Kamikaze Images in Yūshūkan," *Cultural Studies* 29 no. 2 (March 2015): 158–84, https://doi.org/10.1080/09502386.2014.890235; William Connolly, *Neuropolitics: Thinking, Culture, Speed* (Minneapolis: University of Minnesota Press, 2002).

14 Connolly, *Neuropolitics*, 34.

15 Sakamoto, "Mobilizing Affect for Collective War Memory," 175; Ahmed "Affective Economies."

16 Sakamoto, "Mobilizing Affect for Collective War Memory," 175

17 Grant Lafleche, "Winks and Lincs Film Finds Home on U.S. TV," *The Standard* (St. Catharines), 24 February 2004, A1–A2; editorial, "U.S. TV Teaches Us a Lesson," *The Standard*, 25 February 2004, A8.

18 Participants named other sources of knowledge with respect to local history, but they were statistically insignificant. Only five participants identified formal education as a source. Our screen prevented most students from participating in the survey, but we were mildly surprised that more adults did not include it. Other sources named were city council, local churches, and the police.

19 Nigel Taylor, "The Aesthetic Experience of Traffic in the Modern City," *Urban Studies* 40, no. 8 (July 2003): 1609–25, https://doi.org/10.1080/0042098032000094450. See also Lewis Mumford, *The Culture of Cities* (New York: Harcourt, Brace, and World, 1938); Paul C. Adams, "Peripatetic Imagery and Peripatetic Sense of Place," in *Textures of Place: Exploring Humanist Geographies*, ed. Paul C. Adams, Stephen Hoelscher, and Karen E. Till (Minneapolis: University of Minnesota Press, 2001), 186–206; Mimi Sheller and John Urry, "The City and the Car," *International Journal of Urban and Regional Research* 24, no. 4 (December 2000): 737–57, https://doi.org/10.1111/1468-2427.00276.

20 The *Maid of the Mist* was a tour boat on the Niagara River that carried passengers towards the base of Niagara Falls. Boats with this name had operated from the Canadian and American sides of the river since the 1840s, but the Canadian operations stopped using the name in 2012.

21 Erika Doss, *Memorial Mania: Public Feeling in America* (Chicago: University of Chicago Press, 2010); Erika Doss, "Death, Art, and Memory in the Public Sphere: The Visual and Material Culture of Grief in Contemporary America," *Mortality* 7, no. 1 (March 2002): 63–82, https://doi.org/10.1080/13576270120102553.

22 Barry Strauss, "Reflections on the Citizen-Soldier," *Parameters* 33, no. 2 (Summer 2003): 66–78, https://doi.org/10.55540/0031-1723.2151; R. Claire

Snyder, "The Citizen-Soldier Tradition and Gender Integration of the U.S. Military," *Armed Forces and Society* 29, no. 2 (Winter 2003): 185–204, https://doi.org/10.1177/0095327X0302900203; James Wood, *Militia Myths: Ideas of the Canadian Citizen Soldier, 1896–1921* (Vancouver: UBC Press, 2010).

23 City of St. Catharines, General Meeting Minutes, 9 November 2009, 16–20: item no. 547, "Report from the Recreation and Community Services Department," October 27, 2009, re: War Memorial and Outdoor Art Condition Assessments.

24 City of St. Catharines, "Report from the Recreation and Community Services Department," 17.

25 City of St. Catharines, "Report from the Recreation and Community Services Department," 20.

26 City of St. Catharines, Public Art Advisory Committee, Minutes, 18 February 2010, 3.

27 Monique Beech, "Decaying Statue Stirs Debate," *The Standard* (St. Catharines), 16 November 2009, A1.

28 Beech, "Decaying Statue."

29 Wayne Roberts, "Watson Monument in St. Catharines," APTN News (Winnipeg), originally aired on 30 November 2009.

30 Raymond (@Ojibray), "Leave the Watson statue where it is – St. Catharines Standard," Twitter, 23 November 2009, 7:22 a.m., https://twitter.com/Ojibray/status/5974876656. See also "Letters to the editor," *The Standard* (St. Catharines), 25 November 2009), A8.

31 Inside Niagara (@InsideNiagara), "'Mr. Watson – Come here – I want to see you.' Monument in St. Catharines for Pt. Alexander Watson. A real #Canadian hero," Twitter, 27 April 2015, 11:22 a.m., https://twitter.com/InsideNiagara/status/592710350996434947/photo/1. Inside Niagara was the Twitter account for a blog of the same name that featured very occasional news stories about curiosities in Niagara.

32 City of St. Catharines, Public Art Advisory Committee, Minutes, 13 December 2017, 3; City of St. Catharines, Regular Council Minutes, 27 July 2020, 5–9.

33 Li Cohen, "It's Been Over Three Months since George Floyd Was Killed by Police. Police Are Still Killing Black People at Disproportionate Rates," *CBS News*, 10 September 2020, https://www.cbsnews.com/news/george-floyd-killing-police-black-people-killed-164/; Derek Major, "Police Killed at Least 164 Black People in the First Eight Months of 2020," *Black Enterprise*, 16 September 2020, https://www.blackenterprise.com/police-killed-at-least-164-black-people-in-the-first-eight-months-of-2020/.

34 Elizabeth Day, "#BlackLivesMatter: The Birth of New Civil Rights Movement," *The Guardian*, 19 July 2015, https://www.theguardian.com/world/2015/jul/19/blacklivesmatter-birth-civil-rights-movement; Black

Lives Matter, official Twitter feed, accessed 21 September 2020, https://
twitter.com/Blklivesmatter. See also Lynn Schofield Clark, "Participants
on the Margins: #BlackLivesMatter and the Role that Shared Artifacts of
Engagement Played among Minoritized Political Newcomers on Snapchat,
Facebook, and Twitter," *International Journal of Communication* 10 (2016):
235–53; Jelani Incea, Fabio Rojasa, and Clayton A. Davis, "The Social Media
Response to Black Lives Matter: How Twitter Users Interact with Black Lives
Matter through Hashtag Use," *Ethnic and Racial Studies* 40, no. 11 (September
2017): 1814–30, https://doi.org/10.1080/01419870.2017.1334931.

35 Zamira Rahim and Rob Picheta, "Thousands around the World Protest
George Floyd's Death in Global Display of Solidarity," *CNN*, 1 June 2020,
https://www.cnn.com/2020/06/01/world/george-floyd-global-protests
-intl; John Leicester and Frank Jordans, "A Look at Black Lives Matter
Protests from around the World," *Globe and Mail*, 6 June 2020, https://
www.theglobeandmail.com/world/article-a-look-at-black-lives-matter-
protests-from-around-the-world; Thomas Wieder, Cécile Ducourtieux, and
Jean-Pierre Stroobants, "Des manifestations au nom de Black Lives Matter
partout en Europe," *Le Monde*, 8 June 2020, https://www.lemonde.fr
/international/article/2020/06/08/des-manifestations-au-nom-de-black-
lives-matter-partout-en-europe_6042095_3210.html; Alan Taylor, "Images
from a Worldwide Protest Movement," *The Atlantic*, 8 June 2020, https://
www.theatlantic.com/photo/2020/06/images-worldwide-protest-
movement/612811.

36 Mark Spence, *Dispossessing the Wilderness: Indian Removal and the Making
of the National Parks* (New York: Oxford University Press, 1999); Jason E.
Black, "Governmental Colonizing Rhetoric During Indian Removal,"
chap. 2 in *American Indians and the Rhetoric of Removal and Allotment*
(Jackson: University Press of Mississippi, 2015), 37–58; Alfred E. Cave,
Sharp Knife: Andrew Jackson and the American Indians (Santa Barbara, CA:
Praeger, 2017); Jeffrey Ostler, *Surviving Genocide: Native Nations and the
United States from the American Revolution to Bleeding Kansas* (New Haven,
CT: Yale University Press, 2019).

37 Wikipedia, s.v. "List of Monuments and Memorials Removed during
the George Floyd Protests," last modified 12 June 2023, 05:57, https://
en.wikipedia.org/wiki/List_of_monuments_and_memorials_removed
_during_the_George_Floyd_protests. See also "How Statues Are Falling
Around the world," *New York Times*, 24 June 2020, updated 12 September
2020, https://www.nytimes.com/2020/06/24/us/confederate-statues
-photos.html; Jon Quelly, "New Interactive Map Details 67 Confederate
Monuments (and Counting) Removed since George Floyd Murder,"
Common Dreams, 10 July 2020, https://www.commondreams.org
/news/2020/07/10/new-interactive-map-details-67-confederate

-monuments-and-counting-removed-george; Sarah Rankin and David Crary, "Countries Seek 'New History' as Figures Are Reexamined after George Floyd's Death," *Global News*, 12 June 2020, updated 23 June 2020, https://globalnews.ca/news/7058287/us-statues-monuments-racism.

38 "Anti-Black Racism Protests, Vigils Take Place across Canada," *CBC News*, 6 June 2020, https://www.cbc.ca/news/canada/canada-anti-black -racism-protests-solidarity-rallies-1.5601792; "These Windsorites Say the Black Lives Matter Movement Is Empowering, Educating People Locally," *CBC News*, 8 June 2020, https://www.cbc.ca/news/canada/windsor /black-lives-matter-protests-windsor-movement-1.5602902.

39 Allan Benner, "Residents Share Stories of Racism at Niagara-on-the -Lake Rally," *The Standard* (St. Catharines), 6 June 2020, A1; Ray Spiteri, "Racism Is Insidious, It's Systemic," *Niagara Falls Review*, 8 June 2020, A1; Allan Benner, "Black Lives Matter Protests Set to Continue Sunday in St. Catharines," *The Standard*, 12 June 2020, A3; Bob Tymczyszyn, "The Stereotype Is that Black Is Less," *The Standard*, 15 June 2020, A1.

40 Jake Breadman (@jrgbreadman), "If I started a petition or campaign to remove or recontextualize the Watson Monument in downtown St. Catharines, would you sign or join?," Twitter, 9 June 2020, 10:26 p.m., https://twitter.com/hlafweard/status/1270542703806332930; Jake Breadman (@jrgbreadman), "The monument is dedicated to Private Alexander Watson, a Canadian soldier that died at the Battle of Batoche during the Northwest Uprising in 1885. To better understand what the monument commemorates, and why it's problematic, let's go back to 1867," Twitter, 11 June 2020, 5:28 p.m., https://twitter.com/hlafweard /status/1271192521641791488. It should be noted that Breadman also tweeted similar statements at the time from the now defunct @hlafweard account, but these have since been deleted.

41 Gavin Fearon (@madestanding), "Let's talk about the statue at St Catharines City Hall. This is Private Alexander Watson. Watson lived in St Catharines for a time and died in 1885, fighting on the side of Canadian government in the Battle of Batoche, the deciding battle in the North-West Rebellion" and "CC @Wsendzik @karrieporter @MatSiscoe," Twitter, 17 June 2020, 9:06 a.m., https://twitter.com/madestanding/status/1273285841822105602 (last accessed 10 November 2020). The @madestanding account is now defunct but the quoted tweets are archived in the Wayback Machine at https://web. archive.org/web/20200617161047/https://twitter.com/madestanding/ status/1273285835836850178. See also Leanne Kurek (administrative assistant to the mayor, City of St. Catharines), email to Russell Johnston, 2 November 2020; Rhiannon Fleming (digital communications and special events coordinator, Community Care of St. Catharines and Thorold), email to Russell Johnston, 2 November 2020; "Remove the Watson Monument from St. Catharines City Hall," Community Petitions by Avaaz, accessed 17 July

2020, https://secure.avaaz.org/community_petitions/en/st_catharines_city_council_remove_the_watson_monument_from_st_catharines_city_hall.

42 Karrie Porter (@karrieporter), "Gavin, this is great. I just had a discussion with the co-chair of our Anti-Racism Committee suggesting a review of our street and park naming policies. I put in a notice of motion for removing this statue, as well as requesting our city's equity committees review naming policies," Twitter, 17 June 2020, 4:45 p.m., https://twitter.com/madestanding/status/1273285841822105602 (last accessed 10 November 2020). This tweet is no longer available. See also Karrie Porter interviewed by Tom McConnell, *Tom McConnell Show*, 610 CKTB, 17 June 2020.

43 Karena Walter, "Statue May Be Removed from St. Catharines City Hall Lawn," *The Standard* (St. Catharines), 19 June 2020, A1–A2; Bryan Levesque, "Online Petition Calls for Removal of Statue at City Hall," *Niagara This Week*, 25 June 2020, 6.

44 Gavin Fearon interviewed by Tim Denis, *Niagara in the Morning*, 610 CKTB, 18 June 2020; Karrie Porter interviewed by Tom McConnell, *Tom McConnell Show*, 610 CKTB, 17 June 2020; Karrie Porter interviewed by Chrissy Sadowski, *Parental Guidance*, 610 CKTB, 19 June 2020; Kathy Powell interviewed by Matt Holmes, *Weekend Edition*, 610 CKTB, 20 June 2020; Russell Johnston interviewed by Shelby Knox, *Niagara in the Morning*, 610 CKTB, 22 June 2020. The chair of the Niagara Region Anti-Racism Association was interviewed after the decisive meeting took place; Saleh Waziruddin interviewed by Tim Denis, *Niagara in the Morning*, 610 CKTB, 29 July 2020.

45 Jason Gaidola, "Call to Remove the Alexander Watson Statue," *CHCH Evening News*, 18 June 2020.

46 Doug Draper, "City of St. Catharines Undertaking Broad Consultation on Relocation of Contentious Statue," *Niagara at Large*, 29 July 2020, https://niagaraatlarge.com/2020/07/29/city-of-st-catharines-undertaking-broad-consultation-on-relocation-of-contentious-statue; Doug Draper, "Anti-Racism Group Urges Removal of Statue at St. Catharines City Hall," *Niagara at Large*, 27 July 2020, https://niagaraatlarge.com/2020/07/27/anti-racism-group-urges-removal-of-statue-at-st-catharines-city-hall.

47 Kevin Vallier, "A Monument on the Move?," *Niagara Independent*, 31 July 2020, https://niagaraindependent.ca/a-monument-on-the-move.

48 City of St. Catharines, Regular Council Minutes, 22 June 2020, 4–5; City of St. Catharines, Regular Council Minutes, 27 July 2020, 5–9.

49 Alexa Internet's website rankings involved a weighted three-month rolling average of scores for daily views per visitor, daily time spent per visitor, and the number of other websites linking in. "About Us," Alexa Internet, accessed 11 September 2020, https://www.alexa.com/about; "Topsites: Canada," Alexa Internet, accessed 11 November 2020, https://www.alexa.com/topsites/countries/CA. Alexa Internet closed operations

on 1 May 2021. Select ranking data is available through the Wayback Machine at https://archive.org.

50 Deen Freelon, Charlton McIlwain, and Meredith Clark, "Quantifying the Power and Consequences of Social Media Protest," *New Media & Society* 20, no. 3 (March 2018): 990–1011, https://doi.org/10.1177/1461444816676646; W. Carson Byrd, Keon L. Gilbert, and Joseph B. Richardson Jr., "The Vitality of Social Media for Establishing a Research Agenda on Black Lives and the Movement," *Ethnic and Racial Studies* 40, no. 11 (September 2017): 1872–81, https://doi.org/10.1080/01419870.2017.1334937.

51 "About," Reddit, accessed 21 June 2020, https://www.redditinc.com; "Moderator Guidelines for Healthy Communities," Reddit, accessed 21 June 2020, https://www.redditinc.com/policies/moderator-guidelines. Reddit has since revised both of these pages. The relevant pages from June 2020 are available through the Wayback Machine at https://archive.org. See also "Yeah, This Should Probably Happen," Reddit, 18 June 2020, https://www.reddit.com/r/stcatharinesON/comments/hbqguh/yeah_this_should_probably_happen/.

52 One Dish, One Mic, "Trust and Reconciliation Shouldn't Be Easy," Facebook, 28 July 2020, https://www.facebook.com/OneDishOneMic/posts/33036961463220592.

53 John Wylie Sr., "Dear Ms Karrie Porter," Facebook, 26 June 2020, https://www.facebook.com/KarriePorterCouncillor/posts_to_page.

54 Shelby Knox, "Watson Statue and Parking Fees to Be Discussed at Tonight's St. Catharines City Council Meeting," iHeartRadio, 22 June 2020, https://www.iheartradio.ca/610cktb/news/watson-statue-and-parking-fees-to-be-discussed-at-tonight-s-st-catharines-city-council-meeting-1.12768901; Tami Jeanneret, "Private Watson's Future Location to Be Determined," iHeartRadio, 28 July 2020, https://www.iheartradio.ca/610cktb/news/private-watson-s-future-location-to-be-determined-1.13110191.

55 Kathy Powell interviewed by Matt Holmes, *Weekend Edition*, 610 CKTB, 20 June 2020.

56 Vividata, *Newspaper Topline Readership – Average Weekly Audience: Vividata Spring 2020 Adults 18+* (Toronto: Vividata, 2020); Numeris, "St. Catharines-Niagara CTRL," *Top-Line Radio Statistics, Spring 2020* (Toronto: Numeris, 2020); "About Us," *Niagara This Week*, accessed 9 December 2020, https://www.niagarathisweek.com/community-static/2544689-niagarathisweek-com-about-us.

57 Anton Barhan and Andrey Shakhomirov, "Methods for Sentiment Analysis of Twitter Messages," *Proceedings of the 12th Conference of Open Innovations Association FRUCT* 325, no. 12 (2012): 215–22; Paramita Ray, Amlan Chakrabarti, Bhaswati Ganguli, and Pranab Kumar Das, "Demonetization and Its Aftermath: An Analysis Based on Twitter Sentiments," *Sādhanā* 43,

no. 11 (November 2018): 1–10, https://doi.org/10.1007/s12046-018-0949-0; Axel Bruns and Jean Burgess, "Researching News Discussion on Twitter: New Methodologies," *Journalism Studies* 13, nos. 5–6 (October 2012): 801–14, https://doi.org/10.1080/1461670X.2012.664428.

58 Here and elsewhere in this chapter, we use initials to protect the identity of posters.

59 Dan McKnight (@danmcknight) "'To you, from failing hands we throw, the Torch; be it yours to hold high. But, if ye break faith with us who die, We shall not sleep. Though Poppies grow in Flanders Fields.' 🦋 #Remembrance #InFlandersFields #Canada," Twitter, 28 June 2020, 2:29 p.m., https://twitter.com/danmcknight/status/1288179900608917504.

60 Sean Fine, "Nunavut RCMP Accused of Failing in Duty to Protect Inuk man," *Globe and Mail*, 3 June 2020 (updated 4 June 2020), https://www.theglobeandmail.com/canada/article-nunavut-rcmp-accused-of-failing-in-duty-to-protect-inuk-man-knocked; Charles Rusnell and Jennie Russell, "RCMP Dashcam Video Shows Officer Tackling, Punching Chief Allan Adam during Arrest," *CBC News*, 11 June 2020, https://www.cbc.ca/news/canada/edmonton/rcmp-chief-allan-adam-1.5608472; Alexander Quon and Silas Brown, "First Nations Demand Immediate Action in Death of Chantel Moore in Edmundston, N.B.," *Global News*, 5 June 2020, https://globalnews.ca/news/7029776/bc-first-nation-chantel-moore.

61 Bill Jensen, "Save the Statue of Private Alexander Watson," iPetitions, accessed 21 September 2020, https://www.ipetitions.com/petition/save-the-statue-of-private-alexander-watson.

62 Jake Breadman (@jrgbreadman), "I've seen a lot of discussion on monuments recently and I want to write down some of my thoughts on the Watson Monument in downtown St. Catharines and why I think it should be removed or recontextualized," Twitter, 11 June 2020, 5:28 p.m., https://twitter.com/hlafweard/status/1271192520559640579.

63 Gavin Fearon interviewed by Tim Denis, *Niagara in the Morning*, 610 CKTB, 18 June 2020.

64 Karl Dockstader and Sean Vanderklis, *One Dish, One Mic*, 610 CKTB, 21 June 2020.

65 Saleh Waziruddin, Erika Smith, and Marcel Steward for the Niagara Region Anti-Racism Association, "Anti-Racism Group Urges Removal of Statue at St. Catharines City Hall," news release reported in *Niagara at Large*, 27 June 2020, https://niagaraatlarge.com/2020/07/27/anti-racism-group-urges-removal-of-statue-at-st-catharines-city-hall/.

66 Niagara Region Anti-Racism Committee, "Written Submission to St. Catharines City Council," 27 July 2020; Niagara Region Anti-Racism

Association, "For Those Who Missed This Morning's Interview," Facebook, 29 July 2020, https://www.facebook.com/TheNRARA.

67 David Olusoga, "The Toppling of Edward Colston's Statue Is Not an Attack on History," *The Guardian*, 9 June 2020, https://www.theguardian.com/commentisfree/2020/jun/08/edward-colston-statue-history-slave-trader-bristol-protest.

68 James Young, "The Counter-Monument: Memory against Itself in Germany Today," *Critical Inquiry* 18, no. 2 (Winter 1992): 236, https://doi.org/10.1086/448632.

69 Benedict Anderson, *Imagined Communities: Reflections on the Origin and Spread of Nationalism* (London: Verso, 1983); Eric Hobsbawm and Terence Ranger, eds., *The Invention of Tradition* (Cambridge: Cambridge University Press, 1983); Michael Ignatieff, *Blood and Belonging: Journeys into the New Nationalism* (New York: Farrar, Straus and Giroux, 1994).

70 Thomas Janoski, *Citizenship and Society: A Framework of Rights and Obligations in Liberal, Traditional, and Social Democratic Regimes* (Cambridge: Cambridge University Press, 1998); Sara Helman, "Negotiating Obligations, Creating Rights: Conscientious Objection and the Redefinition of Citizenship in Israel," *Citizenship Studies* 3, no. 1 (February 1999): 45–69, https://doi.org/10.1080/13621029908420700; Sonya O. Rose, "Women's Rights, Women's Obligations: Contradictions of Citizenship in World War II Britain," *European Review of History* 7, no. 2 (August 2000): 277–89, https://doi.org/10.1080/713666747.

71 Ignatieff, *Blood and Belonging*; Helman, "Negotiating Obligations, Creating Rights," 49.

72 Janet Donohue, "Dwelling with Monuments," *Philosophy and Geography* 5, no. 2 (2002): 240, https://doi.org/10.1080/10903770220152434.

73 Andrew Butterfield, "Monuments and Memories: What History Can Teach the Architects at Ground Zero," *New Republic*, 3 February 2003, 30.

74 Tracey Raney, "Grassroots Patriotism in Canada: Reconstructing National Identity along the 'Highway of Heroes,'" trans. Ethan Rundell, *Critique Internationale* 58, no. 1 (2013): 19–34, https://doi.org/10.3917/crii.058.0019.

75 Desmond Morton and Glenn Wright, *Winning the Second Battle: Canadian Veterans and the Return to Civilian Life, 1915–1930* (Toronto: University of Toronto Press, 1987); Denise Thomson, "National Sorrow, National Pride: Commemoration of War in Canada, 1918–1945," *Journal of Canadian Studies* 30, no. 4 (Winter 1995–6): 5–27, https://doi.org/10.3138/jcs.30.4.5. For more on the Canadian Legion and the War Amps of Canada campaigns, see https://www.legion.ca and https://www.waramps.ca.

76 Veterans Affairs Canada, *Canada Remembers: The Canadian Armed Forces in Afghanistan* [fact sheet] (Ottawa: Veterans Affairs Canada, 2011); "Canadian Forces' Casualty Statistics (Afghanistan)" [news release: project

no. FS 12.002], Government of Canada, 10 June 2013, https://www
.canada.ca/en/news/archive/2013/06/canadian-armed-forces-casualty
-statistics-afghanistan.html.
77 The roots of this celebration of militarism are debated. Some, like
Kozolanka, contend that this was a top-down process. Others, like Raney,
contend that the Conservatives followed changing popular sentiment.
Kirsten Kozolanka, "The March to Militarism in Canada: Domesticating
the Global Enemy in the Post-9/11, Neo-liberal Nation," *Global Media
Journal: Canadian Edition* 8, no. 1 (2015): 31–51; David Mutimer, "The Road
to Afghanada: Militarization in Canadian Popular Culture during the War
in Afghanistan," *Critical Military Studies* 2, no. 3 (September 2016): 210–25,
https://doi.org/10.1080/23337486.2016.1164982.
78 Kozolanka, "The March to Militarism in Canada."
79 Kozolanka, "The March to Militarism in Canada."
80 Dan Lamothe, "Canada's Highway of Heroes: The Patriotic Tradition
Lives on after Afghanistan," *Washington Post*, 11 March 2015, https://
www.washingtonpost.com/news/checkpoint/wp/2015/03/11/canadas
-highway-of-heroes-the-patriotic-tradition-lives-on-after-afghanistan/.
81 Roger MacIsaac quoted in Norma Ramage, "Advertising Is War,"
Marketing 111, no. 32 (2 October 2006): 10.
82 Janis L. Goldie, "Fighting Change: Representing the Canadian Forces
in the 2006–2008 Fight Recruitment Campaign," *Canadian Journal of
Communication* 39, no. 3 (September 2014): 413–30, https://doi.org
/10.22230/cjc.2014v39n3a2768. See also Ian McKay and Jamie Swift,
Warrior Nation: Rebranding Canada in an Age of Anxiety (Toronto: Between
the Lines, 2012).
83 Rob Gerlsback, "Canadian Forces," *Marketing* 112, no. 22 (26 November
2007): 20–2.
84 Paul Ferriss, "Canada's Top Marketers 2007," *Marketing* 112, no. 22
(26 November 2007): 18; Goldie, "Fighting Change.
85 Goldie, "Fighting Change"; McKay and Swift, *Warrior Nation*; Isabelle
Gusse, "Mythes modernes, propagande et communications publicitaires
de l'armée canadienne en 2010," *Global Media Journal – Canadian Edition*
8, no. 1 (2015): 53–69; Tanner Mirrlees, "The Canadian Armed Forces
'YouTube War': A Cross-Border Military-Social Media Complex," *Global
Media Journal – Canadian Edition* 8, no. 1 (2015): 71–93.

5. Viticulture as a Mnemonic Product

1 United Hotels Company, "Niagara Peninsula Is Full of Beauty: Fine Hotels
at Hamilton and Niagara Falls to Accommodate Motorists," *The Globe*
(Toronto), 27 May 1920, 8.

2   See Michael Ripmeester, Philip Mackintosh, and Christopher Fullerton, eds., *The World of Niagara Wine* (Waterloo: Wilfrid Laurier University Press, 2013).

3   Regina Bendix, *Culture and Value* (Bloomington: Indiana University Press, 2018).

4   Jamie Peck, *Constructions of Neoliberal Reason* (Oxford: Oxford University Press, 2010); John Howkins, *The Creative Economy: How People Make Money from Ideas* (London: Penguin, 2002); Richard Florida, *Cities and the Creative Class* (New York: Routledge, 2005); Richard Florida, *The Rise of the Creative Class, and How It's Transforming Work, Leisure, and Everyday Life* (New York: Basic Books, 2002).

5   Terry N. Clark, *The City as an Entertainment Machine* (Amsterdam: Emerald ePublication, 2004); Organisation for Economic Co-operation and Development, *Tourism and the Creative Economy* (Paris: OECD Publishing, 2014).

6   Niagara Economic Development Corporation, *Energizing Niagara's Wine Country Communities* (St. Catharines: Niagara Economic Development Corporation/Peter J. Smith and Company, 2007), 4.

7   Regional Municipality of Niagara, *Niagara Culture Plan: Creative Niagara: Economy, Places, People, Identity* (Thorold: Regional Municipality of Niagara, 2010). See also J. Michael Robbins, "Southern Ontario Tourism Context and the Challenge for a Sustainable Future," *Environments* 24, no. 3 (1997): 50–60; Chandana Jayawardena, "Tourism in Niagara: Identifying Challenges and Finding Solutions," *International Journal of Contemporary Hospitality Management* 20, no. 3 (2008): 249–58, https://doi.org/10.1108/09596110810866073.

8   Greater Niagara Chamber of Commerce, *Blueprint for Economic Growth and Prosperity: Launching a Report Card for Niagara* (St. Catharines: Greater Niagara Chamber of Commerce, 2013), 16; Paul Forsyth, "Brains over Brawn: Niagara Urged to Pursue Creative Class," *Niagara This Week*, 28 February 2013, 23.

9   Philip T. Kotler, Kevin Lane Keller, Subramanian Sivaramakrishnan, and Peggy H. Cunningham, *Marketing Management*, 14th Canadian ed. (Toronto: Pearson Canada, 2012); Jeannette Hanna and Alan Middleton, *Ikonica: A Field Guide to Canada's Brandscape* (Vancouver: Douglas and McIntyre, 2008).

10 Richard W. Pollay, "Measuring the Cultural Values Manifest in Advertising," in *Current Issues and Research in Advertising*, ed. James H. Leigh and Claude R. Martin Jr. (Ann Arbor: Graduate School of Business, Division of Research, University of Michigan, 1983), 72–92; J.S. Johar and M. Joseph Sirgy, "Value-Expressive versus Utilitarian Advertising Appeals: When and Why to Use Which Appeal," *Journal of Advertising* 20, no. 3 (September 1991): 23–33, https://doi.org/10.1080

/00913367.1991.10673345; Salvador Ruiz and Maria Sicilia, "The Impact of Cognitive and/or Affective Processing Styles on Consumer Response to Advertising Appeals," *Journal of Business Research* 57, no. 6 (June 2004): 657–64, https://doi.org/10.1016/S0148-2963(02)00309-0; Felix Septianto and Loren Pratiwi, "The Moderating role of Construal Level on the Evaluation of Emotional Appeal vs. Cognitive Appeal Advertisements," *Marketing Letters* 27, no. 1 (March 2016): 171–81, https://doi.org/10.1007/s11002-014-9324-z; Loraine G. Lau-Gesk, "Activating Culture through Persuasion Appeals: An Examination of the Bicultural Consumer," *Journal of Consumer Psychology* 13, no. 3 (2003): 301–15, https://psycnet.apa.org/doi/10.1207/S15327663JCP1303_11; Defeng Yang, Yue Lua, Wenting Zhub, and Chenting Sub, "Going Green: How Different Advertising Appeals Impact Green Consumption Behavior," *Journal of Business Research* 68, no. 12 (December 2015): 2663–75, https://doi.org/10.1016/j.jbusres.2015.04.004; Dong Hoo Kim and Doori Song, "Can Brand Experience Shorten Consumers' Psychological Distance toward the Brand? The Effect of Brand Experience on Consumers' Construal Level," *Journal of Brand Management* 26, no. 3 (May 2019): 255–67, https://doi.org/10.1057/s41262-018-0134-0.

11 See Pantea Foroudi, *Place Branding: Connecting Tourist Experiences to Places* (New York: Routledge, 2020); Waldemar Cudny, ed., *Urban Events, Place Branding and Promotion: Place Event Marketing* (Abingdon, UK: Routledge, 2019); Mihalis Kavaratzis and Charles Dennis, "Place Branding Gathering Momentum," *Place Branding and Public Diplomacy* 14, no. 2 (May 2018): 75–7, https://doi.org/10.1057/s41254-018-0098-6.

12 Stephen V. Ward, *Selling Places: The Marketing and Promotion of Towns and Cities, 1850–2000* (London: Spon Press, 1998); Dominic Medway and Gary Warnaby, "What's in a Name? Place Branding and Toponymic Commodification," *Environment and Planning A: Economy and Space* 46, no. 1 (January 2014): 153–67, https://doi.org/10.1068/a45571; Demetris Vrontis and Stanley J. Paliwoda, "Branding and the Cyprus Wine Industry," *Journal of Brand Management* 16, no. 3 (December 2008): 145–59, https://doi.org/10.1057/bm.2008.1; Giannina Warren and Keith Dinnie, "Cultural Intermediaries in Place Branding: Who Are They and How Do They Construct Legitimacy for Their Work and for Themselves?," *Tourism Management* 66 (June 2018): 302–14, https://doi.org/10.1016/j.tourman.2017.12.012.

13 John Urry, *The Tourist Gaze*, 2nd ed. (London: Sage Publications, 2002).

14 Erving Goffman, *Relations in Public: Microstudies of the Public Order* (New York: Basic Books, 1971), 255.

15 Raymond Williams, *Culture* (Glasgow: Collins/Fontana, 1981); Michel Foucault, *Discipline and Punish: The Birth of the Prison*, trans. Alan Sheridan

(New York: Vintage, 1995); Jürgen Habermas, *The Structural Transformation of the Public Sphere: An Inquiry into a Category of Bourgeois Society*, trans. Thomas Burger and Frederick Lawrence (Cambridge, MA: MIT Press, 1991); Allan Pred, *Making Histories and Constructing Human Geographies: The Local Transformation of Practice, Power Relations, and Consciousness* (Boulder, CO: Westview Press, 1990); John Bodnar, *Remaking America: Public Memory, Commemoration, and Patriotism in the Twentieth Century* (Princeton, NJ: Princeton University Press, 1992).

16 Robyn Mayes, "A Place in the Sun: The Politics of Place, Identity, and Branding," *Place Branding and Public Diplomacy* 4, no. 2 (May 2008): 125, https://doi.org/10.1057/pb.2008.1. Emphasis in the original.

17 Jeannette Hannah, "Mapping Community Identity and Place," in *Rediscovering the Wealth of Places*, ed. G. Baeker (St. Thomas: Municipal World, 2010), 91–100; James H. Gilmore and Joseph Pine, *Authenticity: What Consumers Really Want* (Cambridge, MA: Harvard University Press, 2007).

18 Hartmut Berghoff, "From Privilege to Commodity? Modern Tourism and the Rise of the Consumer Society," in *The Making of Modern Tourism: The Cultural History of the British Experience, 1600–2000*, ed. Hartmut Berghoff, Barbara Korte, Ralf Schneider, and Christopher Harvie (London: Palgrave Macmillan, 2002), 159–80; John Walton, "British Tourism between Industrialization and Globalization: An Overview," in Berghoff et al., *The Making of Modern Tourism*; Urry, *The Tourist Gaze*, 109–32; Donald J. MacLaurin and Steve Wolstenholme, "An Analysis of the Gaming Industry in the Niagara Region," *International Journal of Contemporary Hospitality Management* 20, no. 3 (2008): 320–31, https://doi.org/10.1108/09596110810866136.

19 Louis Hennepin, *Nouvelle découverte d'un très grand pays situé dans l'Amerique, entre le Nouveau Mexique, et la Mer Glaciale* (Utrecht: G. Proedelet, 1697); G.M. Davison, *The Fashionable Tour: An Excursion to the Springs, Niagara, Quebec, and through the New-England States*, 3rd ed. (Saratoga Springs, NY: G.M. Davison, 1828), 173–90; Patricia Jasen, *Wild Things: Nature, Culture, and Tourism in Ontario, 1790–1914* (Toronto: University of Toronto Press, 1995), 29–54; Michael J. Broadway, "Urban Tourist Development in the Nineteenth-Century Canadian City," *American Review of Canadian Studies* 26, no. 1 (March 1996): 83–99, https://doi.org/10.1080/02722019609480900.

20 Jasen, *Wild Things*, 29–54.

21 Robert G. Healy, "The Commons Problem and Canada's Niagara Falls," *Annals of Tourism Research* 33, no. 2 (April 2006): 525–44, https://doi.org/10.1016/j.annals.2006.01.003.

22 Ontario, Ministry of Tourism, Culture, and Sport (2014), quoted in *Niagara Connects, Living in Niagara 2017: Critical Indicators for Reflecting on Life in*

Niagara (St. Catharines: Niagara Connects, 2018), 10; "Tourism Research," Niagara Falls Tourism, accessed 16 November 2023, https://www.niagarafallstourism.com/media-kit/contact/tourism-research/.

23 See, for example, Joel Dombrowski, *Niagara Falls* (Berkeley, CA: Avalon/ Perseus, 2014), 8–13; "9 Clifton Hill Attractions to Try This Winter," Niagara Falls Tourism, accessed 9 December 2019, https://www.niagarafallstourism.com/blog/9-clifton-hill-attractions-winter/. The website has been revised but a relevant page is archived in the Wayback Machine at https://web.archive.org/web/20190424195045/www.niagarafallstourism.com/blog/9-clifton-hill-attractions-winter/.

24 Jasen, *Wild Things*; George A. Seibel, *Ontario's Niagara Parks 100 Years* (Niagara Falls: Niagara Parks Commission, 1985); Joan Coutu, "Vehicles of Nationalism: Defining Canada in the 1930s," *Journal of Canadian Studies* 37, no. 1 (Spring 2002): 180–203, https://doi.org/10.3138/jcs.37.1.180.

25 Regional Municipality of Niagara, *Know It* [transportation fact sheet] (Niagara Falls: Niagara Economic Development, 2018).

26 J.B. Jackson, *The Welland Canals and Their Communities* (Toronto: University of Toronto Press, 1997); E.A. Cruikshank, *The Story of Butler's Rangers and the Settlement of Niagara* (Welland: Tribune Printing, 1893); Janet Carnochan, *History of Niagara (in Part)* (Toronto: William Briggs, 1914); Peter J. Stokes, *Old Niagara-on-the-Lake* (Toronto: University of Toronto Press, 1971).

27 Cecilia Morgan, *Creating Colonial Pasts: History, Memory, and Commemoration in Southern Ontario, 1860–1980* (Toronto: University of Toronto Press, 2005); Norman Knowles, *Inventing the Loyalists: The Ontario Loyalist Tradition and the Creation of Usable Pasts* (Toronto: University of Toronto Press, 1997); Elaine Young, "Battlefield to Baseball Diamond: The Niagara Parks Commission and Queenston Heights Park," *London Journal of Canadian Studies* 29 (2014): 47–68.

28 Brian Doherty, *Not Bloody Likely: The Shaw Festival, 1962–1973* (Toronto: J.M. Dent & Sons, 1974), 7–12; Arthur R. Day, "The Shaw Festival at Niagara-on-the-Lake in Ontario, Canada, 1962–1981: A History" (PhD diss., Bowling Green State University, 1982), 1–16; L.W. Conolly, *The Shaw Festival: The First Fifty Years* (Oxford: Oxford University Press, 2011), 9–12; Robin Breon, "A Tale of Two Festivals," *American Theater* 29, no. 5 (May–June 2012): 32–6, 76; Shaw Festival, *Shaw Festival 17: Annual Report* (Niagara-on-the-Lake: Shaw Festival, 2018).

29 Joanna Fountain and Daisy Dawson, "The New Gold: The Role of Place and Heritage in the Marketing of the Central Otago Wine Region," in *Wine and Identity, Branding, Heritage Terroir*, ed. Matt Harvey, Leanne White, and Warwick Frost (London: Routledge, 2014), 106–32; Warwick Frost, Jennifer Frost, Paul Strickland, and Jennifer Smith Maguire, "Seeking a Competitive Advantage in Wine Tourism: Heritage and Storytelling at the

Cellar-Door," *International Journal of Hospitality Management* 87 (May 2020), 1–9, https://doi.org/10.1016/j.ijhm.2020.102460.

30 Vanessa Anne Quintal, Ben Thomas, and Ian Phau, "Incorporating the Winescape into the Theory of Planned Behaviour: Examining 'New World' Wineries," *Tourism Management* 46 (February 2015): 596–609, https://doi.org/10.1016/j.tourman.2014.08.013.

31 Robert Ulin, "Invention and Representation as Cultural Capital: Southwest French Winegrowing History," *American Anthropologist* 97, no. 3 (September 1995): 519–27.

32 Daisy Dawson, Joanna Fountain, and David A. Cohen, "Place-Based Marketing and Wine Tourism: Creating a Point of Difference and Economic Sustainability for Small Wineries," *6th AWBR International Conference, 9–10 June 2011* (Bordeaux: Bordeaux Management School, n.d.), n.p.

33 Quintal, Thomas, and Phau, "Incorporating the Winescape into the Theory of Planned Behaviour," 598.

34 Fountain and Dawson, "The New Gold," 110.

35 Fountain and Dawson, 110. Emphasis in the original.

36 Jules Janick, "The Origins of Fruits, Fruit Growing, and Fruit Breeding," in *Plant Breeding Reviews*, vol. 25, ed. Jules Janick (Hoboken, NJ: John Wiley and Sons, 2005), 255–320; Province of Ontario, Ministry of Agriculture, *Fruits of Ontario* (Toronto: William Briggs, 1914), chap. 3; Elizabeth Simcoe, 3 July 1793, *The Diary of Mrs. John Graves Simcoe* (Toronto: William Briggs, 1911); Robert L. Jones, *History of Agriculture in Ontario, 1613–1880* (Toronto: University of Toronto Press), 17–35; G. Elmore Reaman, *A History of Agriculture in Ontario*, vol. 1 (Toronto: Saunders, 1970), 13–18, 51–4; John N. Jackson, *St. Catharines, Ontario: Its Early Years* (Belleville: Mika, 1976), 61–101; Ross D. Fair, "Gentlemen, Farmers, and Gentlemen Half-Farmers: The Development of Agricultural Societies in Upper Canada, 1792–1846" (PhD diss., Queen's University at Kingston, 1998), 65–6; John McCallum, *Unequal Beginnings: Agriculture and Economic Development in Quebec and Ontario until 1870* (Toronto: University of Toronto Press, 1980), 9–24; R.M. McInnis, "Perspectives on Ontario Agriculture, 1815–1930," *Canadian Papers in Rural History*, vol. 8 (Gananoque: Langdale, 1992), 49–90.

37 Jackson, *St. Catharines*, chap. 1; McCallum, *Unequal Beginnings*, 75–82; Daryll Crewson and Ralph Matthews, "Class Interests in the Emergence of Fruit Growing Cooperation in Lincoln County, Ontario, 1880–1914," in *Canadian Papers in Rural History*, vol. 5 (Gananoque: Langdale, 1986), 23–49. See also R. Janet Powell and Barbara F. Coffman, *Lincoln County 1856–1956* (St. Catharines: Lincoln County Council, 1956), 35–41; William F. Rannie, *Lincoln: The Story of an Ontario Town* (Beamsville: W.F. Rannie, 1974), 55–80.

38 Reaman, *History of Agriculture in Ontario*, 1:146–7; David P.R. Guay, *Great Western Railway of Canada* (Toronto: Dundurn, 2015), 25–81. The Great Western station in Niagara Falls, built in 1879, remained in use with Via Rail in the 2020s.

39 Powell and Coffman, *Lincoln County*, 35–46.

40 "Garden of the Hesperides," *The Globe* (Toronto), 10 July 1902, 8. See also "belt, n.II.8.a." and "belt.n.II.10.b.," *Oxford English Dictionary*, 2 ed., vol. 1 (Oxford: Oxford University Press, 1989); "The Niagara Peninsula," *The Globe*, 22 September 1894, 1; "What Is Toronto Doing?," *Toronto Daily Star*, 8 September 1895, 2.

41 "The Peach Crop," *The Globe* (Toronto), 14 July 1885, 6; "Hard on Peaches," *Toronto Daily Star*, 21 May 1906, 12.

42 Lloyd George Reeds, *Niagara Region Agricultural Research Report: Fruit Belt* (Hamilton: McMaster University, 1969), 1–3; Province of Ontario, Ministry of Agriculture, *Fruits of Ontario*, 107–8.

43 "Why Ontario's Fruit Zone Has Developed," *The Globe* (Toronto), 3 April 1918, 15; George A. Carefoot, *History and Geography of Lincoln County* ([St. Catharines, ON]: s.n., n.d. [1928?]), 10–14; Reaman, *History of Agriculture in Ontario*, 1:146–9.

44 Province of Ontario, Ministry of Agriculture, *Fruits of Ontario*, 5–6; Arthur Loughton, Richard V. Chudyk, and Judy A. Wanner, *Celebrating a Century of Success, 1906–2006* (Guelph: Horticultural Experiment Station, Vineland, University of Guelph, 2006); "Ask Alicia – Buchanan Hall," *Museum Chat* (blog), St. Catharines Museum and Welland Canals Centre, 6 May 2016, https://stcatharinesmuseumblog.com/2016/05/06/ask-alicia-buchanan-hall/.

45 W.H. Upshall, "Ontario," in *History of Fruit Growing and Handling in United States of America and Canada, 1860–1972*, ed. D.V. Fisher and W.H. Upshall (University Park, PA: American Pomological Society/Regatta City, 1976), 172–5; Alan Skeoch, "Changes in Lincoln County Agriculture," in *Agriculture and Farm Life in the Niagara Peninsula*, ed. John Burtniak and Wesley B. Turner (St. Catharines: Brock University, 1983), 37–46; J.H.H. Phillips, "A History of Fruit Crop Research on the Niagara Peninsula," in Burtniak and Turner, *Agriculture and Farm Life in the Niagara Peninsula*; Crewson and Matthews, "Class Interests in the Emergence of Fruit Growing Cooperation in Lincoln County," 23–49; Llewellyn S. Smith with Phyllis Cowan, *The House that Jam Built* (Markham: Baby Boomer, 1995); Paul E. Chapman, "A Field Guide to Niagara s Agriculture: Its History and Present Status," in Burtniak and Turner, *Agriculture and Farm Life in the Niagara Peninsula*, 129–38; Paul Chapman, "Agriculture in Niagara: An Overview," in *Niagara's Changing Landscapes*, ed. Hugh Gayler (Ottawa: Carleton University Press, 1994).

46 See, for example, "The Peach Crop," *The Globe* (Toronto), 14 July 1885, 6; "Crop Bulletin," *The Globe*, 1 June 1895, 7; "Crops in Ontario," *The Globe*, 30 May 1903, 24; "Fruit Crop Outlook Good," *The Globe*, 23 May 1912, 12; "Planting Season Somewhat Early," *The Independent* (Grimsby), 13 April 1921, 1.

47 "Fruit Crop Will Be Bountiful," *The Globe* (Toronto), 24 June 1908, 5. See also "Prosperity in the Fruit Belt," *The Globe*, 10 July 1902, 6; "Hard on Peaches," *Toronto Daily Star*, 21 May 1906, 12; "Niagara Peninsula in Blossom Time," *The Globe*, 24 May 1916, 12.

48 George Altmeyer, "Three Ideas of Nature in Canada, 1893–1914," *Journal of Canadian Studies* 11, no. 3 (August 1976): 21–36, https://doi.org/10.3138/jcs.11.3.21.

49 Jasen, *Wild Things*, 29–54; Jackson Lears, *No Place of Grace: Antimodernism and the Transformation of American Culture, 1880–1920* (New York: Pantheon, 1981); Ian McKay, *The Quest of the Folk: Antimodernism and Cultural Selection in Twentieth Century Nova Scotia* (Montreal: McGill-Queen's University Press, 1994); W. Douglas McCombs, "Therapeutic Rusticity: Antimodernism, Health, and the Wilderness Vacation, 1870–1915," *New York History* 76, no. 4 (1995): 409–28; James Murton, "La « Normandie du Nouveau Monde » : la société Canada Steamship Lines, l'antimodernisme et la promotion du Québec ancien," *Revue d'histoire de l'Amérique française* 55, no. 1 (Summer 2001): 3–44, https://doi.org/10.7202/005554ar; Sharon Wall, *The Nurture of Nature: Childhood, Antimodernism, and Ontario Summer Camps, 1920–55* (Vancouver: UBC Press, 2009); Terrance Young, *Heading Out: A History of American Camping* (Ithaca, NY: Cornell University Press, 2017); Dale Ernest Barbour, "Undressed Toronto: The Transformation of Bathing, 1850 to 1935" (PhD diss., University of Toronto, 2018).

50 William Henry Smith, *Smith's Canadian Gazetteer* (Toronto: H. & W. Rowsell, 1846), 72. See also Ada Bromley, Jean Powell, Phil Dechman, and Linda Coutts, *Once Upon a Little Town … Grimsby: 1876–1976* (Beamsville: Grimsby Historical Society/Rannie Publications, 1979), 36–47.

51 Ronald L. Way, *Ontario's Niagara Parks: A History*, 2nd ed. (Fort Erie: Niagara Parks Commission/Fort Erie Review, 1960).

52 G.H. Hamilton, "Exact Date for Blossom Sunday Hard to Determine," *Globe and Mail*, 7 May 1949, 25.

53 "The Lakeside and Garden City," *Toronto Daily Star*, 20 March 1902, 3.

54 "White and Pink Blossoms," *The Globe* (Toronto), 26 May 1905, 7.

55 Mabel Burkholder, "White and Pink Blossoms," *The Globe* (Toronto), 26 May 1905, 7.

56 "See the Blossoms," *The Globe* (Toronto), 17 May 1910, 2; "See the Blossoms," *The Globe*, 18 May 1910, 7.

57 "Peace Delegates May Ride through Niagara," *The Globe* (Toronto), 13 May 1914, 9.

58 R. Bowen, L. Cox, and F. Fox, "The Interface between Tourism and Agriculture," *Journal of Tourism Studies* 2, no. 2 (1991): 43–54.

59 "Nature Clothed in Loveliest Garb," *The Globe* (Toronto), 22 May 1920, 21; "Few Trips Offer," *The Globe*, 27 May 1920, 8; "Niagara Peninsula," *The Globe*, 29 May 1920, 8; ads for the Clifton, Village Inn, and Royal Connaught hotels, *The Globe*, 10 June 1920, 2.

60 "Peach Blossoms Join in Profusion of Color," *The Globe* (Toronto), 1 May 1922, 3; "View from the Niagara Highway," *The Globe*, 13 May 1922, 27; "Blossom Time in Niagara," *The Globe*, 18 May 1925, 13; "Garden of Canada Bedecked," *The Globe*, 22 May 1926, 1.

61 Bromley et al., *Once Upon a Little Town*, 112–58.

62 "Fruit Belt a Color Riot," *The Independent* (Grimsby), 10 May 1922, 1 and 8.

63 "Heavy Traffic," *The Independent* (Grimsby), 17 May 1922, 1.

64 "Blossom Time Music Festival," *The Independent* (Grimsby), 1 May 1929, 2; "Music Festival Was a Conspicuous Success," *The Independent*, 15 May 1929, 1; "Time Changes," *The Independent*, 15 May 1929, 3.

65 Michelle Nichols, "Fairs and Festivals," in *Grown in the Garden of Canada: The History of the Fruit Industry in Grimsby, Ontario*, Grimsby Museum, Virtual Museum of Canada, accessed 30 December 2021, https://www.communitystories.ca/v1/pm_v2.php?id=story_line&lg=English&fl=0&ex=00000438&sl=4795&pos=1; "Peninsula in Bloom Over the Week-End," *Globe and Mail*, 27 April 1938, 17.

66 "Thousands Swarming Peninsular Highways for Blossom Show," *Globe and Mail*, 7 May 1938, 80.

67 "Carry On," *The Independent* (Grimsby), 1 May 1942, 2; "Blossom Sunday Traffic Drops by 90 Per Cent," *Globe and Mail*, 4 May 1942, 4; "Early Blossom Sunday Unqueened but Superb," *Globe and Mail*, 16 April 1945, 13; "Thousands Visit Our Fairyland," *The Independent*, 9 May 1946, 1 and 3; "Thousands View Fruit Blossoms in Niagara Area," *Globe and Mail*, 3 May 1954, 17; "Blossoms, Canal Jam Create Traffic Tangle in Niagara Peninsula," *Globe and Mail*, 18 May 1959, 1.

68 Powell and Coffman, *Lincoln County 1856–1956*, 46.

69 Jimmy Simpson, "Float Blessed in Blossom Parade," *Globe and Mail*, 9 May 1955, 23; "1955 Blossom Blessing Festival," *Lake Report* (Niagara-on-the-Lake), 29 April 2021, 18.

70 "Niagara Painters Show Work," *Globe and Mail*, 12 May 1962, 17; Karena Walter, "Blossom Parade Was Springtime Tradition," *Niagara Falls Review*, 31 May 2014, A2.

71 "Blossom Time in Niagara," *Museum Chat* (blog), St. Catharines Museum and Welland Canals Centre, 14 May 2021, https://stcatharinesmuseumblog .com/2021/05/14/blossom-time-in-niagara/.

72 "Preparing for Blossom Week," *The Independent* (Grimsby), 10 March 1937, 1 and 5.

73 W.H. Millward letter to Kenneth Armstrong, 10 May 1960; W.J. Berry letter to T.P. McCarthy, 20 June 1960; J. James letter to T.P. McCarthy, 16 June 1960 – all in RG75, box 1, file 5, Ontario Editorial Board fonds, Archives and Special Collections, Brock University.

74 G.W. Hogarth letter to T.P. McCarthy, 3 April 1963, RG75, box 1, file 8, Ontario Editorial Board fonds, Archives and Special Collections, Brock University.

75 Delby Bucknall form letter to convention delegates, 13 October 1960, RG75, box 1, file 5, Ontario Editorial Board fonds, Archives and Special Collections, Brock University.

76 Powell and Coffman, *Lincoln County 1856–1956.*

77 W.H. Millward letter to Donald Holland, 17 March 1960, RG75, box 1, file 5, Ontario Editorial Board fonds, Archives and Special Collections, Brock University. A copy of the film is available in F1741, box 21, County of Lincoln fonds, Public Archives of Ontario.

78 T.P. McCarthy letter to R.H. Rittenhouse, 6 February 1963, RG75, box 1, file 8, Ontario Editorial Board fonds, Archives and Special Collections, Brock University; Francis J. Petrie, *Ball's Falls Conservation Area: Scenic and Historic Heritage* ([Welland, ON]: Niagara Peninsula Conservation Authority, 1972).

79 E.L. Chudleigh with J.R. Rainforth, *Alternatives for the Ontario Tender Fruit Industry* (Toronto: Ontario Food Council, 1972), 3–5.

80 Ontario, Department of Treasury and Economics, *Niagara Escarpment Study Group, Fruit Belt Report* (Toronto: Department of Treasury and Economics, 1968), 8.

81 The pioneering study on Niagara's suburban drift was R.R. Krueger, "Changing Land Use Patterns in the Niagara Fruit Belt," *Transactions of the Royal Canadian Institute* 32, no. 67 (Toronto: Royal Canadian Institute, 1959). See also Lloyd George Reeds, *Niagara Region Agricultural Research Report: Fruit Belt* (Hamilton: McMaster University, 1969); Chudleigh with Rainforth, *Alternatives for the Ontario Tender Fruit Industry*; H.J. Gayler, "Conservation and Development in Urban Growth: The Preservation of Agricultural Land in the Rural-Urban Fringe of Ontario," *Town Planning Review* 53, no. 3 (July 1982): 321–41; Regional Municipality of Niagara, *Niagara Tender Fruit Working Group, the Tender Fruit Industry in Niagara: Issues, Options and Conclusions* ([Thorold]: [Regional Municipality of Niagara], 1990).

82 Chudleigh with Rainforth, *Alternatives for the Ontario Tender Fruit Industry*, 3–5 and 8–9.

83 Tiffany Mayer, *Niagara Food: A Flavoured History of the Peninsula's Bounty* (Charleston, SC: History Press, 2014), 13–24.

84 Ronald C. Moyer, "The Niagara Grape Industry – Evolution to World Status," in Burtniak and Turner, *Agriculture and Farm Life in the Niagara Peninsula*, 1–22; Tony Aspler, *Tony Aspler's Vintage Canada*, 2nd ed. (Toronto: McGraw-Hill Ryerson, 1993); Linda Bramble, *Niagara's Wine Visionaries: Profiles of the Pioneering Winemakers* (Toronto: Lorimer, 2009); Alun Hughes, "Porter Adams and Grape Growing in Niagara," in *History Made in Niagara*, ed. Michael Ripmeester, David Butz, and Loris Gasparotto (St. Catharines: Elbow Island Press, 2019), 281–90.

85 Moyer, "The Niagara Grape Industry"; Hudson Cattell, *Wines of Eastern North America* (Ithaca, NY: Cornell University Press, 2014), 19–22, 85–90, 100–3; Aspler, *Tony Aspler's Vintage Canada*; Bramble, *Niagara's Wine Visionaries*; Linda Bramble, *Touring Niagara's Wine Country*, 2nd ed. (Toronto: Lorimer, 2003); Mayer, *Niagara Food*, 45–53.

86 Gary Pickering and Debbie Inglis, "Vinting on Thin Ice: The Making of Canada's Iconic Dessert Wine," in Ripmeester, Mackintosh, and Fullerton, *The World of Niagara Wine*, 229–48.

87 Donald Ziraldo in Donald Ziraldo and Karl Kaiser, *Icewine: Extreme Winemaking* (Toronto: Key Porter, 2007), 12.

88 Peter G. Mielzynski-Zychlinski, *The Story of Hillebrand Estates Winery* (Toronto: Key Porter, 2001), 29.

89 Robert L. Gluckman, "A Consumer Approach to Branded Wines," *European Journal of Marketing* 24, no. 4 (1990): 27–46, https://doi.org/10.1108/EUM0000000004649; Hassan Sayed, *Vineyard Site Suitability in Ontario* (Toronto: Government of Ontario, Ministry of Agriculture and Food, 1992); Aspler, *Tony Aspler's Vintage Canada*; Mielzynski-Zychlinski, *The Story of Hillebrand Estates*, 9–17, 64–76; Bramble, *Niagara's Wine Visionaries*; Cattell, *Wines of Eastern North America*, 209–11.

90 "Couple Wins Tourism Gamble," *Globe and Mail*, 18 May 1977, 8; "Village Depot" [ad], *Globe and Mail*, 2 November 1977, A17; Bromley et al., *Once Upon a Little Town*, 215. The 1902 station burned down in 1994, but the original Grimsby station of 1853 still stood in the 2020s. See Ron Brown, *The Train Doesn't Stop Here Anymore* (Toronto: Dundurn, 2008), 65–8.

91 "Steam Excursions," *Globe and Mail*, 10 August 1977, A19; "Everyday Outings," *Globe and Mail*, 10 May 1978, A16; "Weekend Getaways," *Globe and Mail*, 22 September 1979, T3.

92 Linda Bramble, "The History of the VQA," in Ripmeester, Mackintosh, and Fullerton, *The World of Niagara Wine*, 67–86. For more on the Niagara Peninsula appellation, see Anthony Shaw, "The Niagara Peninsula

Appellation: A Climactic Analysis of Canada's Largest Wine Region," in Ripmeester, Mackintosh, and Fullerton, *The World of Niagara Wine*, 143–64.

93 Giuseppe Festa, S.M. Riad Shams, Gerardino Metallo, and Maria Teresa Cuomo, "Opportunities and Challenges in the Contribution of Wine Routes to Wine Tourism in Italy – a Stakeholders' Perspective of Development," *Tourism Management Perspectives* 33 (January 2020): 100585, https://doi.org/10.1016/j.tmp.2019.100585; Elena Cruz-Ruiz, Gorka Zamarreño-Aramendia, and Elena Ruiz-Romero de la Cruz, "Key Elements for the Design of a Wine Route: The Case of La Axarquía in Málaga (Spain)," *Sustainability* 12, no. 21 (November [1] 2020): 9242, https://doi.org/10.3390/su12219242; André Lopes, Cláudia Seabra, Carla Silva, and José Luis Abrantes, "Wine Routes: Development of Wine Regions and Local Communities," *International Journal of Multidisciplinarity in Business and Science* 4, no. 5 (2018): 39–44.

94 The Wine Council of Ontario was renamed Ontario Craft Wineries in 2018. Atsuko Hashimoto and David Telfer, "Positioning an Emerging Wine Route in the Niagara Region: Understanding the Wine Tourism Market and Its Implication for Marketing," *Journal of Travel and Tourism Marketing* 14, nos. 3–4 (November 2003): 61–76, https://doi.org/10.1300/J073v14n03_04.

95 "Trip Planner," Wine Country Niagara, accessed 8 September 2021, https://winecountryontario.ca/trip-planner/?msg=existing.

96 David Telfer, "Strategic Alliances along the Niagara Wine Route," *Tourism Management* 22, no. 1 (February 2001): 21–30, https://doi.org/10.1016/S0261-5177(00)00033-9.

97 Insun Lee and Charles Arcodia, "The Role of Regional Food Festivals for Destination Branding," *International Journal of Tourism Research* 13, no. 4 (July/August 2011): 355–67, https://doi.org/10.1002/jtr.852. See also David J. Telfer and Geoffrey Wall, "Linkages between Tourism and Food Production," *Annals of Tourism Research* 23, no. 3 (July 1996): 635–53, https://doi.org/10.1016/0160-7383(95)00087-9.

98 David J. Telfer, "Tastes of Niagara: Building Strategic Alliances between Tourism and Agriculture," *International Journal of Hospitality and Tourism Administration* 1, no. 1 (January 2000): 71–88, https://doi.org/10.1300/J149v01n01_05; Telfer, "Strategic Alliances along the Niagara Wine Route"; Jeffrey W. Stewart, Linda Bramble, and Donald Ziraldo, "Key Challenges in Wine and Culinary Tourism with Practical Recommendations," *International Journal of Contemporary Hospitality Management* 20, no. 3 (2008): 303–13, https://doi.org/10.1108/09596110810866118. For trade books featuring these themes, see, for example, Kathleen Sloane-McIntosh, *A Year in Niagara* (North Vancouver: Whitecap, 2002); Linda Bramble, *Touring Niagara Wine Country*, 2nd ed.

(Toronto: Lorimer, 2003); Walter Sendzik, *Insider Guide to the Niagara Region* (Toronto: CanWest Books, 2005); Tony Aspler and Jean-Franáois Bergeron, *Canadian Wineries* (Richmond Hill: Firefly, 2013); Mayer, *Niagara Food*; Ziraldo and Kaiser, *Icewine*.

99 Regional Municipality of Niagara, *Regional Agricultural Economic Impact Study*, vols. 1–2 (Thorold: Regional Municipality of Niagara, 2003).

100 The NECD was closed in 2017 and replaced by a new agency in 2018 simply called Niagara Economic Development.

101 See, for example, Wine Country Ontario, *Travel Guide 2016* ([Vineland]: Wine Country Ontario, 2020), and other iterations issued annually through the 2010s.

102 Ray Spiteri, "Logo Reflects Region's Diversity; New Brand Is a 'Niagara Original,'" *The Review* (Niagara Falls), 19 November 2008, A1.

103 Regional Municipality of Niagara, *Niagara Culture Plan*.

104 Suzanne A. Hill, "A Serpent in the Garden: Implications of Highway Development in Canada's Niagara Fruit Belt," *Journal of Historical Sociology* 15, no. 4 (December 2002): 495–515, https://doi.org/10.1111/1467-6443.t01-1-00189.

105 Mielzynski-Zychlinski, *The Story of Hillebrand*, 36–44; Aspler and Bergeron, *Canadian Wineries*.

106 Nicholas Baxter-Moore and Caroline Charest, "Constructing Authenticity: Architecture and Landscape in Niagara's Wineries" in Ripmeester, Mackintosh, and Fullerton, *The World of Niagara Wine*, 259–80. See also Aspler and Bergeron, *Canadian Wineries*, 99–169.

107 Sendzik, *Insider Guide*, 20.

108 Tony Spawton, "Of Wine and Live Asses: An Introduction to the Wine Economy and State of Wine Marketing," *European Journal of Marketing* 25, no. 3 (1991): 6–11; Joanna Fountain, Nicola Fish, and Steve Charters, "Making a Connection: Tasting Rooms and Brand Loyalty," *International Journal of Wine Business Research* 20, no. 1 (2008): 8–21, https://doi.org/10.1108/17511060810864589.

109 See Marcel Meler and Marija Ham, "Green Marketing for Green Tourism," *Tourism and Hospitality Management 2012, Conference Proceedings* (2012): 130–9, https://doi.org/10.13140/2.1.3701.5047; Ioannis Papadopoulos, Marios Trigkas, Aikaterini Papadopoulou, Argiro Mallisiova, and Fotini Mpakogiorgou, "Greek Consumers' Awareness and Perceptions for Green Hotels and Green Furniture," in *Strategic Innovative Marketing and Tourism. Springer Proceedings in Business and Economics*, ed. Androniki Kavoura, Efstathios Kefallonitis, and Apostolos Giovanis (Champaign, IL: Springer, 2019), 583–92, https://doi.org/10.1007/978-3-030-12453-3_67; Debbie Gordon, "Wines that Are Better for You and the Environment," *A Must Read Blog*, 28 August 2019,

https://amustreadblog.com/2019/08/28/wines-that-are-better-for-you-the-environment-meet-canadian-environmental-steward-ann-sperling-of-southbrook-vineyards.

110 "VQA: the Sustainable Choice," Wine Country of Ontario, accessed 13 January 2021, https://winecountryontario.ca/sustainability-wco.

111 "Our Green Story," Southbrook Vineyards, accessed 13 January 2021, https://www.southbrook.com/About.

112 Edward Brooker and Jason Burgess, "Marketing Destination Niagara Effectively through the Tourism Life Cycle," *International Journal of Contemporary Hospitality Management* 20, no. 3 (2008): 278–92, https://doi.org/10.1108/09596110810866091; Jayawardena, "Tourism in Niagara," 249–58.

113 Russell Johnston, "Advertising in Canada," in *Mediascapes: New Patterns in Canadian Communication*, 4 ed., ed. Leslie Regan Shade (Scarborough: Nelson, 2013), 150–70; "The Canadian Code of Advertising Standards," Ad Standards Canada, July 2019, https://adstandards.ca/code/the-code-online/.

114 Helen Powell, *Promotional Culture and Convergence: Markets, Methods, Media* (New York: Routledge, 2013).

115 Roland Marchand, *Advertising the American Dream: Making Way for Modernity, 1920–1940* (Los Angeles: University of California Press, 1985); William Leiss, Stephen Kline, Sut Jhally, Jacqueline Botterill, and Kyle Asquith, *Social Communication in Advertising*, 4th ed. (New York: Routledge, 2018).

116 See Quintal, Thomas, and Phau, "Incorporating the Winescape into the Theory of Planned Behaviour"; Fountain and Dawson, "The New Gold"; Frost, Strickland, and Smith Maguire, "Seeking a Competitive Advantage in Wine Tourism."

117 Eastdell Estates, "About Us," accessed 26 January 2010, http://www.eastdell.com/about_us. The website no longer exists.

118 Ravine Vineyard, "Past," accessed 15 March 2010, http://www.ravinevineyard.com/about-ravine/past. The website has been revised, but a relevant page is archived in the Wayback Machine at https://web.archive.org/web/20120226135506/http://www.ravinevineyard.com/about-ravine/past/.

119 Ravine Vineyard, "Tastings," accessed 16 August 2021, https://www.ravinevineyard.com/the-winery/tours–tastings. The website has since been revised. The relevant page from July 2023 is available at https://www.ravinevineyard.com/experiences.

120 "Visit Henry of Pelham Family Estate Winery," Henry of Pelham Winery, accessed 16 August 2021, https://henryofpelham.com/visit-us/#tour.

121 "Reif Estate Winery," Niagara Falls Tourism, accessed 16 August 2021,

https://www.niagarafallstourism.com/play/wineries-breweries/reif-estate-winery/.

122 "About Us," Henry of Pelham Winery, accessed 26 January 2010, http://www.henryofpelham.com/aboutus.php. The website has been revised, but a relevant page is archived in the Wayback Machine at https://web.archive.org/web/20100812041237/http://www.henryofpelham.com/aboutus.php.

123 "1812 Red," Palatine Hills Winery, accessed 10 July 2018, http://www.palatinehillsestatewinery.com/site/wine-shop/detail/6445. The website has been revised, but a relevant page is archived in the Wayback Machine at https://web.archive.org/web/20160805063236/http://www.palatinehillsestatewinery.com/site/wine-shop/detail/851.

124 See, for example, "About Us," Pilliteri Estates Winery, accessed 15 March 2010, http://www.pillitteri.com/pages/about_us.cfm. The website has been revised, but a relevant page is archived in the Wayback Machine at https://web.archive.org/web/20190311211614/https://www.pillitteri.com/our-story/.

125 Fair, "Gentlemen, Farmers, and Gentlemen Half-farmers"; Philip R. Dodds, *The Story of Ontario Agricultural Fairs and Exhibitions, 1792–1967* (Picton: Picton Gazette, 1967); Brian Romagnoli, *Lincoln County Fair, 1857–2007: "A Victorian Country Fair"* (Beamsville: Vintage Arts/Lincoln Agricultural Society, 2007); "About the NRE," Niagara Regional Exhibition, accessed 25 June 2018, https://www.niagaraex.com.

126 Karena Walter, "Wine Festival Has Pressing Need for Space," *The Standard* (St. Catharines), 12 February 2016, A2; Niagara Grape and Wine Festival, accessed 9 September 2020, https://niagarawinefestival.com/.

127 Heather L. Clark, "Residents' Attitudes and Perceptions toward the Niagara Grape and Wine Festival" (honours thesis, Brock University, 1993).

128 Sarah Ackles, "Tickets Selling Fast for Ontario's Biggest VQA Celebration," *Brock News*, 19 March 2018, https://brocku.ca/brock-news/2018/03/tickets-selling-fast-for-ontarios-biggest-vqa-celebration/; Monique Beech, "Cuvée Wine Gala to Mark 20 Years," *The Standard* (St. Catharines), 22 February 2008, D1; Phillip G. Mackintosh, "The Niagara Wine Festival's Grande Parade," in Ripmeester, Macintosh, and Fullerton, *The World of Niagara Wine*, 313–30.

129 "Wine Festivals," Niagara Grape and Wine Festival, accessed 5 September 2020, https://niagarawinefestival.com/. The website has been revised, but a relevant page is archived in the Wayback Machine at https://web.archive.org/web/20200214215236/https://niagarawinefestival.com/.

130 Mackintosh, "The Niagara Wine Festival's Grande Parade."

131 Mackintosh, "The Niagara Wine Festival's Grande Parade."

132 Jim Smelle, "Letter to the Editor," *The Standard* (St. Catharines), 27 August 2009, 8.

133 "Release #20: Is Disaster Fermenting in Ontario's Greenbelt?," Ontario Viniculture Association, accessed 23 March 2010, http://www .realontariowine.ca. The website has been revised, but a relevant page is archived in the Wayback Machine at https://web.archive.org/ web/20110706203231/http://www.realontariowine.ca/?q=OVAR20. See also Hugh Gayler, "Conflict in the Niagara Countryside," in Ripmeester, Macintosh, and Fullerton, *The World of Niagara Wine*, 301–12.

134 Larry Bilkszto, quoted in Tiffany Mayer, "Tilling for Tomorrow," *The Standard* (St. Catharines), 4 March 2008, A1.

135 Susan Cooper, quoted in "Favourite Wine Festival Moments," *The Standard* (St. Catharines), 26 August 2009, A2.

136 Eddie Chau, "Wine Giant Vincor Gobbled Up," *Niagara This Week*, 7 April 2006, NTW00; Christopher Waters, "Purchase of Vincor Resonates in South Africa," *The Standard* (St. Catharines), 8 April 2006, B4; Aspler and Bergeron, *Canadian Wineries*, 99–100.

137 Aspler and Bergeron, *Canadian Wineries*, 100.

138 Roy Rosenzweig and David Thelen, *The Presence of the Past: Popular Uses of History in American Life* (New York: Columbia University Press, 1998); Sanford Levinson, *Written in Stone* (Durham, NC: Duke University Press, 1998); Bodnar, *Remaking America*.

139 Richard Hutton, "Grape and Wine Garden Makes Downtown Debut," *Niagara This Week*, 10 September 2020, 29.

140 Sara Ahmed, "Affective Economies," *Social Text* 22, no. 2 (Summer 2004): 117–39, https://doi.org/10.1215/01642472-22-2_79-117.

6. Residents Engage the Niagara Wine Industry

1 Wine Country Ontario, *Travel Guide 2016* ([Vineland]: Wine Country Ontario, 2016), 71.

 2 Details of the authors' 2012 survey are provided in chapter 1.

 3 Regional Municipality of Niagara, *Niagara COVID-19 Business Impact Survey Part 2* (Thorold: Regional Municipality of Niagara, [2020]), https://niagaracanada.com/wp-content/uploads/sites/7/2020/06/ COVID-2-Report_Final-Draft.pdf.

 4 Sara Nixon, "A Stompin' Good Time: The Niagara Grape and Wine Festival," *Museum Chat* (blog), St. Catharines Museum and Welland Canals Centre, 27 September 2020, https://stcatharinesmuseumblog. com/2020/09/27/a-stompin-good-time-the-niagara-grape-wine-festival/.

 5 Christopher Ray, "Culture, Intellectual Property, and Territorial Rural Development," *Sociologia Ruralis* 38, no. 1 (April 1998): 3–20, https://doi .org/10.1111/1467-9523.00060.

6   Philipp Wassler, Liang Wang, and Kam Hung, "Identity and Destination Branding among Residents," *International Journal of Tourism Research* 21, no. 4 (July/August 2019): 437–46, https://doi.org/10.1002/jtr.2271; Jean-Noël Kapferer, *Strategic Brand Management: Creating and Sustaining Brand Equity Long Term* (London: Kogan Page, 1998); Dmitri Skuras and Efthalia Dimara, "Regional Image and the Consumption of Regionally Denominated Products," *Urban Studies* 41, no. 4 (April 2004): 801–15, https://doi.org/10.1080/0042098042000194115; M. Stephens Balakrishan, "Strategic Branding of Destinations: A Framework," *Emerald Journal of Marketing* 43, nos. 5–6 (2009): 611–29, https://doi.org/10.1108/03090560910946954.

7   Maja Konecnik and Frank Go, "Tourism Destination Brand Identity: The Case of Slovenia," *Brand Management* 15, no. 3 (January 2008): 177, https://doi.org/10.1057/palgrave.bm.2550114.

8   Hugues Seraphin, Stanislav Hristov Ivanov, Frederic Dosquet, and Stéphane Bourliataux-Lajoinie, "Destination Branding and Overtourism," *Journal of Hospitality and Tourism Management* 38 (March 2019): 1–4, https://doi.org/10.1016/j.jhtm.2018.11.003.

9   Noam Shoval, ed., "Tourism, and Urban Planning in European Cities: Overtourism, Placemaking, and Heritage," special issue, *Tourism Geographies* 20, no. 3 (May 2018). See also Biljana Petrevska, Shaul Krakover, and Noga Collins-Kreiner, "Preserving Cultural Assets of Others: Jewish Heritage Sites in Macedonian Cities," 549–72, https://doi.org/10.1080/14616688.2017.1387811; Bálint Kádár, "Hotel Development through Centralized to Liberalized Planning Procedures: Prague Lost in Transition," 461–80, https://doi.org/10.1080/14616688.2017.1375974; and Johannes Novy, "'Destination' Berlin Revisited: From (New) Tourism towards a Pentagon of Mobility and Place Consumption," 418–42, https://doi.org/10.1080/14616688.2017.1357142.

10 Valene L. Smith, *Hosts and Guests: The Anthropology of Tourism* (Philadelphia: University of Pennsylvania Press, 1989); Simon Anholt, "Nation Brands Index: How Does the World See America?," *Journal of Advertising Research* 45, no. 3 (September 2005): 296–304, https://doi.org/10.1017/S0021849905050336.

11 Hyung Min Lee, Jinwoo Park, and Yoonjae Nam, "A Sustainable Solution to Overtourism in the Social Media Era: An Exploratory Analysis on the Roles and Functions of Place-Visitor Relationship (PVR)," *Sustainability* 12, no. 7 (April [1] 2020): 1–15, https://doi.org/10.3390/su12073043.

12 Heather Skinner, "In Search of the Genius Loci: The Essence of a Place Brand," *Marketing Review* 11, no. 3 (Autumn 2011): 281–92, https://doi.org/10.1362/146934711X589471.

13 Richard L. Florida, *The Rise of the Creative Class and How It's Transforming Work, Leisure, Community and Everyday Life* (New York: Basic Books. 2002); Richard Florida, *The Flight of the Creative Class: The New Global Competition for Talent* (New York: HarperCollins, 2005); Richard Florida *Cities and the Creative Class* (New York: Routledge, 2005).

14 Sharon Zukin, *Naked City: The Death and Life of Authentic Urban Places* (Oxford: Oxford University Press, 2010); William Haydock, "The 'Civilising' Effect of a 'Balanced' Night-Time Economy for 'Better People': Class and Cosmopolitan Limit in the Consumption and Regulation of Alcohol in Bournemouth," *Journal of Policy Research in Tourism, Leisure & Events* 6, no. 2 (May 2014): 172–85, https://doi.org/ 10.1080/19407963.2014.900989; Leslie Kern, "Rhythms of Gentrification: Eventfulness and Slow Violence in a Happening Neighbourhood," *Cultural Geographies* 23, no. 3 (July 2016): 441–57, https://doi.org/10.1177 /1474474015591489.

15 Karen Finucan, "What Brand Are You?," *Planning* 68 (August 2002): 10–13; Juergen Gnoth, "The Structure of Destination Brands: Leveraging Values," *Tourism Analysis* 12, nos. 5–6 (2007): 345–58, https://doi.org/10.3727 /108354207783227939.

16 Jeannette Hanna, "Mapping Community and Place," in *Rediscovering the Wealth of Places: A Municipal Cultural Planning Handbook for Canadian Communities*, ed. Greg Baeker (St. Thomas: Municipal World, 2010), 93.

17 Robin Mayes, "A Place in the Sun: The Politics of Place, Identity and Branding," *Place Branding and Public Diplomacy* 4, no. 2 (May 2008): 124–35, https://doi.org/10.1057/pb.2008.1.

18 Fern Willits, Gene Theodori, and Michael Fortunato, "The Rural Mystique in American Society," in *New Realities in an Urbanizing World*, ed. Alexander Thomas and Gregory Fulkerson (Lanham, MD: Lexington Books, 2016), 33–56; Keith Halfacree, "Reading Rural Consumption Practices for Difference: Bolt-Holes, Castles, and Life-Rafts," *Culture Unbound* 2, no. 2 (June 2010): 241–63, https://doi.org/10.3384/cu.2000 .1525.10214241; Brian McLaughlin, "Rural policy in the 1980s: The Revival of the Rural Idyll," *Journal of Rural Studies* 2, no. 2 (1986): 81–90, https:// doi.org/10.1016/0743-0167(86)90047-1.

19 Brian Short, "Idyllic Ruralities," in *The Handbook of Rural Studies*, ed. Paul Cloke, Terry Marsden, and Patrick Mooney (London: Sage Publications, 2006), 133–48; McLaughlin, "Rural Policy in the 1980s"; Lee-Ann Sutherland, "Virtualizing the 'Good Life': Reworking Narratives of Agrarianism and the Rural Idyl in a Computer Game," *Agriculture and Human Values* 37 (2020): 1155–73.

20 Mark Shucksmith, "Re-imagining the Rural: From Rural Idyll to Good Countryside," *Journal of Rural Studies* 59 (April 2018): 163–72, https://doi .org/10.1016/j.jrurstud.2016.07.019.

21 Sutherland, "Virtualizing the 'Good Life'"; Nicola Bishop, "Rural Nostalgia: Revisiting the Lost Idyll in British Library Crime Classics Detective Fiction," *Green Letters* 22, no. 1 (2018): 31–42; Pavel Pospěch, Daniela Spěšná, and Adam Staveník, "Images of a Good Village: A Visual Analysis of the Rural Idyll in the 'Village of the Year' Competition in the Czech Republic," *European Countryside* 7, no. 2 (2015): 68–86; John Horton, "Producing *Postman Pat*: The Popular Cultural Construction of Idyllic Rurality," *Journal of Rural Studies* 24, no. 4 (October 2008): 389–98, https://doi.org/10.1016/j.jrurstud.2008.03.011.

22 Alister Scott, Alana Gilber, and Ayele Gelan, *The Urban-Rural Divide: Myth or Reality* (Aberdeen, UK: Macaulay Institute, 2007); Clemens de Olde and Stijn Oosterlynck, "'The Countryside Starts Here': How the Rural Urban-Rural Divide Continues to Matter in Post-urban Flanders," *European Urban and Regional Studies* 29, no. 3 (July 2021): 1–16, https://doi.org/10.1177/09697764211043448.

23 Valerie du Plessis, Roland Beshiri, Ray D. Bollman, and Heather Clemenson, "Definitions of Rural," *Rural and Small Town Canada Analysis Bulletin* 3, no. 3 (November 2001): 1–17, catalogue no. 21-006-XIE.

24 "Population Centre and Rural Area Classification 2016," Statistics Canada, last modified 8 February 2017, https://www.statcan.gc.ca/en/subjects/standard/pcrac/2016/introduction.

25 A "rural area" is defined as follows: "Rural areas (RAs) include all territory lying outside population centres (POPCTRs). Taken together, population centres and rural areas cover all of Canada. Statistics Canada, "Population Centre and Rural Area Classification 2016."

26 Niagara Peninsula Source Protection Authority, *Updated Assessment Report – Niagara Peninsula Source Protection Area* (Welland: Niagara Peninsula Source Protection Authority, 2013), 16.

27 Hugh Gayler, "Conflict in the Niagara Countryside: Securing the Land Base for the Wine Industry," in *The World of Niagara Wine*, ed. Michael Ripmeester, Philip Mackintosh, and Christopher Fullerton (Waterloo: Wilfrid Laurier University Press, 2013), 301–12.

28 Regional Municipality of Niagara, *Niagara Agriculture Profile* (Thorold: Regional Municipality of Niagara, [2017]), 2–5.

29 Regional Municipality of Niagara, *Niagara Agriculture Profile*, 8; Jill Troyer, "Cannabis Growers in Niagara Double in One Year," *Lake Report* (Niagara-on-the-Lake), 5 September 2019, 1–2.

30 Regional Municipality of Niagara, *Niagara Agriculture Profile*, 8; "PoultryFest Celebrates 10 Years," *Niagara This Week*, 2 July 2009, 1; Scott Rosts, "PoultryFest Cancelled after 18 Years," *Grimsby Lincoln News*, 26 October 2017, 1.

31 Regional Municipality of Niagara, *Niagara Agriculture Profile*, 8.

32 Regional Municipality of Niagara, "Appendix 3: Economic Analysis and Forecasting Draft Discussion Paper, Global Investment Attraction Group," *Niagara Economic Development Strategy 2019–2024* (Thorold: Regional Municipality of Niagara, [2019]), 23.

33 Geoff A. Wilson, "From Productivism to Post-productivism ... and Back Again? Exploring the (Un)changed Natural and Mental Landscapes of European Agriculture," *Transactions – Institute of British Geographers* 26, no. 1 (March 2001): 77–102, https://doi.org/10.1111/1475-5661.00007; Lone Søderkvist Kristensen, Claudine Thenail, and Søren Pilgaard Kristensen, "Landscape Changes in Agrarian Landscapes in the 1990s: The Interaction between Farmers and the Farmed Landscape: A Case Study from Jutland, Denmark," *Journal of Environmental Management* 71, no. 3 (July 2004): 231–44, https://doi.org/10.1016/j.jenvman.2004.03.003.

34 Regional Municipality of Niagara, *Niagara Agriculture Profile and Economic Impact* (Thorold: Regional Municipality of Niagara, [2018]), 2–5.

35 Carla Barbieri, "An Important-Performance Analysis of the Motivation behind Agritourism and Other Farm Enterprise Developments in Canada," *Journal of Rural and Community Development* 5, nos. 1–2 (2010): 1–20; Jie Gao, Carla Barbieri, and Corinne Valdivia, "Agricultural Landscape Preferences: Implications for Agritourism Development," *Journal of Travel Research* 53, no. 3 (May 2014): 366–79, https://doi.org/10.1177/0047287513496471.

36 Hilary Dawson, Mark Holmes, Hersch Jacobs, and Richard I. Wade, "Wine Tourism: Winery Visitation in the Wine Appellations of Ontario," *Journal of Vacation Marketing* 17, no. 3 (July 2011): 237–46, https://doi.org/10.1177/1356766711409185; Bruce McAdams, Statia Elliot, and Joshua LeBlanc, "Drive by My Cellar Door: Rethinking the Benefits of Wine Tourism in Niagara, *Tourism Analysis* 26, nos. 2–3 (2021): 225–36, https://doi.org/10.3727/108354221X16079839951484.

37 Francesco Vistentin and Francesco Vallerani, "A Countryside to Sip: Venice Inland and the Prosecco's Uneasy Relationship with Wine Tourism and Rural Exploitation," *Sustainability* 10, no. 7 (July 2018): 1–18, https://doi.org/10.3390/su10072195; Marjorie Jones, Neha Singh, and Yvonne Hsiung, "Determining the Critical Success Factors of the Wine Tourism Region of Napa from a Supply Perspective," *International Journal of Tourism Research* 7, no. 3 (May/June 2015): 261–71, https://doi.org/10.1002/jtr.1984; John Holmes and Kate Hartig, "Metropolitan Colonization and the Reinvention of Place: Class Polarization along the Cessnock-Pokolbin Fault-Line," *Geographical Research* 45, no. 1 (March 2007): 54–70, https://doi.org/10.1111/j.1745-5871.2007.00404.x; John Overton, "Rural Idylls and Urban Economies: The Making of Metropolitan Wine Regions," *Journal of Wine Research* 30, no. 3 (July 2019): 238–58, https://doi.org/10.1080/0957

1264.2019.1652150; Peter Howland, "Martinborough's Wine Tourists and the Metro-Rural Idyll," *Journal of New Zealand Studies*, nos. 6–7 (October 2008): 77–100, https://doi.org/10.26686/jnzs.v0i6/7.135.

38 Howland, "Martinborough's Wine Tourists and the Metro-Rural Idyll"; Overton, "Rural Idylls and Urban Economies."

39 Barbara Carmichael, "Understanding the Wine Tourism Experience for Winery Visitors in the Niagara Region, Ontario, Canada," *Tourism Geographies* 7, no. 2 (2005): 185–204.

40 Carman Cullen, Eugene Kaciak, and Linda Bramble, "Segmentation of the Off-Peak Wine Tourist in Canada's Niagara Region," *International Business and Economic Research Journal* 4, no. 8 (2005): 13–24, https://doi .org/10.19030/iber.v4i8.3608.

41 McAdams, Elliot, and LeBlanc, "Drive by My Cellar Door," 231.

42 Carmichael, "Understanding the Wine Tourism Experience," 185–204; Cullen, Kaciak, and Bramble, "Segmentation of the Off-Peak Wine Tourist in Canada's Niagara Region," 13–24; Dawson et al., "Wine Tourism," 237–46; McAdams, Elliot, and LeBlanc, "Drive by My Cellar Door," 231.

43 Tourism Research Unit, *The Niagara Region's Tourism Opportunities: The U.S. and Ontario Markets* (Toronto: Ontario, Ministry of Tourism, 2008), 27.

44 Carmichael, "Understanding the Wine Tourism Experience," 196. See also Dawson et al., "Wine Tourism"; Hwansuk Chris Choi, Shuyue Huang, Joan Flaherty, and Anahita Khazaei, "Segmenting Wine Tourists in Niagara, Using Motivation and Involvement," *International Journal of Tourism Sciences* 17, no. 3 (July 2017): 198–212, https://doi.org/10.1080 /15980634.2017.1351083.

45 Donald Getz and Graham Brown, "Critical Success Factors for Wine Tourism Regions: A Demand Analysis," *Tourism Management* 27, no. 1 (February 2006): 155, https://doi.org/10.1016/j.tourman.2004.08.002.

46 Carmichael, "Understanding the Wine Tourism Experience," 185–204; Eli Cohen and Livnat Ben-Nun, "The Important Dimensions of Tourism Experience from Potential Visitors' Perception," *Tourism and Hospitality Management* 9, no. 1 (January 2009): 20–31, https://doi.org/10.1057 /thr.2008.42; Choi et al., "Segmenting Wine Tourists in Niagara."

47 Overton, "Rural Idylls and Urban Economies."

48 Halfacree, "Reading Rural Consumption Practices for Difference."

49 See Careoline Ritchie, "Beyond Drinking: The Role of Wine in the Life of the UK Consumer," *International Journal of Consumer Studies* 31, no. 5 (September 2007): 534–40, https://doi.org/10.1111/j.1470-6431.2007 .00610.x; Johan Bruwer and Karin Alant, "The Hedonic Nature of Wine Tourism Consumption: An Experiential View," *International Journal of Wine Business Research* 21, no. 3 (2009): 235–57, https://doi.org/10.1108 /17511060910985962; John Overton and Warwick E. Murray, "Class in

a Glass: Capital, Neoliberalism, and Social Space in the Global Wine Industry," *Antipode* 45, no. 3 (June 2013): 702–18, https://doi.org/10.1111/j.1467-8330.2012.01042.x.

50 Overton and Murray, "Class in a Glass," 711.

51 Niagara Economic Development and Ownera Media, "Eleven Reasons to Move to Niagara Right Now," *Toronto Life*, accessed 19 April 2022, https://torontolife.com/real-estate/eleven-reasons-to-move-to-niagara/.

52 L. Anders Sandberg and Gerda R. Wekerle, "Reaping Nature's Dividends: The Neoliberalization and Gentrification of Nature on the Oak Ridges Moraine," *Journal of Environmental Policy and Planning* 12, no. 1 (March 2010): 41–57, https://doi.org/10.1080/15239080903371915; Sara Epp, "Competing and Conflicting Land Uses at the Rural-Urban Interface: Understanding the Impacts of Residential Development on Agricultural Land" (master's thesis, Brock University, 2013); Peter B. Nelson and J. Dwight Hines, "Rural Gentrification and Networks of Capital Accumulation – a Case Study of Jackson, Wyoming," *Environment and Planning A: Economy and Space* 50, no. 7 (October 2018): 1473–95, https://doi.org/10.1177/0308518X18778595.

53 Wassler, Wang, and Hung, "Identity and Destination Branding among Residents," 437–46; Paul Bergeron III, "App Brings New Dynamic to Resident Referrals: Residents Just Might Be Your Best Brand Ambassadors," *Units* 43, no. 3 (March 2019): 47.

54 Canada, Census Profiles, *2016 Census*, catalogue no. 98–316-X2016001, Town of Grimsby. On Virgil, see David F. Hemmings, *The Cross Roads: Fortune Favours the Strong* (Niagara-on-the-Lake: Bygones, 2015).

55 Canada, Public Works Canada, *Area Screening Canada: Niagara Region, Ontario* (Ottawa: Ministry of Supply and Services, 1982), 4–5.

56 John N. Jackson, *The Welland Canals and Their Communities: Engineering, Industrial, and Urban Transformation* (Toronto: University of Toronto Press, 1997).

57 Wassler, Wang, and Hung, "Identity and Destination Branding among Residents," 437–46.

58 See, for example, Johan Bruwer and Isabelle Lesschaeve, "Sources of Information Used by Tourists Travelling to Visit Canadian Winery Tasting Rooms," *Tourism Planning & Development* 9, no. 3 (2012): 269–89, https://doi.org/10.1080/21568316.2012.672452.

59 The *Fort Erie Times* closed in 2017.

60 George Lipsitz, *Time Passages: Collective Memory and American Popular Culture* (Minneapolis: University of Minnesota Press, 1990).

61 Barbara Zelizer, "The Voice of the Visual in Memory," in *Framing Public Memory*, ed. Kendall R. Phillips (Tuscaloosa: University of Alabama Press, 2004), 157–86; Barbara A. Biesecker, "Renovating the National Imagery: A

Prolegomenon on Contemporary Paregoric Rhetoric," in Phillips, *Framing Public Memory*, 212–47; Sabine Marschall, "The *Sunday Times* Heritage Project: Heritage, the Media, and the Formation of National Consciousness," *Social Dynamics* 37, no. 3 (September 2011): 409–23, https://doi.org/10.1 080/02533952.2011.656430; Daniel Michon and Ahmed El Antably, "It's Hard to Be Down When You're Up: Interpreting Cultural Heritage through Alternative Media," *International Journal of Heritage Studies* 19, no. 1 (January 2013): 16–40, https://doi.org/10.1080/13527258.2011.633539.

62 Alison Landsberg, *Prosthetic Memory: The Transformation of American Remembrance in the Age of Mass Culture* (New York: Columbia University Press, 2004), 19.

63 Landsberg, 18.

64 Maurice Halbwachs, *On Collective Memory*, ed. and trans. Lewis A. Coser (Chicago: University of Chicago Press, 1992).

65 Ray, "Culture, Intellectual Property and Territorial Rural Development," 3–20.

66 Regional Municipality of Niagara, *Regional Agricultural Economic Impact Study*, vol. 1 (Thorold: Regional Municipality of Niagara, 2003), 7.2. See also Andrea Scarantino, "Affordances Explained," *Philosophy of Science* 70, no. 5 (December 2003): 949–61, https://doi.org/10.1086/377380.

67 Ontario, Department of Municipal Affairs, *The Niagara Area: Changing Land Uses* (Toronto: Government of Ontario, 1961); Ontario, Department of Agriculture and Food, W.J. Dillon, *Grape Production in the Niagara Peninsula* (Toronto: Government of Ontario, 1968); Ontario, Department of Treasury and Economics *Niagara Escarpment Study Group, Fruit Belt Report* (Toronto: Department of Treasury and Economics, 1968); Ralph R. Krueger, "Urbanization of the Niagara Fruit Belt," *Canadian Geographer* 22, no. 3 (September 1978): 179–94, https://doi.org/10.1111/j.1541-0064.1978. tb01011.x; Hugh J. Gayler, "Urban Sprawl and the Decline of the Niagara Fruit Belt," in *Geographical Snapshots of North America*, ed. D.G. Janelle (New York: Guidlford, 1992), 128–32; A. Suzanne Hill, "A Serpent in the Garden: Implications of Highway Development in the Niagara Fruit Belt," *Journal of Historical Sociology* 15, no. 4 (December 2002): 495–509, https:// doi.org/10.1111/1467-6443.t01-1-00189.

68 These observations echo research regarding Canadians' food choices. See Brenda L. Beagan, Gwen E. Chapman, Josée Johnston, Deborah McPhail, Elaine M. Power, and Helen Vallianatos, *Acquired Tastes: Why Families Eat the Way They Do* (Vancouver: UBC Press, 2015).

69 See Sebastian Zenker, Erik Braun, and Sibylle Petersen, "Branding the Destination Versus the Place: The Effects of Brand Complexity and Identification for Residents and Visitors," *Tourism Management* 58 (February 2017): 15–27, https://doi.org/10.1016/j.tourman.2016.10.008.

70 Natália Lozano, "Recommendation Culture's Influence on the Promotional Communication of Spanish Autonomous Community Brands," *Catalan Journal of Communication & Cultural Studies* 3 (2003): 289–96, https://doi.org/10.1386/cjcs.3.2.149_2; Natàlia Lozano-Monterrubio and Assumpció Huertas, "The Image of Barcelona in Online Travel Reviews during 2017 Catalan Independence Process," *Communication & Society* 33, no. 3 (June 2020): 33–49, https://doi.org/10.15581/003.33.3.33-49.

71 João Ricardo Freire, "'Local People': A Critical Dimension for Place Brands," *Journal of Brand Management* 16, no. 7 (June 2007): 420–38, https://doi.org/10.1057/palgrave.bm.2550097; Robert Govers, "From Place Marketing to Place Branding and Back," *Place Branding and Public Diplomacy* 7, no. 4 (November 2011): 227–31, https://doi.org/10.1057/pb.2011.28; Weilin Lu and Svetlana Stepchenkova, "User-Generated Content as a Research Mode in Tourism and Hospitality Applications: Topics, Methods, and Software," *Journal of Hospitality Marketing & Management* 24, no. 2 (February 2015): 119–54, https://doi.org/10.1080/19368623.2014.907758; Maria-Irina Ana and Laura-Gabriela Istudor, "The Role of Social Media and User-Generated-Content in Millennials' Travel Behavior," *Management Dynamics in the Knowledge Economy* 7, no. 1 (March 2019): 87–104, http://dx.doi.org/10.25019/MDKE/7.1.05.

7. Conclusion

1 "KNOW IT," Niagara Economic Development, accessed 15 July 2018, https://niagaracanada.com. The website has been revised, but a relevant page is archived in the Wayback Machine at https://web.archive.org/web/20190623131013/https://niagaracanada.com/.

2 Eviatar Zerubavel, *Time Maps: Collective Memory and the Social Shape of the Past* (Chicago: University of Chicago Press, 2003).

3 Tim Cresswell, *Maxwell Street: Writing and Thinking Place* (Chicago: University of Chicago Press, 2019).

4 Andrew Baldwin, Laura Cameron, and Audrey Kobayashi, *Rethinking the Great White North: Race, Nature and Historical Geographies of Whiteness in Canada* (Vancouver: UBC Press, 2012).

5 Tony Aspler, *The Wine Atlas of Canada* (Toronto: Random House Canada, 2006).

6 Niagara Economic Development Corporation, *Energizing Niagara's Wine Country Communities* (St. Catharines: Niagara Economic Development Corporation/Peter J. Smith and Company, 2007).

7 Christian Schmid, "Henri Lefebvre, the Right to the City, and the New Metropolitan Mainstream," in *Cities for People, Not for Profit: Critical Urban Theory and the Right to the City*, ed. Neil Brenner, Peter Marcuse, and Margit Mayer (New York: Routledge, 2012), 56.

8 Sharon Zukin, *Naked City: The Death and Life of Authentic Urban Places* (Oxford: Oxford University Press, 2010); William Haydock, "The 'Civilising' Effect of a 'Balanced' Night-Time Economy for 'Better People': Class and Cosmopolitan Limit in the Consumption and Regulation of Alcohol in Bournemouth," *Journal of Policy Research In Tourism, Leisure & Events* 6, no. 2 (May 2014): 172–85, https://doi.org/10.1080/19407963.2014.900989; Leslie Kern, "Rhythms of Gentrification: Eventfulness and Slow Violence in a Happening Neighbourhood," *Cultural Geographies* 23, no. 3 (July 2016): 441–57, https://doi.org/10.1177/1474474015591489.

9 Sharon Zukin, "Consuming Authenticity," *Cultural Studies* 22, no. 5 (September 2008): 724–48, https://doi.org/10.1080/09502380802245985; Robyn Mayes, "A Place in the Sun: The Politics of Place, Identity, and Branding," *Place Branding and Public Diplomacy* 4, no. 2 (May 2008): 124–35, https://doi.org/10.1057/pb.2008.1.

10 Sara Epp, "Competing and Conflicting Land Uses at the Rural-Urban Fringe Interface: Understanding the Impacts of Residential Development on Agricultural Landscapes" (master's thesis, Brock University, 2013).

11 Larry Davis, "OFA Opposes Residential Severances on Prime Ag Land," *Nation Valley News*, 14 February 2020, https://nationvalleynews.com/2020/02/14/ofa-opposes-residential-severances-prime-ag-land/.

12 Regional Municipality of Niagara, *Niagara Agriculture Profile* (Thorold: Regional Municipality of Niagara, [2017]), 2–5.

13 Epp, "Competing and Conflicting Land Uses at the Rural-Urban Fringe Interface," 3.

14 Allan Pred, *Making Histories and Constructing Human Geographies: The Local Transformation of Practice, Power Relations, and Consciousness* (Boulder, CO: Westview Press, 1990); Ruth Finnegan, *Tales of the City: A Study of Narrative and Urban Life* (Cambridge: Cambridge University Press, 1998).

15 Peter G. Prins, "Group Preferences for Rural Amenities and Farmland Preservation in the Niagara Fruit Belt" (master's thesis, University of Waterloo, 2005).

16 John Bodnar, *Remaking America: Public Memory, Commemoration, and Patriotism in the Twentieth Century* (Princeton, NJ: Princeton University Press, 1992).

17 Anthony Giddens, *Modernity and Self-Identity: Self and Society in the Late Modern Age* (Stanford, CA: Stanford University Press, 1991), 128. See also Michael N. Billig, *Banal Nationalism* (London: Sage, 1995).

18 Bonnie Honig, *Public Democracy in Disrepair* (New York: Fordham University Press, 2017).

19 Honig, *Public Democracy*, 5. See also Don Mitchell, "The End of Public Space? People's Park, Definitions of the Public and Democracy," *Annals of the Association of American Geographers* 85, no. 1 (March 1995): 108–33; Richard Sennett, *The Fall of Public Man* (New York: Knopf, 1977); Mike

Davis, "Fortress Los Angeles: The Militarization of Urban Space," in *Variations on a Theme Park: The New American City and the End of Public Space*, ed. Michael Sorkin (New York: Hill and Wang, 1992), 154–80; Sharon Zukin, *Cultures of Cities* (Cambridge: Blackwell, 1995).

20 Edit András, "Public Monuments in Changing Societies," *ARS* 43, no. 1 (2010): 40.

21 Alex Barker, "In Whose Honor/In Whose Time? Regimes of Historicity and the Debate over Confederate Monuments," *Museum Anthropology* 41, no. 2 (Fall 2018): 125–8, https://doi.org/10.1111/muan.12179.

22 Honig, *Public Democracy*.

23 Barker, "In Whose Honor."

24 Joanne Heritz, *Looking Ahead and Looking Up: Affordable Housing in Niagara*, Niagara Community Observatory Policy Brief #48 (St. Catharines: Niagara Community Observatory, 2020).

25 Erika Doss, *Memorial Mania: Public Feeling in America* (Chicago: University of Chicago Press, 2016); Quentin Stevens, Karen A. Franck, and Ruth Fazakerley, "Countermonuments: The Anti-monumental and the Dialogic," *Journal of Architecture* 23, no. 5 (July 2018): 718–39, https://doi.org/10.1080/13602365.2018.1495914.

26 James Young, "The Counter-Monument: Memory against Itself in Germany Today," *Critical Inquiry* 18, no. 2 (Winter 1992): 271, https://doi.org/10.1086/448632.

27 Stevens, Franck, and Fazakerley, "Countermonuments."

28 Doss, *Memorial Mania*; William Logan and Keir Reeves, eds., *Places of Pain and Shame: Dealing with Difficult Heritage* (London: Routledge, 2009); Benjamin Cohen Rossi, "False Exemplars: Admiration and the Ethics of Public Monuments," *Journal of Ethics and Social Philosophy* 18, no. 1 (2020): 49–83, https://doi.org/10.26556/jesp.v18i1.696.

29 Shoshana Felman, "In an Era of Testimony: Claude Lanzmann's Shoah," *Yale French Studies*, no. 97 (2000): 103–50, https://doi.org/10.2307/2903217; Annie Coombes, *History after Apartheid: Visual Culture and Public Memory in a Democratic South Africa* (Durham, NC: Duke University Press, 2003); Kelly Oliver, "Witnessing, Recognition, and Response Ethics," *Philosophy and Rhetoric* 48, no. 4 (November 2015): 473–93, https://doi.org/10.5325/philrhet.48.4.0473; Doss, *Memorial Mania*.

30 Young, "The Counter-Monument," 270.

31 See András, "Public Monuments in Changing Societies."

Index

Notes: The letter *f* following a page number denotes a figure; the letter *m*, a map; and the letter *t*, a table.

Adams, George K.B., 96
Adamson, Anthony, 161
advertising. *See* marketing, wine industry
affect, 10–12
Afghanistan, War in, 102, 142–3, 209
agriculture: agritourism, 184–5; blossoms, 103, 156, 157–60, 164; fruit and fruit belt, 154–6, 161–2, 178, 199–200; in Niagara, 22, 166, 178, 184, 193, 197. *See also* wine industry
Ahmed, Sara, 11, 12, 179
Allward, Walter, 82, 239n99
American Civil War, 65–6, 68, 84, 88
Andrés Wines, 164
Anishinaabe peoples, 23–4. *See also* Chippewa peoples
A-pis-chas-koos, 82
Arcodia, Charles, 165
Armistice Day. *See* Remembrance Day
Arnold, Henry M., 93, 109
Ashton, Paul, 14
Aspler, Tony, 178
Assiniboine Nation, 59–60, 71, 130, 136

authority (authoritative sponsorship), 8–10, 85, 113, 141–2, 210–11, 212. *See also* power
Azaryahu, Maoz, 9

Baltimore: Battle Monument, 68
Barker, Alex, 211
Barnes, George, 162
Barthes, Roland, 10
Batoche, Battle of (1885), 58, 60, 61, 63, 77, 82–3, 104, 106
Batoche National Historic Site, 106
Baxter-Moore, Nicholas, 167
Beamsville (ON), 159
Beech, Monique, 122
Big Bear (Mistahi-maskwa), 60
Billig, Michael, 13, 51, 52
Black Lives Matter movement, 13, 22, 123–4, 125
Blais, Achilles, 70, 72
blossoms (Blossom Time tourism), 103, 156, 157–60, 164. *See also* fruit and fruit belt
Bodnar, John, 13, 211
Boerstler, Charles, 47
Boer War, Second, 87, 91–3, 102, 107, 137

Booth, Oliver J., 76
Boyer, M. Christine, 113
Boy Scouts, 93, 97, 98
branding and place branding: about, 150–1; comparison to memory entrepreneurship, 198–9; local residents and, 204; Niagara Falls, 151–2; Niagara-on-the-Lake, 152–3; regional economic development and, 181–3; wine industry, 153–4, 163–9, 181, 185–6, 203–4, 208–9
Breadman, Jake, 124, 132
Brights Wines, 155, 162
Brock, Isaac, 19–20, 45, 47, 67, 84
Brock's Monument (Queenston Heights), 25, 30, 173
Brock University, 45, 175
Brock University Student Social Justice Centre, 126
Broughall, Lewis W., 97
Brown, Chester: Louis Riel, 106
Brown, Graham, 185
Burgoyne, William B., 29, 226n56
burial practices. See funerary practices
Burkholder, Mabel, 158
Butterfield, Andrew, 142

Cahill, Louis J., 160
Campbell, Gary, 13
Canada: A People's History (CBC), 106
Canadian Armed Forces, 143
Canadian Militia Gazette (newspaper), 57, 71
Canadian Volunteers Monument (Queen's Park, Toronto), 70, 88
Carlisle, George C., 77
Carmichael, Barbara, 185
Caron, Adolphe, 70
Carousel Players, 108–9
Castells, Manuel, 12
Cenotaph (London), 98, 101

Cenotaph (St. Catharines), 100–3, 107
Charest, Caroline, 167
Château des Charmes (winery), 174
Chateau-Gai Wines, 162
Chippewa peoples, 23, 24
Christianity, 64–5
citizen-soldier: Canadian Volunteers Monument and, 70; cultural capital of, 142–3; Soldiers' Cross and, 96; symbolic resonance of, 11, 66, 67–8; Watson Monument and, 78–9, 83–4, 87, 120–1, 144, 207. See also soldiers
CKTB, 127
collective memory, 6–7, 8
colonialism, 23–4, 59
commemograms, 14, 26–7, 205
Commonwealth War Graves Commission (formerly Imperial War Graves Commission), 94–5, 233n35, 242n38
Connolly, William, 114
Conrad, Margaret, 14
constructivism, 33–4, 206
Cook, Wayne, 27, 225n53
Coombs, Albert E., 102
Cornwallis, Edward, 112
counter-monuments, 212
COVID-19 pandemic, 125, 180, 186
Cowan, David L., 70, 72
creative economy, 25–6, 148, 182
Cree Nation, 59–60, 71, 82, 130, 136
Cresswell, Tim, 11
Crooke, Elizabeth, 9
Crystal Beach (ON), 157
Cuc, Alexandru, 8
culture and cultural mapping, 150–1, 182
Cuvée Weekend, 174, 175

Dawson, Daisy, 154
death, good, 64–5

DeCew House, 45, 47–8
Decoration Day, 88–90, 95, 97, 98, 99, 100–2, 103. *See also* Memorial Day; Remembrance Day (Armistice Day)
Demitrova, Snezhana, 58
Denison, George T., III, 70, 89
Devonshire, Victor Cavendish, 9th Duke, 98
Dockstader, Karl, 138
Donohue, Janet, 142
Doss, Erika, 20, 212
Dreschel, Andrew, 108

EastDell Estates Winery, 171
E.D. Smith (jam maker), 155
Energizing Niagara's Wine Communities (2007 report), 148–9, 165
enthymemes, 51
environmental sustainability, 168, 178

Facebook, 126–7
Fearon, Gavin, 124, 125, 132, 136, 141
Fenian Raids, 25, 68, 87–8
festivals, wine, 174–6, 177, 208. *See also* Grape and Wine Festival
First World War: Armistice Day and, 99–100; Canadian National Vimy Memorial, 82; North-West Resistance and Watson Monument displaced by, 100, 101–2, 109–10, 113; public commemorations of, 28, 30, 86, 94–7, 101
FitzGibbon, James, 45, 47
Florida, Richard, 182
Floyd, George, 123
Folk Arts Festival, 36
Food and Drink (magazine), 196
Foote, James G., 76
Foote, Kenneth, 9
Fort Erie (ON), 22–3, 24, 30, 47, 157, 166

Fort Erie Times (newspaper), 196
Fort George, 30, 47, 107, 152
Fort Mississauga, 152
Fort Niagara, 23
44th Regiment, 74, 76, 107. *See also* Lincoln and Welland Regiment
Foucault, Michel, 8
Fountain, Joanna, 154
Freedom Seekers (Underground Railroad), 3, 6, 11, 24, 28, 29, 38, 214n2
fruit and fruit belt, 35, 154–6, 161–2, 178, 199–200. *See also* blossoms
Fruitland (ON), 161
funerary practices: Commonwealth War Graves Commission, 94–5, 233n35, 242n38; development of, 64–5, 66–7; North-West Resistance, 70–6; Second Boer War, 92–3; for soldiers, 57–8, 63–4, 65–6, 87

Garagozov, Rauf, 8
General Motors, 25, 35, 38, 178
Getz, Donald, 185
Giddens, Anthony, 12, 14, 211
Girl Guides, 93, 158
Gladstone, William, 67, 234n39
The Globe (Toronto), 67, 71–2, 87, 89, 98, 155, 157–8
Globe and Mail (newspaper), 159, 197
Go, Frank, 182
Goffman, Erving, 150
Goodman, Edwin, 77
Gordon, Alan, 30
Gowanlock, Theresa, 60, 71–2
Grape and Wine Festival (now Niagara Wine Festival), 36, 39, 103, 174–5, 177, 180–1, 199, 208, 209
grape and wine industry. *See* wine industry
Great War. *See* First World War
Grech, Caroline, 111

Greenbelt Act, 183
Grimsby (ON): about, 41, 188;
   Blossom Time tourism, 158–9;
   Indigenous archaeological sites,
   22; on local agriculture, 193*t*;
   on Niagara identity, 42*t*, 189*t*;
   rural idyll reputation, 156–7;
   urban development, 200; on wine
   industry, 41–2, 42*t*, 190*t*, 192,
   192*t*, 194, 195*t*, 197, 197*t*; wine
   industry experiential marketing,
   163–4
Grimsby Chamber of Commerce, 160
Grip (magazine), 89
Grote, J.W., 89–90
Guild of Loyal Women of South
   Africa, 92–3
Gulf War, 142

Habermas, Tilmann, 8
Halbwachs, Maurice, 6, 8
Halifax: Welsford-Parker Arch, 68
Hamilton, G.H., 157
Hamilton, Paula, 14
Hamilton Spectator (newspaper), 74,
   196
Hands, Harry, 97
Hanna, Jeannette, 182
Harding, Robert J., 86
Harper, Stephen (Harper
   government), 43–4, 46–7
Haudenosaunee Confederacy, 23, 24, 45
Hay, Iain, 85
Henley Regatta, 36
Henry of Pelham Family Estate
   Winery, 172, 173
heritage: basis for, 6–7; and branding
   and economic development, 180–2;
   local understandings of, 200–1;
   methods for learning local history,
   115–16, 249n18; residents on wine
   industry and, 42, 42*t*, 187, 193–4;

wine industry branding and,
   153–4, 171–4, 178–9
"Highway of Heroes," 143
Hill, A.S., 166
Hillebrand Estates Winery, 173
Historical Council (Lincoln County),
   160–1
Holmes, Matt, 127
Homegrown Festival, 174, 175
Honig, Bonnie, 211
Howland, Peter, 185
Hughes, Andrew, 85
Hutchison, James, 72

icewine, 163, 164
Icewine Festival, 174, 175, 192
Ignatieff, Michael, 142
Imperial Order Daughters of the
   Empire (IODE), 92–3, 97, 98, 101
Imperial War Graves Commission
   (now Commonwealth War Graves
   Commission), 94–5, 233n35, 242n38
The Independent (Grimsby
   newspaper), 159
Indigenous Peoples: critique of
   mnemonic products and, 112;
   fruits and, 154; lack of plaques
   commemorating, 28; in Niagara
   region, 7, 22–3, 24, 212; North-West
   Resistance and, 59–60, 82, 105;
   Watson Monument debate and,
   137–8
Inniskillin Wines, 162, 163
intertextuality: identities and, 141;
   memorial commemorations
   and, 21, 85, 207, 211; mnemonic
   products and, 113–14, 210, 212–13;
   and resonance and resilience,
   10–11, 46; wine industry and, 196,
   203
Irwin, Robert, 93
Itka, 82

Jelin, Elizabeth, 9
Johansson Bar, 29, 226n55
Jordan (ON), 156
Jordan Wines, 162

Kah-paypamhchukwao, 82
Key, Francis Scott, 47
King, Henry, 77
Kingston: McBurney Park (Skeleton
    Park), 65
Kippen, A.W., grave monument, 77,
    78f
Kit-awah-ke-ni, 82
Klerides, Eleftherios, 9–10
Köber, Christin, 8
Konecnik, Maja, 182
Konzelmann Estate Winery, 174
Korean War, 87, 102
Kozolanka, Kirsten, 257n77

Lähdesmäki, Tuuli, 11
Landsberg, Alison, 198
Langewin, Hector-Louis, 112
Laurier, Wilfrid (Laurier
    government), 92
learning, passive, 52
Lee, Insun, 165
Lewis, Paul, 122
Lincoln, Abraham, 66, 69
Lincoln and Welland Regiment,
    57, 58, 103, 107–8. See also 19th
    Regiment; 44th Regiment
Lincoln County, 160–1
Lipsitz, George, 197
Litt, Paul, 30
Llewellyn, Nigel, 64–5
Local Architectural Conservation
    Advisory Committee, 108
local history, methods for learning,
    115–16, 249n18
Logan, William, 212
Lozano-Monterrubio, Natàlia, 204

Lundy's Lane, Battle of (1814), 44, 47
L'Union Nationale Métisse, 82–3

Macdonald, John A. (Macdonald
    government), 61, 62, 71, 83, 105,
    112, 131
Macdonald, Norman, 104–5
MacDonald, Ronald I., 23
MacIsaac, Roger, 143
MacKenzie, David, 109
Mackenzie, William Lyon, 24–5
Maid of the Mist (Niagara River tour
    boat), 118, 249n20
Manchoose, 82
Manitoba, 59, 62, 105–6
marketing, wine industry, 169–74,
    170t, 171t, 181, 196–7, 197t
McCrae, John: "In Flanders Fields,"
    97, 131
McGee, Thomas D'Arcy, 67
media, mass, 52, 115–16, 196–8.
    See also newspapers
Memorial Day, 88. See also Decoration
    Day; Remembrance Day (Armistice
    Day)
memorials. See mnemonic products;
    monuments and memorials
memory, 6–7, 8–9
memory entrepreneurs, 9, 30, 198–9,
    205–6
Merritt, William H., 25
Merritton (ON), 62
Métis Nation, 59–60, 61, 71, 82–3,
    122
Métis Nation of Ontario, 86
Mewburn, John H., 69–70, 74, 84, 88,
    235n50, 240n11
Michael, Moina, 97
Michalski, Sergiusz, 20
Middleton, Frederick, 70, 71, 80
military. See citizen-soldier; soldiers
militia myth, 67–8, 69–70, 76–7, 84

Miller, E.W.: "On the Erection of a Monument on the Battlefield of Lundy's Lane," 19
Mires, Charlene, 86
Mississaugas of the Credit First Nation, 23, 24
Mistahi-maskwa (Big Bear), 60
mnemonic products: affect and, 10–12; approach to, 7–8, 14, 20; authority and, 9–10, 85, 113, 141–2, 210–11, 212; Black Lives Matter and other activist critiques, 112, 123–4; collective memory and shared narratives, 6–7, 20–1, 53, 85–6, 112–13, 213; commemogram analysis, 26–7; constructivist approach, 33–4, 206; democratic importance of, 211; national identity and war memorials, 141–3, 257n77; passive learning and, 52; plaques and other projects in Niagara region, 26, 27–30, 28f, 29f, 31m, 32f, 205–6; reinterpretation and change over time, 21–2, 86, 113, 141; removal of and sustainable social change, 211–13; residents' engagement with, 12–14, 30, 32, 50–3, 206–7, 209–10, 211; resilience and, 10–11; resonance, 114; resonance and, 11–12, 21, 52–3, 85, 210–11, 212; somatic markers and, 113–14; ubiquity (repetition) and, 10–11, 85, 203, 210–11, 212; War of 1812 bicentennial case study, 42–8; wine industry and, 178–9, 199, 207–8. *See also* funerary practices; Niagara, Regional Municipality of – mnemonic products engagement surveys; Watson Monument
Montreal: Dorchester Square, 65
Montreal Witness (newspaper), 70–1

monuments and memorials, 20–1, 53, 85–6, 90, 207. *See also* mnemonic products; Watson Monument
Moore, James, 46–7
Moose Jaw (SK), 71
Morgan, Cecilia, 153
Morton, Desmond, 76
Munro, James, 78, 87
Murphy, Kaitlin, 10
Murray, Warwick E., 186
Musil, Robert, 22
Muziani, Hamza, 21

Nahpase, 82
Napier of Magdala, Robert Napier, 1st Baron, 67, 234n39
narratives. *See* mnemonic products
Nelson, Horatio Nelson, 1st Viscount, 67, 234n39
Neutral Nation, 23, 24
Newspaper Audience Databank (NADbank), 45
newspapers, 44, 45–8, 115, 196, 228n89. *See also* media, mass
Nguyen, C. Thi, 11–12
Niagara, Regional Municipality of: agriculture, 22, 166, 178, 184, 193, 197; approach to, 7–8, 14–15; Blossom Time tourism, 103, 156, 157–60, 164; branding, 151–3, 205; commemorative efforts and mnemonic products, 3, 6; COVID-19 pandemic and, 180; economy, 25–6; Fenian Raids memorials, 68–70; fruit belt, 154–6, 161–2; gentrification of, 186; history and foundational narratives, 23–5, 147; Indigenous Peoples, 7, 22–3, 24, 212; map, 5m; mnemonic products and plaque inventory, 26, 27–30, 28f, 29f, 31m, 32f, 205–6; municipalities and land

distribution, 183; North-West Resistance (1885) and, 61–2, 71–2; population, 7; residents' engagement with mnemonic products, 50–3, 206–7; tourism, 26, 156–7, 160–1; TV stations, 228n88; War of 1812 bicentennial, 42–8. *See also* Watson Monument; wine industry; *specific municipalities*
– mnemonic products engagement surveys: 2005 survey, 34–6, 35*f*, 50–2, 114–21, 141, 143–4, 147, 186–7, 207; 2008 survey, 36–9, 38*f*, 111, 186–7; 2009 survey, 39–41, 40*t*, 187, 189*t*; 2012 survey, 41–2, 42*t*, 180, 188–97, 189*t*, 199–203; 2016 survey, 49–50, 49*f*; research methodology, 32–4
Niagara at Large (blog), 124
Niagara Chamber of Commerce, 149
Niagara Culture Plan (2010), 149
Niagara District Volunteer Veterans' Association (NDVVA), 89–90, 97, 98–9, 100
Niagara Economic Development (formerly Niagara Economic Development Corporation [NECD]), 166, 205, 269n100
Niagara Escarpment, 25, 187
Niagara Falls, 25, 35, 38, 39, 41, 151–2, 187, 188
Niagara Falls (ON), 44–5, 124, 152, 159–60, 170, 187
Niagara Falls Review (newspaper), 48, 227n79
Niagara Independent (news site), 124
Niagara-on-the-Lake (ON): about, 187, 188; association with local identity, 38, 39; Blossom Time tourism, 159; branding and, 152–3, 156; establishment of, 24; War of 1812 bicentennial and, 45

Niagara Parks Commission, 157, 159
Niagara Regional Exhibition, 174
Niagara Region Anti-Racism Association (NRARA), 124, 126, 140
Niagara River, 23–4, 30
Niagara This Week (newspaper), 48, 124, 127, 196
Niagara Wine Festival. *See* Grape and Wine Festival
19th Regiment, 74, 76, 80, 90, 92, 93, 94, 95–6, 97, 98, 107. *See also* Lincoln and Welland Regiment
90th Battalion (Royal Winnipeg Rifles), 58, 61, 62–3, 70–1, 72–3, 77, 82, 89, 93
The 90th on Active Service (music hall farce), 62–3
Nixon, Sara, 180–1
North-West Mounted Police (now Royal Canadian Mounted Police), 60, 70, 106, 132–3
North-West Resistance (1885): about, 59–62; changing views of, 105–6; commemorative activities and funerals, 62–3, 70–6, 82–3, 239n102; displaced by First World War, 102, 106–7, 109–10; Watson's funeral, 58, 73–6, 75*f*. *See also* Watson Monument
Norton, John, 47

Olick, Jeffrey, 6
One Dish, One Mic (radio show), 126, 138
Ontario Craft Wineries (formerly Wine Council of Ontario), 165, 166, 174, 268n94; *Wine Country Ontario Travel Guide*, 196
Ontario Editorial Bureau, 160–1
Ontario Motor League, 158
Orange Order, 59, 61–2, 74, 75–6

Ord, Lewis R., 57
Osborne, Brian, 21–2
Ottawa: McDonald Park, 65
Overton, John, 185, 186

Pahpah-me-kee-sick, 82
Palatine Hills Estate Winery, 173
pamphlets, 169–70, 170*t*
Paquette, Rick, 122
passive learning, 52
peacekeeping, 43, 102, 143
Perales-García, Cristina, 46
Perth (ON), 77
Pîhtokahanapiwiyin (Poundmaker), 60
Pilliteri Estates Winery, 174
place branding. *See* branding and place branding
Places to Grow Act, 183
plaques, 27–30, 28*f*, 29*f*, 31*m*, 32*f*
poppies, 97, 142
Port Colborne (ON): about, 25, 41, 188; on local agriculture, 193*t*; on Niagara identity, 42*t*, 189*t*; on wine industry, 41–2, 42*t*, 190, 190*t*, 192*t*, 193–4, 195*t*, 197, 197*t*
Port Dalhousie (St. Catharines), 34, 88, 103, 114, 156
Porter, Karrie, 124, 126, 140
Postmedia, 44
Poundmaker (Pîhtokahanapiwiyin), 60
power, 8–9, 12, 66, 151, 207, 211. *See also* authority
Print Measurement Bureau, 45
Puntscher, Sibylle, 9

Queen's Own Rifles, 70, 235n50
Queenston Heights, 19, 45, 47, 107, 157

Raney, Tracey, 257n77
Ratcliffe, John H., 76
Ravine Vineyard, 172

Ray, Christopher, 181, 198
Rebellion of 1837, 24–5
Reddit, 126
"Red Fridays," 143
Red River Resistance (1870), 59, 88, 105
Reeves, Keir, 212
Remembrance Day (Armistice Day): comparison to wine industry, 199, 208; development of, 97–8; in mass media, 46, 107; as mnemonic practice, 207, 209; state promotion of, 142; in St. Catharines, 98, 99–100, 101, 102–3; Watson Monument and, 101, 103, 142. *See also* Decoration Day; Memorial Day
repetition (ubiquity), 10–11, 85, 203, 210–11, 212. *See also* intertextuality
resilience, 10–11
resonance, 11–12, 21, 52–3, 85, 114, 210–11, 212
The Retrial of Louis Riel (CBC), 106
rhetorical topoi, 113
Rib-Fest, 36
Ridgeway, Battle of (1866), 68–9, 87–8, 89–91, 91*f*, 93, 97, 100–1
Rief Estate Winery, 172
Riel, Louis, 59, 60, 61–2, 70, 83, 105–7, 136
Roberts, Wayne, 122
Robertson, Gail, 108
Rogers, R.L., 108
Roll of Honour (St. Catharines City Hall), 102
Rosenzweig, Roy, 13–14
Royal Canadian Legion, 98–9, 99*f*, 100–2, 103, 142
Royal Canadian Mounted Police (formerly North-West Mounted Police), 60, 70, 106, 132–3
rural idylls and rural areas, 166, 181, 183, 209, 275n25

Said, Edward, 8–9
Sainte-Marie, Buffy: "The Universal
    Soldier," 109, 247n95
Sakamoto, Rumi, 11, 114
Salem Chapel (St. Catharines), 3, 4*f*,
    6, 29
Savage, Kirk, 21
Scott, Thomas, 59
Second Boer War, 87, 91–3, 102, 107,
    137
Second World War, 28, 30, 102, 107
Secord, Laura, 45, 47, 211
Séraphin, Hugues, 182
settler colonialism, 23–4, 59
Shaw Festival, 35, 38, 153, 165
Sheaffe, Roger Hale, 47
Shucksmith, Mark, 183
silence, two-minute, 97–8
Sjolandaer, Claire, 43
Skinner, Heather, 182
Smith, Laurajane, 6–7, 13
social change, 211–13
social media: brand marketing and,
    204; Watson Monument debates,
    125–41, 128*t*; wine industry and,
    195
soldiers: commemoration of and
    national identity, 141–3, 257n77;
    Commonwealth War Graves
    Commission, 94–5, 233n35, 242n38;
    Decoration Day, 88–90, 95, 97, 98,
    99, 100–2, 103; funerals for Fenian
    Raids militia volunteers, 68–70;
    IODE monuments for Second
    Boer War dead, 92–3; Memorial
    Day, 88; North-West Resistance
    funerals, 70–6; public burial of
    dead soldiers, 57–8, 63–4, 65–6, 84;
    recruitment by Canadian Armed
    Forces, 143; service memorials,
    207. *See also* citizen-soldier;
    Remembrance Day

Soldiers' Cross (St. Catharines), 94–7,
    94*f*, 98, 99, 101, 102, 242n38
somatic markers, 113–14
Southbrook Vineyards, 168
Spring Sparkles Festival, 175
Stamford (ON), 69
The Standard (St. Catharines): about,
    227n79; on Decoration Day, 89–90,
    100, 101–2; on loss of North-
    West Resistance and Boer War
    memories, 109; on need for First
    World War memorial, 101; on
    Soldiers' Cross, 98; on War of 1812
    bicentennial, 46–7, 48; on Watson
    Monument, 87, 104, 122, 124, 127;
    William Burgoyne and, 226n56;
    on wine festivals, 176, 177; wine
    industry and, 196
statues. *See* mnemonic products;
    monuments and memorials
St. Catharines (ON): about, 25,
    152, 188–9; Black Lives Matter
    and, 124; Cenotaph, 100–3, 107;
    and changing views of Riel and
    North-West Resistance, 106–7; city
    hall rebuilt, 103–4; Decoration Day,
    89–91, 95, 97, 98, 99, 100–2, 103;
    Fenian Raids and, 88; horticultural
    research facility, 156; Legion's
    War Shrine, 99, 99*f*; on Niagara
    identity, 40–1, 40*t*, 189*t*; plaques in,
    30; Remembrance Day (Armistice
    Day), 97, 98, 99–100, 101, 102–3,
    142; Salem Chapel, 3, 4*f*, 6, 29;
    Second Boer War and, 92, 93;
    Soldiers' Cross, 94–7, 94*f*, 98, 99,
    101, 102, 242n38; tourism, 103, 158,
    160; Victoria Lawn Cemetery, 65,
    89–90, 97, 102, 103, 140; Watson's
    funeral, 58, 73–6, 75*f*. *See also*
    Watson Monument
St. Catharines Historical Society, 111

St. Catharines Journal (newspaper), 69, 76
St. David's (ON), 22
Steward, Kathleen, 10
stickiness, 11, 21, 114, 179. *See also* resonance
Storm, Albert, 57, 84, 209
Stupples, Peter, 113
Sydenham, Charles Thomson, 1st Baron, 67

Taché, Alexandre-Antonin, 72
Taché, Étienne-Paschal, 67
"Tastes of Niagara: A Quality Food Alliance," 165
Taylor, Nigel, 117
Taylor, Robert R.: *Discovering St. Catharines' Heritage*, 108, 247n92
Tecumseh (Shawnee chief), 47
terroir, 153–4
Tesla, Nikola, 25
Thelen, David, 13–14
Thorold (ON), 44
Thorold Post (newspaper), 61, 62, 90
time, 21–2
Toronto: Canadian Volunteers Monument, 70, 88; Decoration Day, 89; North-West Resistance memorial, 82, 86; Victoria Memorial Square, 65
Toronto Star (formerly *Toronto Daily Star*), 89, 156, 157
Toronto World (newspaper), 159
tourism: agritourism, 184; Blossom Time, 103, 156, 157–60, 164; creative economy and, 25–6, 148, 182; engagement with mnemonic products, 30, 32; Niagara identity and, 38, 39; in Niagara region, 26, 156–7, 160–1; overtourism, 182; wine industry and, 26, 163–9, 184–5
tourist gaze, 150

Truth and Reconciliation Commission, 112, 212
Tubman, Harriet, 3, 4f, 6, 29, 140
Tutton, Mark, 85
Tweedsmuir, John Buchan, Lord, 101
Twitter, 123, 125–6

ubiquity (repetition), 10–11, 85, 203, 210–11, 212. *See also* intertextuality
Underground Railroad (Freedom Seekers), 3, 6, 11, 24, 28, 29, 38, 214n2
United Hotels Company, 158
United States of America: Civil War, 65–6, 68, 84, 88; Vietnam Veterans Memorial, 212
urbanization, 117, 183, 200
Urry, John, 150

Vanderklis, Sean, 138
Van Rensselaer, Stephen, 47
Vietnam Veterans Memorial, 212
Vineland (ON), 168
Vineland Research Station, 156, 162
Vintners Quality Alliance (VQA), 164, 174
Vintners Quality Alliance Act, 164
Virgil (ON): about, 24, 41, 188; on local agriculture, 193t; on Niagara identity, 42t, 189t; on wine industry, 41–2, 42t, 190t, 192, 192t, 194, 195t, 197, 197t
visibility, 194
Vividata, 45, 127
Les Voltigeurs de Québec, 70–1, 73, 237n83

War Amps of Canada, 142
War of 1812: association with Niagara, 11, 24; bicentennial as mnemonic products case study, 42–8, 206; Brock and, 19; burial

of dead soldiers, 63; destruction from and rebuilding, 152; plaques commemorating, 28, 29, 30

Washington, DC: Vietnam Veterans Memorial, 212

Watson, Alexander, 58, 60–1, 63, 72, 73–5, 75f. *See also* Watson Monument

Watson, David, 60, 87

Watson, John, 63

Watson Monument: approach to, 15, 58–9, 85, 86, 111–12; assessment and public consultation (2009–19), 121–2; Black Lives Matter and petition to remove, 124–5; construction and unveiling, 77–8, 80–2, 83–4, 238n91; Decoration Day celebrations and, 90–1, 91f, 97; eclipsed by Cenotaph (1927–71), 100–3, 107; message of, 78; move to new location, 103–4; North-West Resistance context, 59–62; photographs and depictions, 79f, 81f, 91f, 112f; as quaint anachronism (1971–2009), 106–10; residents' engagement with, 114, 117–21, 141, 143–4; restoration of, 104–5, 110; Second Boer War dead and, 93; as service memorial (1886–1927), 87–100, 207; social media discussions, on petition to remove, 129–41, 206–7; social media discussions, research methodology, 125–9, 128t, 253n49; vs. Soldiers' Cross, 95–6, 98

Waywahnitch, 82

websites, for wineries, 170, 171–2, 171t

Welland (ON): about, 25, 41, 188; on local agriculture, 193t; on Niagara identity, 42t, 189t; on wine industry, 41–2, 42t, 190t, 192t, 193–4, 195t, 197, 197t

Welland Canal, 25, 36, 38, 152

Welland Rose Festival, 160

Welland Telegraph (newspaper), 74

Welland Tribune (newspaper), 61, 62, 196, 227n79

Wellington, Arthur Wellesley, 1st Duke, 67, 234n39

white settlement, 24, 28, 29, 83, 165–6

Williamson, Ronald F., 23

Wine Council of Ontario (now Ontario Craft Wineries), 165, 166, 174, 268n94; *Wine Country Ontario Travel Guide*, 196

wine industry: approach to, 15, 147–8; branding, 153–4, 163–9, 181, 185–6, 203–4, 208–9; challenges facing, 176–7; development of, 148–9, 162–3; festivals, 174–6, 177, 208; green marketing, 168; icewine, 163, 164; local media coverage, 177–8; location of Niagara wineries, 191m; marketing, 169–74, 170t, 171t, 181, 196–7; as mnemonic product, 178–9, 199, 207–8; within Niagara agriculture, 184; tourism and, 26, 163–9, 184–5
– residents' engagement with: 2009 survey, 40–1, 187; 2012 survey, 41–2, 188–97, 199–203; on advertising, 197, 197t; on branding, 201–3; on heritage and wine industry, 192–4, 193t, 200–1; on identification with Niagara, 35, 38, 209–10; on proximity to and role in daily routines, 189–92, 190t, 192t, 210–11; on sources of wine industry information, 194–6, 195t; on wine industry's place within Niagara, 199–200

wine routes, 164–5

winescapes, 153

Winnipeg, 82, 88–9, 212, 239n102
Winnipeg Free Press (newspaper), 63, 72, 79
Winnipeg Times (newspaper), 63
Winona (ON), 155, 161, 164
Winona Peach Festival, 160
Withers, Charles, 9
Wolfe, James, 21
World War I. *See* First World War
World War II. *See* Second World War

Xambó, Rafael, 46
Xicoy, Enric, 46

Yeoh, Brenda, 21
Young, James E., 21, 141, 212

Zembylas, Michalinos, 9–10
Zerubavel, Eviatar, 9, 11, 14, 26, 205
Ziraldo, Donald, 163

Printed in the USA
CPSIA information can be obtained
at www.ICGtesting.com
CBHW031205040524
7895CB00022B/216